VARIORUM COLLECTED STUDIES SERIES

The Identity of the History of Science and Medicine

Andrew Cunningham

Andrew Cunningham

The Identity of the History of Science and Medicine

Routledge
Taylor & Francis Group

LONDON AND NEW YORK

First published 2012 by Ashgate Publishing

Published 2016 by Routledge
2 Park Square, Milton Park, Abingdon, Oxon OX14 4RN
711 Third Avenue, New York, NY 10017, USA

First issued in paperback 2017

Routledge is an imprint of the Taylor & Francis Group, an informa business

British Library Cataloguing in Publication Data
Cunningham, Andrew, Dr.
 The identity of the history of science and medicine.
 – (Variorum collected studies series ; CS1003)
 1. Science – History. 2. Science – Philosophy.
 3. Medicine – Research – History. 4. Knowledge, Theory of.
 I. Title II. Series
 501–dc23

 ISBN 978–1–4094–4024–6

Library of Congress Control Number: 2011945181

VARIORUM COLLECTED STUDIES SERIES CS1003

ISBN 13: 978-1-138-11032-8 (pbk)
ISBN 13: 978-1-4094-4024-6 (hbk)

CONTENTS

This volume contains xii + 270 pages

PUBLISHER'S NOTE

The articles in this volume, as in all others in the Variorum Collected Studies Series, have not been given a new, continuous pagination. In order to avoid confusion, and to facilitate their use where these same studies have been referred to elsewhere, the original pagination has been maintained wherever possible.

Each article has been given a Roman number in order of appearance, as listed in the Contents. This number is repeated on each page and is quoted in the index entries.

INTRODUCTION

'We will not anticipate the past, our retrospection will now be all to the future'

It's Mrs Malaprop, of course, forgiving and forgetting in Sheridan's play of 1775, *The Rivals*. Had she not been Mrs Malaprop and got her particles in a twist, she would probably have said something like 'We will not dwell upon the past, our anticipation now will all be for the future'. But in fact the Malapropism contains much excellent advice for the historian of ideas, including the historian of science or medicine. We actually have a spontaneous tendency to anticipate the past – to assume that the past was like the present, only a bit earlier, or to assume that the people of the past were trying to bring about our present ways of thinking and behaving, or that they did their thinking within the same categories and disciplines as we do today.

Looking back on my work in the history of science and medicine, I am struck by how consistently and persistently I have been concerned with issues of *identity* – what was the identity of topics, disciplines, arguments, diseases in the past, and are they identical with (more usually, how they are *not* identical with) topics, disciplines, arguments or diseases in the present. Historians usually tend to assume continuous identities of present attitudes and activities with past ones, and rarely question them. In my view this often gives us a false image of the very things in the past that we went to look for.

I believe that if we can stop anticipating the past then we can write much better history, that is to say, history of the past, of past people and past ways of looking, thinking and acting. This is what these essays have been about, so they are as historiographical as they are historical, and of concern to historians of ideas in general.

Historians are historical characters just as much as the past people they study, and their lives, thoughts, beliefs and actions are shaped by the circumstances in which they find themselves. My concern with the identity of the activities of past people, especially their thinking activities, arose when I was working in the history of medicine at Cambridge in my early thirties, in the middle of that period between the Sixties, when the feminists taught us that 'the personal is political', and the fall of the Berlin Wall in 1989, after which all properly political perspectives unfortunately seem to have disappeared from our history-writing. In this context those of us of a general Leftist leaning

– typical young products of the Sixties – were dissatisfied with the kinds of history being written by our teachers. In the history of science and medicine this meant that we were no longer satisfied by an account which was very positivist, which glorified and never criticised the achievements of science and medicine, and which was peopled predominantly by heroic male thinkers struggling against the irrationality of their contemporaries.

When I encountered the discussion it was, on the surface, about 'internal' and 'external' factors in the historical explanations we gave in the history of science and medicine. The 'internalists' were our mostly positivist teachers, who thought that concentration on the text as a memorial of the life of the mind of the individual was what was most important. We 'externalists' by contrast were beginning to believe that the social, political and economic context of past discovery and practice in science and medicine were indispensable factors that had to be taken into account if we were properly to understand not just the past but the very words and terms used by past people to express their ideas.

But it was not at all clear how we could and should do this. We had earnest discussions about quite how much external 'context' was required to explain any past moment in the history of science or medicine: did you need none, a little, a lot, or what, and of what kind? Framed like that, the issue looks very silly now. But where you stood on this internalist-externalist spectrum was important at the time because it really affected job prospects, publication possibilities and indeed the future direction of the subject.

My personal route out of these dilemmas was worked out primarily in a very happy collaboration with a friend and colleague, Adrian Wilson, which lasted several years, and was the most important intellectual adventure of my life. Our joint discussions, for some time carried on day after day,[1] were put to the test by us working through practical research exercises in our respective fields.[2]

Among the writings we read which helped us out of these dilemmas were R.G. Collingwood's *An Autobiography* of 1939, and two then-recently published papers by precocious young scholars, one by Quentin Skinner, 'Meaning and Understanding in the History of Ideas' of 1969, and John Dunn's very dense companion paper of 1968, 'The Identity of the History of Ideas',

---

[1] It will come as no surprise to readers of the first piece below, 'Getting the game right', that we were also playing a lot of bridge at the time with our girlfriends.

[2] Other friends and colleagues, especially John Gabbay and Perry Williams, also shared their thinking on these areas with me, which I found very helpful. Yoko Mitsui helped, then as now, to explain the philosophical implications of my own thinking to me, for which I am most grateful, as I am not the most gracious pupil for philosophy.

to which the title of the present volume alludes.[3] All of these, in their different ways, pointed towards the importance of studying past thought – in the case of Skinner and Dunn, this was 'political thought' – as a human activity, and as part of human activity, rather than as a sequence of disembodied texts or doctrines.

But our solutions were not the same as any of these writers. I applied our joint thinking particularly to my own subject areas, the history of science and medicine.[4] For me, the answer now lay in relocating ideas back to the place where they happened and existed, viz. people's heads. So I reformulated the question 'Why do new ideas arise?', and put it my way: 'Why do *particular people* have particular new ideas, reinterpret old views, etc.?' I began to interpret changes in ideas, whether they were scientific, medical, or whatever, as being intimately bound up with people's plans, intentions, projects, programmes of research, all within the context (taken in the widest sense) of their lives as they were living them. As a historical approach it all sounds commonsensical and straightforward now, but it was hard to work out, and continues to be challenging to put into practice effectively.

My approach looked relativist to some people, as if I was saying that all ideas and truths are only – merely – time-specific and contingent. Given the long-established positivist tradition in writing the history of science and medicine, this was not perhaps surprising. But I am not an in-principle relativist, just an (I hope) open-minded historian trying to be alert to what past people wrote, what they meant, and what they said to their contemporaries. If they used vocabulary or disciplinary categories different from our own, then our duty as historians (I believe) is to try and understand that vocabulary and those categories, rather than just elide them with our own vocabulary and categories, and thus make any differences disappear.

The following essays, I hope, taken together, will show some of the advantages of following Mrs Malaprop's advice and not anticipating the past, if for no other reason than that the past that happened is more interesting than any past that we project back from the present.

Now a few brief introductory comments on the essays.

First, the group I have labelled 'Natural philosophy/ science identity issues'. In pursuing my new method of interpretation and applying it to the history of science and medicine, my first major finding was that we historians had been

---

[3]   Quentin Skinner, 'Meaning and Understanding in the History of Ideas', *History and Theory* 8, 1969, 3–53; John Dunn, 'The Identity of the History of Ideas', *Philosophy* 43, 1968, 85–104.

[4]   Adrian's primary historiographic concerns in these areas resulted in two articles on Whig history, present-centred history, and the whole problem of historical knowledge, which he wrote with Timothy Ashplant, and which were published in *The Historical Journal* in 1988.

blithely ignoring the existence of the major discipline for studying the universe and its contents, which was in practice in Christian Europe from the early 13th century to the end of the 18th century: *natural philosophy*. Hitherto, those texts or doctrines of natural philosophers that we approved of, in our positivist judgements, we had just turned seamlessly into science. And that means that most of the people we used to put in our traditional history of science weren't doing science at all!

We historians had also for a generation or more in the mid-twentieth century put much effort into delineating and portraying in detail what we took to be the major change from pre-scientific ways of exploring the universe and its contents, to the scientific way. Historians had called this 'The scientific revolution of the 17th century', culminating in the work and achievements of Isaac Newton at the end of that century. But if natural philosophy was now to be taken seriously by us as an important actors' category, then this neat picture of the development of science and scientific ways of thought needed radical rethinking. This is the theme pursued in the first three essays. Incidentally, I chose Newton's famous book as my focus in 'How the *Principia* got its name', in order to show that the greatest investigator of all was working within the natural philosophy paradigm – just like everyone else.

If I wanted to get other historians in my field to take natural philosophy seriously as a past category of knowledge, then I felt it was important to look at the origin of some forms of it. With the late Roger French I wrote a book called *Before Science: The Invention of the Friars' Natural Philosophy* (1996), which does this and which was a great revelation to both of us as we worked on the material. The next two essays on 'The identity of "medieval science"' – the quote marks are essential here – build on that work. Amongst modern-day historians of 'medieval science' it is an unquestioned practice to include as instances of 'medieval science' what seem to be the pertinent writings of bishops, friars and other medieval professional religious men. In these two papers I go as deeply as anyone has ever gone into the life and attitudes of St Francis and his immediate followers, both to dispose of any bizarre attempt to make him a modern scientist (in this case a supposed ecologist) before his time, and also to discover what his typically medieval attitudes to nature really were, if indeed he had any. The distinction I call attention to in these papers between 'creature' and 'nature' epitomises the difference between natural philosophy and science. The natural philosopher was concerned with creatures, as the creation of the Creator, while the scientist is concerned with investigating nature in a way which owes nothing to any religious dogma or Bronze Age creation myths.

The third set of essays deals with the identity of the investigative projects of certain individuals, each of whom has a reputation today for having been a modern before his time. But no-one is before their time – it is a logical and practical impossibility. So these reputations are ripe for reassessment. The first is Aristotle, still widely held today to have been a biologist in Antiquity – a claim which is both unsustainable and absurd, as I try to show. The second is Paracelsus, whose supposed modern achievements all turn out to be projections back of differing interests of modern historians and other interested parties. Thomas Sydenham, too, the supposed great modern reformer of medicine in 17th century England, was a man of his own time, and neither ahead of his time, nor a man of our time. I try to show that his investigative programme on fevers in the Restoration period was a product of his political commitments, and methodologically inspired by the philosophical concerns of his friends and contemporaries and the recent political traumas through which they had all lived.

Finally I don my historian of medicine hat and present two pieces on disease identity. Medical doctors, when they turn their hand to doing history – like historians in general – tend to try and make retrospective diagnoses. In my view this is not only logically impossible in the case of infectious diseases, but also undesirable as history. In the first piece I try to show how infectious diseases in general, and plague in particular, have had their identities changed with the invention of the bacteriological laboratory in the late 19th century. This means that it is logically impossible for us to compare the identity of (say) plague before and after: the two constructs are simply incommensurable. Understanding the crucial role of the laboratory in structuring our own thinking about infectious disease, opens the way to investigating past illnesses in the way in which they were experienced and theorised by people in the pre-laboratory world. The second piece offers some ways to do this, by taking seriously both past experiences of disease and past diagnoses.

ANDREW CUNNINGHAM

Cambridge
January 2012

ACKNOWLEDGEMENTS

Grateful acknowledgement is made to the following persons, institutions, journals and publishers for their kind permission to reproduce the papers included in this volume: Elsevier (for articles I, IV and V); Science History Publications Ltd (II), Perry Williams, and the *British Journal for the History of Science* (III); Cambridge University Press (III, VIII, IX); University of Arkansas Press (VI); Koninklijke Brill N.V., Leiden (VII); and the Spanish National Research Council (*Asclepio*) (X).

I

GETTING THE GAME RIGHT:
SOME PLAIN WORDS ON
THE IDENTITY AND INVENTION OF SCIENCE

1. Identifying science

ARE WE studying the right subject? I ask this question as a historian of science, and my question concerns how we do, and how we should, go about our self-imposed task of writing the history of science. In asking whether we are studying the right subject I do not mean to suggest that we should abandon our discipline for another. I mean: when we set out to study the history of science, are we properly equipped to identify *science* in the past in order to study it? When we decide to put some episode into our account of the history of science, have we identified it correctly: *was it science* in any meaningful sense? To be able to answer this question in the affirmative we obviously need to be in possession of criteria which enable us to identify science in the past correctly every time. For our story properly to be about science, it has to be not about non-science: we have to be able to distinguish science in the past from all the things that it was not and that were not it. It lies at the very heart of our self-imposed enterprise as historians of science, this issue as to the nature and identity of science itself. Behind everything we do lies the assumption that we are getting this identification right, that we can indeed identify science in the past without any problem or doubt. And it follows that if we get it wrong — if we are identifying the wrong thing as science — we will be writing myths, hallucinations and romances which can only purport to be a history of science: we will be writing accounts of events which may not have happened, and of the adventures of a something which may well not have existed.

And yet the tools we currently use to make this identification are not very sophisticated. For in practice all we do is simply assume that the presence of science in the past *identifies itself* to our eyes as enquiring historians. We read old documents, and it comes leaping off the page into our eyes: here is some science, this should be in our history of science! That is what it feels like. There may be the odd marginal case (such as Paracelsus), but in general this way of proceeding seems to work: it gives us a history of science which includes Copernicus, Kepler and Newton as practitioners, but which excludes

This article was published in *Studies in History and Philosophy of Science* 19 (1988), pp. 365–389. © 1988 by Pergamon Press. Reprinted with permission of Elsevier.

Shakespeare, Columbus and Luther. And very rarely does it lead to any problems, for there is broad (almost total) agreement amongst historians of science as to what does, and what does not, 'count as' past science and who does and who does not qualify as a past scientist.

Yet it should be obvious that, although it may indeed feel as though science in the past is identifying itself to us as we do our research, what is actually happening is that we — the historians — are the active agents: that we are actually taking to our investigation a ready-prepared set of finding guides to identify past science. If it 'looks like science' to us, then we believe that it will 'look like science' to our colleagues and readers; indeed, if it 'looks like science' to us, then for us it *is* (or more properly, *was*) science! That is to say, what in practice determines whether some past episode, piece of work or whatever was or was not science, is whether it fits, to the satisfaction of the historian, some template which the historian carries around in his mind. Yet for some reason or other we are not usually aware that this is happening, and if pressed we would probably point to something other than ourselves as providing the criteria for positive identification: we would say that the thing or event in question is 'self-evidently' an instance of science, or we would point to old documents as being (or containing) the 'evidence' for the presence of a given instance of science in the past. To be able to point to a physical thing such as an old document feels conclusive: we can hold in our hands the evidence which 'speaks for itself', evidence both tangible and audible. And yet the real arbiter of 'what counts as science' in the past for us to study, is in fact not an old document, nor some passage in an old document. Nor are instances of past science remotely 'self-evident'. The arbiter is us: the historian in the present. And this is the problem.

For whether our account of the history of science can claim to be referring to anything meaningful at all depends on whether *we* have got an appropriate concept of 'what science is' — a concept appropriate, that is, to the past that we are seeking to explore. It is our concept of 'what science is' which is going to determine all the history that we write in our discipline. It determines 'what counts' as an appropriate question for us to ask or topic for us to study, 'what counts' as an appropriate answer or account for us to give, and what is the proper way for us to do the research. It is thus the major determinant of the account we can give of the history — of 'what happened in the past' with respect to science. That is why we need to investigate this matter of the identity of science — what kind of thing it is — for only then will we be able to know whether the history of science that we write bears any determinable relation to the past, the human past, it is supposedly about.

So this is where we need to call on an important distinction. We need to distinguish between on the one hand that set of *assumptions and techniques by which* we find out about the past (let us call it our 'historiography') and, on the other hand, what we use them to find out, i.e. *what happened in the past*. Now, if we are to get a valid picture of the past, and of science in the past, then our 'historiography' must be a suitable tool for the job, it must be as a key for a lock: our assumptions and techniques must be appropriate for our goal. Yet currently it is the historian *in the present* who (whether aware of it or otherwise) sets the criteria for 'what counts as science' *in the past* he or she studies. This can hardly be appropriate: uninspected concepts of and about the present can hardly legitimately be used as a measure either of what happened in the past or of whether something or other happened in the past. In other words, 'what looks like science' to us now in the present is simply that: our view from the present. It cannot in itself be taken as a legitimate criterion for identifying what was science *in the past*. This misapprehension arises from our 'present-centredness' as it has been termed: from the fact that we look at the past with both eyes in the present.[1] I shall be arguing that until the historian can, in some way, 'get out of the present', then the history he or she produces can never be (or be known to be) authentically about the past and true to what happened in the past. But if he or she can do this, then the picture that the history of science will then present will be radically different from the one we are familiar with.

2. The specialness of science

The first and major obstacle standing in our way and hindering us from discovering what kind of thing that science is of which we want to write the history, is the apparent specialness of science. For if science is different in kind and quality from all other subjects, from all other pursuits, and from all other truths, then presumably different criteria would apply to the writing of its history (and perhaps even for its identification) than the historian would apply elsewhere. That science is indeed special is something we are led (and lead each other) to believe, and it is this very specialness which attracted many of us to study the history of science in the first place. But it is in this apparent specialness of science that our present-centredness as historians of science finds its source. So we need to start with that very specialness, lest it bedevil the rest of our enterprise.

---

[1]For the coining and meaning of this term see Adrian Wilson and T. G. Ashplant, 'Whig History and Present-centred History', and T. G. Ashplant and Adrian Wilson, 'Present-centred History and the Problem of Historical Knowledge', both in *The Historical Journal* Vol. **31** (1988). I am in agreement with everything they write there, and the present article should be seen as complementary to theirs. Their discussion of, and references to, the historiographic and historical literature should also be considered as pertinent to my present argument. Quentin Skinner's characterisation of 'the mythology of doctrines' ['Meaning and Understanding in the History of Ideas', *History and Theory* **8** (1969), 3 – 53] bears some resemblance to my critique of present-centredness in the history of science; however, my argument goes beyond his in significant ways.

To do this we need to start with a fact about the present, the present predicament, a fact which as historians of science we quite simply take for granted. This is the enormous importance that science has for our society today. Science, its claims and achievements, totally dominate our modern outlook. The world we live in, the physical, the technological, and the intellectual world, is deeply pervaded and affected by the presence of science and scientists. We take this fact for granted: that is to say, it is because we live in a world which has been profoundly structured by the achievements of science and scientists, that we treat it as 'natural' that science is the authentic way of looking at the world. And this is to be expected. In the same way we tend to take the dominant moral values of western society for granted too, and treat them as 'natural', as somehow acultural.

In taking science for granted because it is dominant in our own society today, we also take for granted a certain kind of claim about the nature of science: that it produces 'objective' findings, and hence that it is itself 'objective'.[2] It is in this that its specialness supposedly consists. Most historians of science never in fact mention whether or not they think that science produces 'objective' findings; nor do they bother their heads with what the philosophers of science have to say about this. But our actual practice reveals that we do think this: we act and write as if we believe (in my formulation) that science unproblematically 'reads off the truth about Nature from Nature, and in the categories of Nature'. Treating this as a 'natural' attitude to hold, we historians of science set out to write a history of science which matches it: a history about how man learnt to read off the truth about Nature from Nature and in the categories of Nature. It is our untroubled acceptance of the 'naturalness' of science and of the 'objective' nature of science, that leads us to write like this; to put it another way, the objectivity of science seems to demand that we write a history of the emergence of its objectivity. That we take this attitude to the history of science is due to our 'present-centredness'.

Now the problem with this is that if (as most of us customarily do) we take the objectivity of science and its findings for granted in our work on the history of science, then we do not allow ourselves to raise a single question about the nature of science, or about the appropriate shape of a valid history of science. For we are taking for granted that science is special — rather than asking historical questions about how and why people in the past (and hence we ourselves) came to realise that it was special, or came to see it as special, or

---

[2]I do not want to enter here on a specification of what 'objective' means when in common speech we apply it to science. But for my purposes here we can take it that it refers to those truths which are considered to be the same for any time and place and independent of personal whim, cultural context, political outlook, social position, race, gender, creed, or colour. More will be said about 'objectivity' later in the paper.

perhaps came to make it special. Our present-centredness has already settled our views about the nature of science, and hence of the appropriate shape of a valid history of science, long before we start our research. And because our minds are already made up, the history we write gives us the illusion that it proves that the history of science actually consisted of a sequence of people learning, with increasing accuracy and sophistication over time, how to read off the truth about Nature from Nature and in the categories of Nature. But here we are going round in a circle. First we shape the past of science to our own preconceptions about the nature and importance of science — preconceptions which are derived from the present. And then we take the past of science (as now shaped by us) as confirming the appropriateness of the history that we write about it!

To get out of this dilemma we need to do two things. Firstly we need to see the specialness of science as an issue which the historian needs to investigate. Secondly, having separated the subject (science) from our evaluation of it, we need to put aside that specialness from our considerations for the time being. Once we do this we can see that the same criteria will apply to writing the history of science as apply to the histories of other subjects. But if we do not do this, we will simply be writing self-serving and self-confirming history, from which all properly historical questions have been refused application. I appreciate that this is going to be a difficult matter, and may seem like a call to repudiate the whole point of our enterprise. But to put the specialness, the 'objectivity', of science temporarily aside is not to deny that science may be unique in certain respects; nor is it to deny that science produces findings with a special ('objective') status. It is simply the crucial first move for us to make in order for us to be able to understand what kind of thing science is, and thus ask properly historical questions about science and its history.

3. Science as human activity

Our assumption about the specialness of science (I have been arguing) has stood in our way of treating the history of science historically: our assumption that it has (or our attribution to it of) a unique specialness has inhibited us from asking (for instance) how it got that specialness — perhaps the most interesting historical question that could be asked about it. But having now put aside for the moment our concern with the specialness, the 'objectivity' of science, we are now in a position to look at science just like anything else, and to ask historical questions about it, just as we might about anything else. Thus we can now turn to the question of what kind of thing that 'science' is of which we want to write the history, without begging the question as to its status and how it acquired that status.

My contention is that science is a human activity, a human practice. It is an activity, and it is an activity engaged in only by humans. As a claim it should be pretty uncontroversial, and I hope that no historian of science would want to deny it. For the histories of science we produce normally include accounts of people in the past (named individuals, usually) engaged in those uniquely human activities of 'thinking', 'investigating', 'drawing conclusions' and so forth which are central and essential to the practice of science: 'thinking' and 'investigating' are activities which demand the use of the human mind or brain and of human hands as tools of the thinking mind, and cannot be carried out without them.

But if I can gain ready acquiescence to my contention that the practice and practising of science is a human activity, is that all it is? Is there something about the practice or the practising of science which is more than human activity, which lies outside of human activity? The practising scientist of course takes as his raw material the world 'out there'; and it is the business of the practising scientist to produce *products* of a certain kind: laws, statements, judgements, findings, conclusions, theories, knowledge, truths, things to believe about the world 'out there'. But although we may want to ascribe a special kind of status to these products (to these laws, findings and so on), yet the actual production or producing of them lies wholly within the human activity of practising science. In this sense, then, everything about the doing of science, everything about its practice, is a human activity, wholly a human activity, and nothing but a human activity.[3]

Now, if the practice of science is a human activity, wholly a human activity, and nothing but a human activity, then we should hardly be satisfied with setting out to produce history of science which merely makes just some concession to this, which by lucky chance just happens to include some account or other of people engaged in those human activities essential to science, such as thinking or investigating. For, given my contention, if an historian is claiming to offer an account of science in the past, then it is essential that he or she regard the reconstruction of it *as a human activity* as one of his or her central aims, and indeed the fundamental one.

But what we write are not histories of 'science as a human activity', but histories of something else entirely. In this we are commiting a logical error, a category error: we mistake one kind of thing for another.[4] The core of our mistake lies in the fact that as historians of science we (quite unconsciously)

---

[3] It is tempting to assume that science is simply a set of discrete statements (together constituting a 'body of knowledge'), with an identity independent of their status as products of a particular human activity. But in fact such statements acquire or are given their status as 'scientific' as a consequence of how they are produced and who produces them: simply as a discrete statement (i.e. abstracted from its status as a product of the human practice of science) even $e = mc^2$ is simply incoherent, it has neither status nor meaning. This issue is discussed further in the last section below.

[4] For the classic account of category mistakes see Gilbert Ryle, *The Concept of Mind* (London: Hutchison, 1949), Chapter 1.

turn the human activity of 'science' into a thing other than human activity, and it is of this non-human-activity thing that we attempt to write the history. The thing of which we write is science — taken not as a human practice but as something else, such as a series of discrete 'ideas', or as a 'body of knowledge'. Our usual historical theme is the career of science over time. Our particular topic may be science as a whole; or it can be a scientific discipline, such as physiology, physics or astronomy; or it can be something believed to be true about science, such as its 'objectivity' or its 'rationality'; or it can be one of the things which supposedly are the components of science such as its 'method' or 'the elements of physics'; or again it can be a thing which is the product of scientific activity, such as a 'law' about the behaviour of the natural world; or it can simply be a scientific 'idea' or 'concept'. More particularly, we choose to write accounts of the birth, development or progress of that something, or of its status at a certain period, or of its relations to other things, such as 'society' or 'the Enlightenment'. Thus the topic ('science' or one of its relatives) comes first in our choice of things in the past to study; the *people* who were engaged in the activity of practising that science come to our attention simply because we believe we have 'found' them engaged in practising the topic we chose to study in the past. This way of choosing topics to study, a choosing which takes place without reflection, seems natural to us because it matches the way of conceptualising that brought us into the subject in the first place. What could be more natural to study than such topics? After all we are historians of science first, and specialists in Newton, Pasteur or Einstein second. But such a way of perceiving and of delimiting topics to study comes from our position, from our present-centredness: it does not come from — and it bears no necessary relation to — the history, to what happened in the past. If we recognise that science is a human activity, wholly a human activity, and nothing but a human activity, yet we set out to study it as if it were something other than a human activity, then it is most unlikely that we will produce history of science which corresponds to what happened in the past! (And even if, by chance, we do produce accounts which correspond to what happened in the past, we will have no ways of recognising that they do.)

There is nothing in principle wrong with this turning of human activities into things other than human activities. The problem comes when we mistake the one for the other: when the reification we have made leads us to give the thing we are talking about the wrong identity, turning it into something quite independent of the human activity which actually constitutes it. Conventionally we have treated science as if it is some kind of material object, and we have given it the history over time appropriate to a material object. It is as if we have been writing a history of the earth: here is a material object (not made by man) whose identity has always been the same (it is always the earth), but whose form and fortunes have changed over time. It is man's role as historian of the earth to decipher what changes the earth has undergone over time: to wonder about the beginnings, to read the strata, to identify the erratics, to chart the shifts of the continents, in short to investigate how that object with a persistent identity came to be the way it is today. This is the kind of history that we usually give to science. But science is not a material object. It is a human activity. It therefore deserves the history appropriate to a human activity. That history will be centrally about people, about people engaged (or not) in that activity, about how and why they started that activity for themselves to engage in, about how they pursued, changed or abandoned that activity over time, about how their pursuit of that activity affected the way they pursued other activities.

This shows us, then, what ought to be our subject of study. For if the practice of science is in fact a human activity, and nothing but a human activity, then the historian of science should hardly be exploring how 'the Truth' became visible to men. For the products of scientific activity — the laws, truths — should be seen as just that: products of a human activity. Thus what we should be exploring and trying to reconstruct is human activity in investigating Nature — even including asking why man should ever have seen (or demarcated off) 'Nature' as a thing separate from himself and needing investigation. It is a matter of seeing the history of science as an active human process: of people doing things, not of things doing things. Does science 'relate to' society? Did religion and science 'conflict' with each other in the nineteenth century? Of course not. What happens is that *people* believe, argue, investigate, seek to achieve human goals: people do things. Science is one of the things that people do.

The customary focus of our attention as historians of science has not primarily been on people in the practice of this human activity 'science', but on one or other abstraction of a different kind — abstracted, that is, from the human activity which constitutes it. The position may perhaps be defended by arguing that within such history-writing produced by historians of science, it is often the case that the thinking or investigative activity of the people in the past will have been reconstructed satisfactorily to show them doing science. Surely (one might argue) this is good enough, surely this sort of enterprise takes sufficient account of the fact that the practice and practising of science is a human activity? Unfortunately not: for to write the history of science in this way is quite simply to beg the question. For if we engage in this form of history of science we never ask ourselves what is perhaps the most important question about the people we are dealing with: were these people doing science at all? If we are concerned with science as the activity of these people, surely we ought to be raising this question first of all.

So the problem arising from this mistake in the object of study is that we may well be ascribing to people in the past activities in which they were not engaged, in which they did not claim (and would not have claimed) to have been engaged, and in which they could not possibly have been engaged. The activity we ascribe to them will have been wrongly identified. To see the force of this, we need to turn to the intentionality of the practice of science.

4. Science as intentional human activity

We have established that the practice of science ('doing science' in my phrase) is a human activity, wholly a human activity and nothing but a human activity. But it is something more precise than this: it is a human activity of a very particular kind. It is an intentional human activity. What is it for an activity to be an intentional one?[5] If we were to go up to someone who was performing some action — let it be something as simple and non-specific as moving her arm — and we were to ask her why she was performing that action, then there are a range of possible kinds of reply. If our person looked up with surprise and said 'I had no idea I was doing that; thank you for telling me, I shall stop at once', then (assuming she was not lying) what she was doing can definitely be said to be not an intentional activity. If however she were to reply 'Because I am climbing a mountain/going for a walk/doing the shopping', then we can see that the particular *action* that we asked about is part of some *activity* which that person intended and intends to engage in. The answer we are given thus serves to explain both why the action was called into being, and also what it means: the separate actions get their meaning from the person's engagement in that particular activity. And as that person *intended* to engage in that activity, she must have a concept of it as an activity to engage in (climbing mountains, doing the shopping). For we cannot intend to engage in an activity of which we have no concept; for instance, we could not intend to 'go for a walk' if we had no idea what 'going for a walk' is. Such activities constitute *intentional* activities, and science is one of them.

It is however very difficult, perhaps impossible, actually to demonstrate that science is an intentional activity. So I will show what I mean by reference to an activity which we all understand and which we will all spontaneously agree is intentional, and show that science has the same form and structure as it. Then it should be clear what I mean when I say that science is intentional. The intentional activity which will serve as our model is that of a *game*. I will be claiming that the practice of science is just like a game. In the same breath I must add that in making this parallel I am not seeking to disparage science, its practitioners or its findings in any way: my claim is that science is 'just like' a game, not that it is 'just' a game (deadly serious though many games are).

---

[5] I am indebted here to Gillian E. M. Anscombe, *Intention* (Oxford: Basil Blackwell, 1957). Professor Anscombe is concerned with intentional *actions* and does not (as far as I can see) distinguish between them and intentional *activities* as I have chosen to do.

Games are intentional. They may indeed be taken as the paradigm of intentional activities in that the fact that one is playing a given game is so clearly what both *calls forth* and *gives coherence to* the actions that one makes in the playing of it. Let us take card games as our instance. If a certain card game is being played, then particular simple actions (such as putting a card on a table) are called into being and derive their meaning from the fact that the person who performs them is engaged in playing this particular game: this person would not be making these actions here, now and thus if he were not currently engaged in this particular game. (And if he did happen to perform those actions, outside the framework of the game, then those actions would have either no meaning or a different meaning.)

Most important to note with respect to the intentionality of games is that if someone is playing a particular game, then this game is what he means to play, it is what he knows he is playing: and he has to have a concept of it — both of what a game is, and of what this particular game is — for him to be able to engage in playing it. One cannot be playing a game by accident. Nor can one sensibly be credited with playing one game when one is in fact playing another game — or with *failing* to play one game when one is in fact playing a different game! Games are just not like that — because intentional activities are not like that. And it will be noticed that the question of skill or competence on the part of the players is irrelevant to whether games are intentional activities, or to the identity of the game being played. I can be playing a game so badly that my partner seriously contemplates murder, yet I am nevertheless still engaged on it as a game, and it is still the game it is.

So much for the intentionality of games from the participant's point of view. The innocent observer of a game is in a completely different position. When a player he is observing puts (say) an ace on the table, this action is meaningless to the observer unless he knows what game it is, and the state of play when the ace is played. But if he has the patience our observer can, of course, discover the nature of the game being played. Indeed, if he persists long enough in his observing he could become a competent player or judge of the game himself (even if he were never to learn the name of the game). All that is required is that he should recognise what kind of activity he is watching (i.e. that it is a game), and then watch, watch, watch. In working out the point and rules of this unfamiliar game he is observing, our watcher might well compare it to other games he is already familiar with. But, unless he is very careful, his knowledge of other games might be a serious handicap to him. Imagine, for instance, that he knows the game of whist: he sees a card game going on in which tricks are being made, which uses a standard pack of cards and is for four players. He watches for some time and, struck by how much of the play he recognises, he pronounces that it is whist. If the game was in fact bridge, then our observer would be wrong. There is only one correct answer to the question 'what game is it?', and our observer has not got it. It is no defence for him that it is possible for the same players, at different times, to play both whist and bridge, and with the same cards. For it is in the nature of games that it is not possible to play both at once! Thus his position as being detached from the game does not give our friend the observer any privileged viewpoint: indeed, until proved otherwise, he must be expected to be more ignorant of the game in question than the people he is watching.

Handily for us, games, as intentional activities, have histories too. Often they are created out of one another, the creators of a new game taking certain features of an old game and modifying them. This has happened for instance in the case of whist: out of whist was created auction bridge, and out of auction bridge was created contract bridge. But we can expect the origin of any particular game to be very intricate and messy, even the apparently straightforward examples being revealed as very complex in origin on closer inspection.[6] Games also have different careers and survival rates: whist survived the invention of auction bridge, but auction bridge did not survive the invention of contract bridge, auction bridge being a largely forgotten game now. Thus in some cases one game totally replaces another out of which it was created, and in other cases the two games may exist, and be played, alongside each other. Finally, and perhaps most important, we can see that in the history of any game, the game itself does not do anything: it is people who create it, it is people who play it, it is people who change or abandon it. The history of a game is the history of the people playing it, discussing it, watching it, modifying it — as a game, and as the game that it is.

Games, then, are intentional activities. If we wanted to draw up a simple check-list of some of the features typical of games, in order then to compare them to science, we might settle for the following:

(i) They are structured and disciplined ways for man to behave in, invented by man, and played only by man.

(ii) Every game has a point, it has rules, it has procedures and conventions. In playing a game we put all of these into practice at the same time.

(iii) A given game is something that (at a given moment) one either is playing or one is not; and if you are playing a particular game you know you are doing so — because that is part of what playing a game is.

(iv) The identity of any game is not affected by which particular individuals are playing it at any time, nor how well or badly.

(v) The experts on the identity and rules of a game, and the judges of competence at it, are those skilled at that game.

(vi) Any game must have been invented before it can be played.

---

[6]For a view of the complexity of games and their histories see the exhaustive *The Game of Tarot from Ferrara to Salt Lake City*, by Michael Dummett with the assistance of Sylvia Mann (London: Duckworth, 1980).

5. The game of science

Is science like a game in this sense? If we take the model seriously, we can see that if science is the same kind of human activity as a game, then according to the first set of criteria above, it would be a structured and disciplined way for humans to behave in, invented by man, and played only by man. I have, I hope, already established that the practice of science is a human activity and nothing but a human activity, and it is self-evident that it is a structured and disciplined way for man to behave in, engaged in only by man. The only initial criterion not yet satisfied is whether or not the practice of science was 'invented' by man, and this remains so far unproven; let us put this aside for the moment, and proceed to the other distinguishing features of games to see whether they have legitimate parallels in the practice of science.

Starting then from the next set of criteria (every game has a point, it has rules, it has procedures and conventions), let us think about the daily practice of science to see what parallels can readily be drawn. Every day thousands of people who would describe themselves as scientists go off to the laboratory to do science: presumably they know what sort of thing it is that they are trying to achieve, they will know when they have achieved it, they have been educated to try to achieve it in certain ways, and know that certain other ways are illegitimate. The point of the daily doing of science, we might say, for the regular practitioner, is to discover something further about 'how Nature is'. (We may take it that, no matter what he may think or claim when talking philosophy in his spare time, while the workaday scientist is carrying out his work measuring, for instance, wavelengths, he takes it that such things are out there in Nature to be measured; in his practice therefore we may treat him as being, at least pragmatically, what the philosophers of science would call a 'realist'.) If he is lucky or inspired or particularly skilled, the scientist might discover some new phenomenon, or might even get to frame a new law. These may not be events of the day-to-day round, but if they happen then the scientist certainly recognises them to be the sort of thing he is after in his practice: he knows that they are one of the points of the activity he is engaged in. What is more, the scientist does not make mistakes about what the material of his activity consists of: he does not, for instance, expect to frame a law about the behaviour of God, or even to find new ways of measuring God. The material of his activity is the phenomena of the world 'out there'; the aim of the enterprise consists in the better elucidation of these phenomena in certain terms (mathematical, for instance), but not others (not theological, for instance). The activity is competitive, between individuals or between teams, but there is a strong sense of community between those engaged in it. There are ways in which the activity must be conducted (it must be possible, for instance, for someone to follow the procedure you took and reach the same position that you reached); and there are ways in which it must not be conducted (you must not fudge your experimental results). If you make a good move, it is other members of the community of scientists who judge that it is a good one. There is a language of the activity opaque to outsiders: the technical jargon used among the participants. There are 'good' (successful) scientists, and there are 'poor' (unsuccessful) scientists.

It is also clear that at a given moment a person either is doing science, or he is not: it is something which, if he is doing it, he intended to do. Although such a person may be doing science poorly or half-heartedly, he is not 'half-following' scientific procedures: his engagement is with this particular activity, not 'half' with it, nor with 'half' of this activity and 'half' of another. And moreover, if you are doing science, then you know that you are doing it — you do not do science by accident; nor do you find that you thought you had been doing one activity, only later to discover that you had 'really' been doing science. Doing science is not like speaking prose!

I trust that I need not go on through the other criteria. The extensive parallels between games and the practice of science do indeed show that the practice of science may legitimately be taken to have the same type of nature, characteristics and structure as a game.[7] And that brings us back to the central point: that as science is just like a game in nature and structure, so it is like it in being an exclusively intentional human activity. No matter if we think that science is by far the most complicated and sophisticated activity yet engaged in by man, its complexity does not disqualify it from being an intentional activity; it is precisely what qualifies it as such. For science is not just a complex activity, it is equally a coherent one, and that coherence is what identifies it as intentional. To engage in science you must intend to do so; you must have a concept of it as something to engage in. What you engage in is an activity which is highly complex but which is also fully coherent.

But it must be admitted that the thought that science is like a game is somewhat disturbing: it feels like a comparison of the sublime with the ridiculous. Why should this be? The answer seems to lie with our perfectly understandable concern with the 'objective' status of the products of this activity: we tend to treat the products of science as if they were not directly

---

[7]This may also help account for why it is impossible to give an exhaustive specification of what science is or comprises. For though many features of a game can be specified (such as the 'rules'), it is in the nature of games that exhaustive specification is not possible: there are always tacit 'competences' governing proper application of the rules. For the moment I am concerned with what all the games have in common, and thus treat science as if it was just one game. As my argument proceeds it will be apparent that equivalents could be seen like this:

(1) science = a family of games (say, card games),
(2) a particular discipline within science = a particular card-game (such as whist),
(3) a particular scientific discipline as pursued at a particular time and place = a particular variety or variation of whist (e.g. solo).

produced by human activity at all — as if they were free from human contact, as if man's role consisted merely in learning to read them off from Nature and in the categories of Nature. Hence we may feel that the comparison of science with a game is almost sacrilegious. But we can in fact recognise the intentionality of the practice of science without prejudice to the 'objective' status we wish to accord to its products. To do this we simply have to recognise that the 'point' of the intentional activity science may be characterised as 'the production of findings-about-how-Nature-is'. The actual production or producing of these findings, however, will lie within the intentional activity (within the 'game'). What will also lie within the 'game' of science are the rules about what counts as such a product, and how (in what terms) it should be conceived and expressed. The judges of 'what counts' and of 'how it should be expressed' are, of course, skilled and competent participants; and this is just as we would expect with any game too.

6. What game were they playing?

If the practice of science really is an intentional activity (like a game) then some most important consequences follow, with respect both to the activity of people in the past and to our practice as historians of science today. In the first place, people in the past could only have intended to engage in science if they had the concept of science as an activity they could engage in. People can only intend to do what they know, what they have a concept of. If they did not have the concept of 'science' then they could not possibly have been undertaking science. Furthermore, if people in the past were engaged in science then they would have known that they were so engaged, for they would have intended to engage in it. And if they did not have the same concept of 'science' that we have, they could not possibly have been undertaking science in the sense that we normally mean it (as when we call ourselves 'historians of science'). This means that to describe them as having been engaged in science can only be a true description of their activity if it could have been *their own* description of their activity. With respect to our practice as historians of science, it thus follows that when we look at the intentional activity of people in the past, their description and their account of their own activity must be taken seriously by us as historians, and must take primacy over our own preconceptions about their activity.

This is the sort of tool we have been needing: a specification of the identity of science and of the conditions for science being practised in the past. And it comes down to this: if a given person in the past *did not have* or *could not have had* the concept of science as something to engage in, then he could not possibly have been doing science. And if he did have the *concept*, then it is necessary that he also had the *intention* to engage in science, before we can validly judge him to have been engaged in science. Unless a given person in the past satisfies both of these conditions, then he does not qualify as someone who was practising science or making scientific findings.

Now, what were the people we conventionally study in the history of science actually doing? We can find out what they were doing by taking a double approach: first, by asking *what was their description of their own activity*, if they gave one? And second, by then trying to *reconstruct that activity* with the extension, boundaries, aims, typical products — in short with the 'wholeness' — that that activity had for its practitioners. This second route is always available to us, whether or not the people in question mentioned the title of their activity. Until we have done this we can hardly claim to know what that activity comprised; and we must certainly not take the apparent short-cut of simply dividing up that activity into parts that we think we can recognise, for this is to impose our boundaries on their activity. When we have discovered what they called their activity, and when we have reconstructed what that activity was to them, we must take seriously what we have then discovered: for it was their engagement in that precise intentional activity which led them to produce the 'texts' or documents or books containing what we initially thought were 'scientific findings'.

The general enterprise that most people in the past would have usually described themselves as engaged in when they produced what we normally take to be their 'scientific' findings, was either *Philosophy* or *Natural Philosophy* (which for many centuries was treated as a large branch of Philosophy).[8] More particularly they might have described their enterprise as 'anatomy', 'chemistry', or whatever — *as a branch or sub-discipline of Philosophy or Natural Philosophy*.[9] This is what they called it. We must now take this

[8]See for instance the titles they gave those of their books where we look to find them reporting science: even in the seventeenth century (where we customarily identify a 'Scientific Revolution'), people consistently use the terms *philosophy* and *natural philosophy* as the appropriate terms to specify the identity of what they are doing and saying. For instance, Newton's major work is called *Philosophiae naturalis principia mathematica* (1687) — an assertion that the proper principles of Natural Philosophy are mathematical; Boyle's is called *Some Considerations Touching the Usefulness of Experimental Natural Philosophy* (1663). Similarly the full title of The Royal Society (in its second charter) is 'Regalis Societas Londini pro scientia naturali promovenda', always correctly translated as 'for improving natural knowledge', and its journal was called the *Philosophical Transactions*. Thousands of other instances could be cited.

[9]Some of the disciplines we now count under science would have been placed under Mathematics (for instance, optics, astronomy, mechanics). The relation that people in the past believed to hold between Mathematics and Natural Philosophy, and between Mathematics and Philosophy, would bear further investigation; it would seem that different views on the matter were held at different times and places. It is certainly the case that mathematics was sometimes placed under Natural Philosophy, and sometimes studied primarily in order to pursue Natural Philosophy — as at Cambridge University in the eighteenth and early-nineteenth century. Here the curriculum and examination consisted almost entirely of mathematics, but the original intention of this was to equip students to be able to read and understand Newton's *Principia*, a work of Natural Philosophy [see Harvey W. Becher, 'William Whewell and Cambridge Mathematics', *Historical Studies in the Physical Sciences* 11 (1980), 1 – 48].

description seriously, and regard it as a description of an intentional activity, as a 'game' in its own right. We do not as yet know what this activity in its wholeness consisted in because, as historians of *science*, we have never hitherto asked. But did these people in the past perhaps *also* describe this self-same activity of theirs as 'science'? The answer must be that until at least 1750, and possibly until as late as 1800, *no-one at all* described their activity like this. They did of course have a Latin word *scientia* (Englished as 'science'), which is the Latin equivalent of the Greek *episteme*. But it referred to speculative sub-disciplines of Philosophy or Natural Philosophy, and to knowledge of a certain status acquired by the practitioners of those disciplines. It did not refer to the same set or 'family' of disciplines and activities that our modern word 'science' does. Indeed, into the late eighteenth century and beyond, all sorts of studies that we would exclude from science were described quite validly as 'sciences'; such were logic, grammar, theology and ethics.

Thus we customarily take people who, by their own accounts, were engaged in intentional activities *other than* science, and treat them as having been engaged in science. We mistake one activity for another. As a consequence we also give the wrong identity to what these people said and did. For us to ascribe the activity 'science' to people who were not only not engaged in science but who were actively engaged in another activity altogether, is for us to hijack their actions and statements into our context, a modern-day context, and give them a *post factum* identity. We turn their acts and statements into new acts, new statements — acts or statements they never were. If, however, we are trying to give an account of 'what happened in *the past*', then we have to respect intentional activities for what they were; and the authorities on the identity of particular intentional activities in the past are the people who were engaged in them.

Whatever happened to Natural Philosophy? For no-one today plays this 'game'; I do not know of anyone this century who would have claimed to have been engaged in it. Natural Philosophy, as such, has fallen out of practice. (And the fact that no-one today pursues it is an additional reason for us to take care that we do not assume that we know what the 'game' was until we have investigated how it was played.) We have this potentially odd situation: that perhaps a couple of thousand people across Europe would have claimed in, say the 1740s, that they were engaged in the activity of 'Natural Philosophy', while none of them would have claimed they were engaged in the practice of 'science' in our sense. Then today, there are hundreds of thousands of people who would claim that they are engaged in the practice of 'science' in our sense, but none of them would claim they were engaged in the practice of 'Natural Philosophy'. It is beginning to look as though the practice of that intentional human activity Natural Philosophy got replaced or superseded; and that it was was replaced by the practice of the intentional human activity that we know as

Science. It is also beginning to look as though we can with reasonable assurance point to the period when this happened.

We must remember that if Science is an intentional activity, as I have been arguing, and that if Natural Philosophy was equally an intentional activity for its practitioners, then what we have here is *two* activities (two 'games'). Not taking seriously the description that people in the past applied to their own activities, historians have thought, written and acted as if Natural Philosophy was simply renamed as Science: as if the two terms are simply alternatives for an earlier and later practice of the same activity. Hence we have tended to slip happily back and forth between the two terms.

Thus Science appears to have been introduced as a new activity, which eventually displaced Natural Philosophy. But, even if it is accepted that they are two activities, it is still true that, even on a fairly detailed inspection, Natural Philosophy seems to have a great many similarities to Science. Our discussion of games helps us see why this can be the case, without obliging us to revert to our assumption that they were one activity under two names. For there is no reason why a new game should not be created out of an old one, as auction bridge was created out of whist. Equally there is no reason why a new game should not have great similarities to an old one, as contract bridge is very similar to auction bridge. Further, there is no reason why the people who played the earlier game (whist) should not also have played the later game (bridge) once it existed. But nevertheless the new game will be a *different* game from the old one, however great the similarities and however direct the lineage. That is how games are, as intentional activities. A new game will have differences in the point, in the play, in the rules, in the conventions, in the procedures. Thus in the 'games' of Natural Philosophy and Science, although both deal with the natural world, and both produce a 'product' (i.e. findings or statements about Nature), yet what *counts as* an appropriate product in the one may well differ from what counts as an appropriate product in the other. This will be the case no matter how similar the two activities may appear to someone acquainted with only one of them. We may again take the parallel of whist and bridge: in both games the object is to make tricks of a certain kind, but in whist it is to make as many as possible, while in bridge it is to contract for the number of tricks you can make and then make them. If one thinks that 'what counts' in bridge is the absolute number of tricks (as it is in whist), then one is misunderstanding the game, no matter how much it seems to resemble whist in the fact that it produces tricks, and no matter how similar those tricks look. And, as has been said before, the arbiters of what does and what does not 'count' in a given game are not the innocent outsiders, nor even those expert at a different game, but the practitioners of the game in question. It was the practitioners of Natural Philosophy who were the arbiters of 'what counted' as authentic Natural Philosophy, not the present-day historians of

Science — historians, that is, of a different intentional activity, a different 'game'.

7. The importance of being curious

It is perhaps now possible to see why it is insufficient for us historians of science to try to write either the history of 'science' (the 'family' of intentional activities), or the history of 'physiology', 'astronomy' or other members of that family, or of 'force' or 'nature' or 'life' or 'objectivity' or some other category of thing detached from any intentional activity, and why we must instead *set out to* write the history of human activity. For if we pursue the history of 'science', and only subsequently come to reconstruct the activity of the people concerned, we shall probably be reconstructing the *wrong* activity. We shall be foisting upon people of the past an intentional activity in which they were not engaged. And we shall have no means of checking whether or not they were engaged in the activity we thus ascribe to them, since we shall not have directly asked what their activity was. Thus as historians of science we are not faced with an open choice about which of these two ways to take when setting out to write our history, either of which will eventually lead to the same result: for if we choose to write the history of something *other than* human activity, we will close off from ourselves the possibility of discovering what activity in the past people were actually pursuing. The choice of initial question will shape and constrain the result the historian will arrive at. Of course, looking for 'science' in the past greatly simplifies our work as historians of science, but it systematically distorts our understanding of what was actually happening. Our traditional way of doing the history of science systematically conceals from us the possibility of seeing that the studies conducted by people in the past may well have been different studies than those we pursue, with different goals, approaches and conclusions. To quote someone's happy phrase (about something else), science 'is like a pair of glasses on our nose through which we see whatever we look at. It never occurs to us to take them off'. Taking those glasses off is a liberating experience which will for the first time allow us to comprehend and reconstruct the actual intentional intellectual activities of people in the past in their investigation of Nature. And once those glasses are off, we will no longer be writing a history of science over the centuries when it was not being practised, but we will instead be writing a history of humans in the practice of the activities of investigating Nature which they were actually pursuing.

To 'ask' of people in the past (in the way that I have been suggesting) what their own description and understanding of their own intentional activity was, and then to take seriously what we learn (i.e. to set out to reconstruct that activity in its wholeness), is our means of 'getting out of the present', of transcending our present-centredness as historians of science. It may turn out, unfortunately for us, that what people in the past were actually doing does not coincide with what we wanted to find them doing. But *our* intentions in wanting to write history of *science* nevertheless give us no warrant for distorting and misrepresenting the activities that people in the past were actually engaged in. To do our history historically it may therefore be necessary for us to give up our initial intentions (at least with respect to people who flourished before about 1770). However, if we do so we will find that what people in the past were actually doing in their investigations into Nature was at least as interesting, at least as significant as we take science to be: and what they were actually doing helped both to make and change the world of their day. Our present world is the outcome of the activities our ancestors were actually engaged in.

If we are going to comment on a game (as in writing its history) it is important that we have the game right. Eternal curiosity is the only way: starting constantly from a position of honest ignorance rather than supposed knowledge about what the game was. But of course if we want to be particularly obtuse we can carry on taking some dead person who was actually playing whist, and pronounce him to have been 'really' playing bridge, and then cast judgement on his lack of expertise as a bridge player. And we can, if we want, carry on taking an intentional activity of (for instance) Aristotle, an activity which Aristotle claimed to be, and knew to be 'philosophy', and pronounce that it was 'really' science, and then pass judgement on Aristotle as a good or bad practitioner of science. But if we do these things, then we are simply failing to do our own job properly. However, if our first enquiry is (as it ought to be) 'What were people X and Y *doing* in their investigation of Nature in the past? In what intentional activity were they engaged in investigating Nature?', we will find that for two millenia their activity was not that of practising Science, but of practising something else: usually something they called 'Philosophy' or something they called 'Natural Philosophy'. And these are not the same as Science. Men did not start doing Science until they stopped doing Natural Philosophy.[10] They did not start doing Science until they had invented it as an activity to do.

We do not yet know much about the enterprise of Natural Philosophy, and much work will be needed before we do. But perhaps the single greatest

---

[10] I am not implying by this that the new Science comprised (new versions of) all and only the disciplines which had constituted Natural Philosophy, for it is obvious that several disciplines which had been considered part of Mathematics were also now put under the umbrella of Science. But it would also be premature of us to assume that pre-Science mathematics and its disciplines had the same identity as the later Science versions.

difference between Natural Philosophy and Science is that Natural Philosophy was an enterprise which was about God; Science by contrast is an enterprise which (virtually by definition) is *not* about God. God, His existence and attributes are taken to be *irrelevant* to science and the practising scientist (he or she may believe what they like, however, outside science). Natural Philosophy, however, had Nature as its subject-matter, but not as its goal. For Nature was the book of God's works. Thus Natural Philosophy could be an exploration of God's creation and an admiration of His wisdom and foresight, or it could be an attempt to discover God's laws, or an attempt to penetrate the mind of God: it was about God's achievements, God's intentions, God's purposes, God's messages to man. Nature was explored, but not for itself. Certainly over the centuries a few people may have tried to minimise God's role as the efficient cause of why things are as they are; a few may even have repudiated God as the First Cause of things. But it was only when men stopped looking for God in Nature that they stopped doing Natural Philosophy. The God-less activity they started to do was Science.[11]

What then have we learnt? We have learnt firstly that science is an intentional human activity, which deserves a history appropriate to the kind of thing it is. Second, we have learnt that our usual approach as historians of science is to give science the kind of history appropriate to a material object (or to a mysterious thing as it 'becomes visible'). In doing this we have mistaken the identity of science — and hence looked for the wrong thing. Our capacity to find it, whether it was there or not, is what has been distracting us from noticing our error. Third, we have learnt that science, as an intentional human activity, could not have been practised before it was invented. Hence investigations of Nature undertaken before science was invented must have been as part of *some other* intentional activity, one yet to be investigated in its wholeness by us. Here is a new field of study for us. And another new field is opened up by our discovery of the 'invention' of science: where, when and how was it invented as a practice, by whom and why? What a tale that would be.

These points have such large implications for our practice as historians of science that I want to stress that my whole argument here is built on something other (and stronger) than on the usage of particular *words*: my case does not stand or fall on whether particular past people used the term 'science' (or *scientia* or an equivalent). My argument is about certain human *practices* and

---

[11] Certainly a distinction was long recognised between Theology (rather than religion) and Natural Philosophy; but nevertheless the first dealt with *God's* book, the second with *God's* works. Science, by contrast, does not deal with God's anything. The many works by historians on the supposed age-old conflict between science and religion should not blind us to the fact that such a conflict could not have predated the existence of science as a practice.

their intentionality. And in exploring these we can, on occasion, take the word-use of past people and use it as a clue in order to make proper identifications of their intentional practices. Of course many past people used the term *scientia*, and used it often, sometimes in its English form *science*. But what they meant by it was different from what we mean by science: both the getting of *scientia* and its deployment once got, were part of an intentional activity different from our science.

8. 'Ideas' and the whiggish historian

Our historiographic concern with the identity of the actual activities of people in the past in investigating Nature has brought to light an historical event of the period c1780 – c1850, which I am calling 'the invention of science'.[12] But is it an artefact or a fact: is it merely a mirage given visibility by this historian's procedure, or did it actually happen? The 'evidence' of the past will not decide for us, for there is no such thing without the intrusion of an historian. Whether or not science was 'invented' is as incapable of proof or disproof as is our customary view of science as either having been practised somewhat intermittently for centuries but in a cumulative and progressive way, or as something (or things) which became increasingly and progressively visible over time. All one can present — for this as for any other claim about what happened in the past — is a historical story presented to be as convincing as possible (like a legal case or argument), and in the course of which one brings forward or produces things to be 'evidence' for one's story and make it more plausible. Presumably an historical story is the better for explaining or accounting for more. Now, the finding that science was 'invented', and in this particular period, cannot be explained within our conventional portrayal of the past of science: indeed, it contradicts it. But on the other hand our normal way of seeing the past of science can itself be accounted for historically if we can accept that science was invented as an intentional activity in a particular

---

[12] In the light of the writings of Michel Foucault and others which deal with or touch on the origin of disciplines, it may seem no great novelty to claim that the enterprise of science was invented in the decades around 1800. Foucault, for instance, writes of biology that in the eighteenth century it 'did not exist then, and that the pattern of knowledge that has been familiar to us for a hundred and fifty years is not valid for a previous period' [*The Order of Things: an Archeology of the Human Sciences* (London: Tavistock Publications, 1974; 1982 edn), pp. 127 – 128; *Les mots et les choses* (Paris: Gallimard, 1966), p. 139]. But neither Foucault nor anyone else (as far as I can see) would want to claim that science, or biology as a particular science of science, was actively — and possibly also deliberately — brought into existence as a new practice, as I do here. The intended force of my use of the term *invention* is to stress the fact that science as a practice was a creation of men: I do not regard expressions such as 'came into existence', 'emerged' or 'was born' as referring to the kind of event I have in mind.

period and location: it can be explained as a side-effect of the invention of science itself.

Let us now indulge in a little bit of historical speculation, and for this purpose let us here take it as fact that science was indeed invented in this period, and that a particular claim — that it produced 'objective' knowledge, knowledge independent of men — was made about it. Let us also make the quite reasonable supposition that the invention of science on the one hand, and the massive political, social, intellectual and economic changes of the period on the other, would have been causally inter-related. It should now be possible to see why people might first have introduced the 'whiggish' tradition of conceiving and writing the history of science, and also why the invention of science might have become historically 'invisible'.[13]

Writing the history of science was an early nineteenth century innovation (as it could only be, if the practice itself had only just been invented). The inventors of science and their immediate successors unselfconsciously rewrote the past in a way which showed themselves to be the heirs to a grand tradition. But one might say that they *had to* give science this supposititious lineage in order for their invention of this novel discipline, and the claim they were making about it, to go unchallenged. The most general form that this history-of-science writing took was to write histories which were in practice actually making novel *assertions* about where the 'natural' subject-boundaries of knowledge now lay: for instance, histories of the 'inductive' sciences, of the 'exact' sciences, or histories of 'biology', of 'geology', or of 'physics';[14] sometimes too people wrote histories about the development over time of one or other aspect of the 'method' of science. But, in the very process of writing such histories of science, our nineteenth century historians gave science itself a new identity. That is to say, they presented what was a *human practice* as if it was a series of *autonomous concepts* — 'ideas' each with a career of their own, which are taken to 'become clear', be 'contributed to', or whose elements or ingredients 'come together' over time. They separated the thought — the 'idea' — from the thinker. And while the historians took it as their duty to trace the 'idea', the concept, they treated the thinkers of these 'ideas' or concepts as if they were simply the temporary guardians or nurses of the 'ideas': the best guardians were portrayed as Geniuses, the worst as 'blind'.

By their own lights, these historians were giving a causal account of why science developed. The story is not of people X and Y struggling with all the difficulties of living and thinking in order to achieve particular human goal A or B. Instead the story is of elements of the Truth (discrete 'ideas') dawning

---

[13] On revolutions being made historically 'invisible', see Thomas S. Kuhn, *The Structure of Scientific Revolutions* (Chicago: University of Chicago Press, 1962), Chapter XI.

[14] Valuable here is the chapter "The invention of physics" in Susan Faye Cannon's *Science in Culture* (New York: Science History Publications, 1978).

on, being revealed to, becoming visible to the Human Mind over time; segments of the Truth become available at particular times and places either because the Time Is Ripe (the Spirit of the Age is moving), and/or because of the presence of some specially gifted individual, a Genius (and preferably triumphing over enormous odds). The validity of this causal story depends on the implicit assumption that the Truth which science is about is *eternal*; given this, then it is consistent to see that Truth as something which was there all the time, gradually coming into view. And hence it follows that there is Progress in science: the making visible over time of more and more of the Truth. This in crude outline is 'whiggish' history as these nineteenth century historians started it (and as we still know it and sometimes still practise it today), both in its topics and its mode of explanation. More Truth is visible today, *ergo* there has been progress; thus, science is progressive.[15]

When these nineteenth century historians of science took 'ideas' as the subject-matter of their history-of-science writing, they stripped the 'ideas' of their human genesis and of their roots in the world of human experience, beliefs, interests and thinking. Why might it have felt 'natural' to these early historians of science to perform this transformation of a *human practice* into discrete autonomous '*idea*' units with a truth-content 'independent of' men? Why should certain people have wanted, or felt impelled, to present one kind of knowledge as 'objective', as 'independent of' men? Why did they not see (or not want to see) that it is *men who make the ideas*? Here I draw attention to the circumstances and locale of the invention of science — in deist and post-deist France and in protestant Germany, and perhaps Scotland and England — and to the political, intellectual and industrial revolutions in the course of which the invention happened, and to the economic revolution which underlay everything else. Science was the prime knowledge-claim of the people who were now in power. Naturally, it would have matched and served their interests.

Putting two and two together: science, the supreme form of knowledge because it is supposedly independent of men, was invented at the very same time and places in which the bourgeoisie triumphed politically and where industrial capitalism first became the dominant mode of economic production: capitalism which (in Marx's analysis) separates the *product* of man's labour from the *human process* of its production and treats it as an autonomous thing with a status and value independent of the conditions of its production. Science is its corresponding form of knowledge: and it came to be presented as

---

[15] I take this to be the underlying assumption of Prof. A. R. Hall's argument in 'On Whiggism', *History of Science* **xxi** (1983), 45 – 59, which allows him to take it as given that the history of science *shows* that there has been progress.

I

the supreme kind of knowledge.[16] And the history of discrete 'ideas', separated from their human producers, is its appropriate kind of history.

It is no coincidence that this usage (our present-day one) of the term 'idea' was itself an innovation of this same period. Hitherto, an 'idea' had been taken to be *an image in the mind* cast there by objects *outside* it. In the new usage, an idea became something quite different — something *originated in* the mind, and with a possible subsequent career (or 'realisation') *outside* it and independent of it. This new usage of the term 'idea', and the concomitant separation of thought from thinker, was characteristic also in that parallel — that mirror-image — invention, Art, and hence also in histories of art or artists.[17] There is equal novelty in the use of the term 'genius': hitherto 'genius' was something you did (or did not) *have*; henceforth it was something you could *be*. No wonder we eventually reached the position where we historians of science could treat it as one of our central duties to ensure that a given 'idea' is correctly attributed to the 'Genius' who first formulated or discovered it and made it available to his fellow men. It is as if he was an intellectual *entrepreneur*, engaged in a risky enterprise against great odds; we are in his *debt*, and hence his 'originality' (another term now given a new meaning) deserves the proper *credit*. In this sense we treat 'ideas' like items of private *property*: as if they are objects which are rightfully owned by those who are or were 'original' enough to 'discover' and exploit them. And in doing this we are simply following in the footsteps (and interests) of the people who were involved in the invention of science itself. As Goethe was heard to remark in 1823:

> Questions of science are very frequently career questions. A single discovery may make a man famous and lay the foundation of his fortunes as a citizen...Every

---

[16] The *laboratory* was the special place put aside as the *factory* for the production of scientific knowledge. For fascinating accounts of what goes on in the laboratory see Karin D. Knorr-Cetina, *The Manufacture of Knowledge: An Essay in the Constructivist and Contextual Nature of Science* (Oxford: Pergamon Press, 1981), and Bruno Latour and Steve Woolgar, *Laboratory Life: The Construction of Scientific Facts* (1979; 2nd edn, Princeton University Press, 1986). For a politically Marxist account of the historical relation of the production of scientific knowledge to the economic structure of the society which produces it, see Alfred Sohn-Rethel, *Intellectual Labour and Manual Labour: A Critique of Epistemology* (London: Macmillan, 1978). I differ from him on several matters, especially in my claim about the time in the past at which science *per se* was invented as a practice; but his is nevertheless far and away the most serious and informed exploration of these issues known to me.

[17] On the invention of the category, values and enterprise of Art as we know it, the starting-point must be Raymond Williams, *Culture and Society, 1780–1950* (London: Chatto & Windus, 1958), especially the Introduction and first two chapters. See also M. H. Abrams, *The Mirror and the Lamp: Romantic Theory and the Critical Tradition* (Oxford: Oxford University Press, 1953), and P. O. Kristeller, 'The Modern System of the Arts: A Study in the History of Aesthetics', *Journal of the History of Ideas* 12 (1951), 496–527 and 13 (1952), 17–46. Most books on Romanticism comment on how radical a change was brought about in this period in people's understandings of what Art was 'about', what it was 'for', what 'counted as' Art, and what qualities the proper Artist was supposed to have.

newly observed phenomenon is a discovery, every discovery is property. Touch a man's property and his passions are immediately aroused.[18]

Such considerations can give us some inkling as to why the invention of science may have been, quite unconsciously, rendered historically invisible by its inventors: to have made it visible (to themselves or others) would have made it questionable. And in our chosen role as historians of science we put ourselves in a similar position today: as long as we write the history of science as the history of discrete 'ideas' we not only continue to misrepresent the identity of the subject whose history we claim to be studying, but we are also perpetuating the illusions and values that were built into the invention of science itself.

Acknowledgements — Dr Adrian Wilson gave me the encouragement to develop these thoughts and Dr Yoko Mitsui taught me a vocabulary in which to express them: my deepest thanks to them both. I am also grateful to Drs Nick Jardine, Perry Williams, Simon Schaffer and Timothy Ashplant for their valuable comments.

[18] This translation is as given in E. J. Hobsbawm, *The Age of Revolution 1789–1848* (London: Wiedenfeld & Nicholson, 1962), motto for Chapter 15. The original German comes from the third (1848) volume of Johann Peter Eckermann's 'Conversations with Goethe'; see Fritz Bergemann (ed.) *Gespräche mit Goethe in den letzten Jahren seines Lebens* (Wiesbaden: Insel-Verlag 1955), pp. 483–484, 30 December 1823: "Wir sprachen sodann über Naturwissenschaften...'Die Fragen der Wissenschaft', versetzte Goethe, 'sind sehr häufig Fragen der Existenz. Eine einzige Entdeckung kann einen Mann berühmt machen und sein bürgerliches Glück begründen...Jedes wahrgenommene neue Phänomen ist eine Entdeckung, jede Entdeckung ein Eigentum. Taste aber nur einer das Eigentum an, und der Mensch mit seinen Leidenshaften wird sogleich da sein.'"

II

HOW THE *PRINCIPIA* GOT ITS NAME; OR, TAKING NATURAL PHILOSOPHY SERIOUSLY

I

The first part of my title — How the *Principia* got its name — will, I hope, immediately recall to mind the "Just So" stories of Rudyard Kipling: "How the camel got his hump", "How the leopard got his spots", and most important of all, the question of How the elephant got its trunk. In these stories Kipling took what is the most typical and distinctive thing about certain animals, the feature which more than any other gives them their *identity* in our eyes, and imagines a state of affairs when that animal did not have that feature, and then tells a fantastic story about how it acquired it. That is to say, he makes the basic distinguishing, definitional, identity-giving characteristic first look odd, and then asks about its origin.

I want to do something like this with respect to Newton's most famous book. I want to draw my reader's attention to something we all know well — the title of Newton's great work, that title which announces its *identity* — then suggest that we might try and look at this as odd or unexpected, and then ask about why it merited that title. While I hope that my account will not be fantastic in the fictional sense, yet it has important implications for our whole subject of 'history of science'.

II

"How the *Principia* got its name": what is that name? It is *Philosophiae naturalis principia mathematica*, The mathematical principles of natural philosophy (see Figure 1). As far as anyone knows, this title was of Newton's own choosing, and we may therefore assume that it was the title he felt was appropriate for the work, and which he felt would convey the nature of the work to his intended audience.

The only time that Newton mentioned the title, as any kind of issue or problem, in anything he wrote was in a letter to Edmond Halley of June 1686. Halley was seeing the text through the press at his own expense, and Robert Hooke was threatening to claim priority over the inverse-square relation. Halley possessed the first part of the text and was urging Newton on, "for the

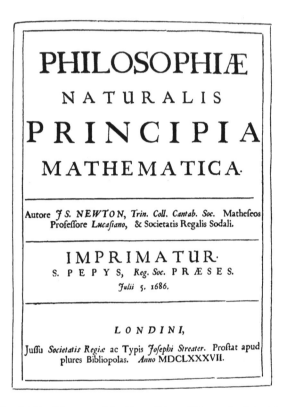

FIG. 1. The title page of the first edition of Newton's *Principia*.

application of this Mathematical part, to the System of the world; is what will render it acceptable to all Naturalists, as well as Mathematiciens; and much advance the sale of ye book".[1] This is how Newton responded:

> The Proof you sent me I like very well. I designed ye whole to consist of three books.... The third I now designe to suppress. Philosophy is such an impertinently litigious Lady that a man had as good be engaged in Law suits as have to do with her. I found it so formerly & now I no sooner come near her again but she gives me warning. The two first books without the third will not so well-beare ye title of *Philosophiae naturalis Principia Mathematica* & therefore I had altered it to this *De motu corporum libri duo:* but upon second thoughts I retain ye former title. Twill help ye sale of ye book wch I ought not to diminish now tis yours.[2]

In the event, of course, Newton did include the third book. It is interesting to note the importance of market considerations in both Halley's and Newton's thinking: a volume called *The mathematical principles of natural philosophy* and which gave a mathematical treatment to the System of the World would, they believed, sell better than one called *Two books on the motion of bodies*. Thus Newton believed the title would speak to its intended audience. And none of Newton's contemporaries seemed to think that the book had an inappropriate name: at least, I know of no-one complaining that the book was about something other than its title announced, or of anyone criticizing it for being about this subject. As the work received a lot of criticism, this fact is noteworthy in itself. So our starting-point is that this is Newton's own title, and that the terms it used were, in themselves, non-controversial to his intended audience.

Of the three elements in the title — Principles, Mathematical, Natural Philosophy — I want to concentrate on the last: *natural philosophy*. I want to ask (in familiar language) "What kind of beast was this?"

Some modern scholars have dealt with the *extension* of natural philosophy in the seventeenth and other centuries. For instance Alan Gabbey has recently written (and this is uncontroversial) that natural philosophy was "the science of natural bodies, of their powers, natures, operations and interactions",[3] and that this was the case before, during and after Newton's time. Thus, Gabbey maintains, only *part* of the range of subjects that conventionally constituted natural philosophy was actually represented in the topics dealt with by Newton in the published versions of his *Principia*.[4] But Newton worked on other areas of natural philosophy; for instance he did anatomising and experiments on the senses, and these were conventionally seen as falling under natural philosophy (as any historian of anatomy knows).[5] Such other areas of natural philosophy Newton covered in his *Opticks*, especially in the Queries to that work.

Newton's concept of natural philosophy corresponded with the contents of Magirus's old textbook which Newton read and took notes on as a student, the *Physiologia peripatetica* of 1597. This, as Gabbey and also Wallace have recently shown,[6] followed the range of topics of those of Aristotle's books which had long before been adopted as the texts of natural philosophy: the *Physica*, the *De caelo et mundo*; the *De generatione et corruptione*; the *Meteorologia*; the *De mineralibus*, the *De plantis* and the *De animalibus*; and finally the *De anima* and the *Parva naturalia* — which last work includes a book on dreams. This was Newton's (and, it seems, everyone else's) understanding of the range of natural philosophy, and is thus unremarkable. We also have to note that, as Betty Jo Teeter Dobbs has shown (and Gabbey follows her here), for Newton alchemy was also amongst the range of disciplines which were part of natural philosophy, though on this point not all of his contemporaries may have agreed with him.[7]

From a modern point of view, however, there is indeed something remarkable about a discipline (natural philosophy) which included both physics *and* the soul — and everything in between. Such an extension is not shared by any modern division of knowledge.

The transformation that Newton wrought in natural philosophy was not to change its extension, but to insist that it had to be treated — and could only properly be understood — via mathematical principles, and this is one of the messages that his title was meant to convey. This fact is now very familiar. Newton can be quoted on this from several places; I choose a (draft) letter of 1712 to Roger Cotes who was editing the second edition of the *Principia*, and had asked Newton for "something more particularly concerning the manner of Philosophizing made use of & wherein it differs from that of De Cartes & others. I mean in first demonstrating the Principles it imploys." Newton's draft reply ran (in part):

> I like your designe of adding something more particularly concerning the manner of Philosophizing made use of in the Principia & wherein it differs from the method of others, vizt by deducing things mathematically from principles derived from Phaenomena by Induction. These Principles are the 3 laws of motion. And these Laws in being deduced from Phaenomena by Induction & backt with reason & the three general Rules of philoso-phizing[,] are distinguished from Hypotheses & considered as Axioms. Upon these are founded all the Propositions in the first & second Book. And these Propositions are in the third Book applied to the motions of ye heavenly Bodies.... Experimental Philosophy reduces Phaenomena to general Rules & looks upon the Rules to be general when they hold generally in Phaenomena.... Hypothetical Philosophy consists in imagin-ary explications of things & imaginary arguments for or against such explications, or against the arguments of Experimental Philosophers founded upon Induction. The first sort of Philosophy is followed by me, the latter too much by Cartes, Leibnitz & some others....[8]

This makes it clear what Newton meant by introducing 'mathematical principles' for natural philosophy.

So much for the moment for the *extension* of natural philosophy, as understood by its practitioners. Central to it was physics, but physics was not the whole of it.

But what about what we might, by contrast, call the *intension* of natural philosophy? What was its identity? Why, that is, was it *this* set of topics — extending from physics to the soul — which constituted 'natural philosophy', and what was it that they were taken to have in common, such that they constituted 'natural philosophy'? To put it bluntly: what was 'natural philoso-phy' about: what was the point of it, both the point it was meant to achieve, and the point of engaging in it? What was its essence? Did natural philosophy differ from modern science in its point? And if so, in what particular way or ways? What Newton referred to (in the quotation above) as his 'experimental philosophy' falls under this question, since this expression was shorthand he and others sometimes used for 'experimental natural philosophy', just as 'philosophy' in his usage was often shorthand for 'natural philosophy'.

This is where we might care to recall the "Just So" stories — and try and make natural philosophy look *un*familiar, rather than assuming its familiarity. For, as McGuire and Rattansi reminded us as long ago as 1966, Newton "was not a 'scientist' but a Philosopher of Nature".[9] And the same was true for all other practitioners of natural philosophy: they were philosophers of nature, not 'scientists'. Given this difference in self-description, the identity of the natural philosophy that natural philosophers practised (in all its different varieties) is likely also to have differed from modern science.[10]

To my knowledge only one modern scholar, John Schuster, has dealt with this, and almost as a throwaway line in just one paragraph of the recent *Companion to the history of modern science*. Talking about the seventeenth century he writes — correctly, to my mind:

> every system of natural philosophy, whether of a generally Aristotelian, mechanistic or Neo-Platonic magical/alchemical type, purported to describe and explain the entire universe and *the relation of that universe to God*, however conceived. The enterprise also involved, explicitly, a concern with the place of human beings and society in that universe [my italics].[11]

This is what I want to bring to general attention: that over and above any other defining feature which marks natural philosophy off from modern science — for instance, that it was *philosophy*, and therefore primarily for the purposes of contemplation rather than action — natural philosophy was *about God* and *about God's universe*. Indeed, this was the central pillar of its identity as a discipline, both with respect to its subject-matter and to its goals, its purposes, and the functions it served. This is what, more than anything else, distinguishes it from our modern science.

Now, I and John Schuster are hardly the first to suggest that Newton and other natural philosophers were obsessed with God and His relation to His world, nor to point out that their natural philosophy was obsessed with God. It was, for instance, a central theme of Alexandre Koyré's interpretation in *From the closed world to the infinite universe* of 1957, which has had such an influence on our discipline; everyone Koyré deals with was, in his account of them, totally obsessed with God and with the universe as God's creation. Since then it has become basic to historians' approaches to studying investigation into nature in the early modern period — as witness for instance Amos Funkenstein's work on *Theology and the scientific imagination from the Middle*

Ages to the seventeenth century, or the recent volume edited by Professors Lindberg and Numbers called *God and nature,* or a recent issue of the journal *Science in context,* every article of which deals with this relationship.[12]

Newton has not been left out of this reassessment, and God has largely been put back into our understanding of Newton and his work. The old view was the one first hinted at as early as 1821 by the French physicist Jean Baptiste Biot, who portrayed a Newton who first did his astonishing work on nature, *then* went mad, and *then* became interested in theology. So Newton's work on nature, with which Biot sympathized, and Newton's obsession with God, with which Biot did not sympathize, thus turned out to have been separated in a way that Biot felt appropriate: by Newton suffering a bout of madness![13] But much work has been done by more sympathetic historians in recent decades to bridge this artificial divide in Newton's interests, and to look anew at the relationship between Newton's theology and his natural philosophy, most notably perhaps by Professor McGuire, and we now have a subtle and nuanced picture of their relationship in Newton's work and interests.[14]

Nevertheless almost all our discussions of the importance of religion or theology to natural philosophers — including even Newton — still characterize them as a lot of religious men in a religious age doing *science* as well as they could under the circumstances! This is particularly evident when we get ourselves involved (as we sometimes still do) in fruitless questions about how far their 'science' (or do we mean their natural philosophy?) 'influenced' for good or ill their religious views, or vice versa; these discussions are pointless on our part because they depend for their validity on religion and natural philosophy having been seen as being *as separate* from each other in the past as religion and science are seen as being today. The point that I am trying to make here, by contrast, is that the project(s) of exploring nature pursued by natural philosophers — their natural philosophies, that is — were themselves about God and His creation, because that is what the point of natural philosophy as a discipline and subject was. Hence each and every variety of natural philosophy that was put forward was an argument for particular and specific views of God. In other words, what we should be looking at (in my view) is not the 'influence' of their religion on their natural philosophies (or vice versa), but on how their particular religious outlooks and goals *were constitutive* of their particular natural philosophies.

When I say that natural philosophy was 'about God' and about His creation, I mean that this was so fundamentally the case that it hardly ever got mentioned. This may sound like something of a blank cheque for me to make any claim I like without having to produce any evidence, but I mean it in the same sense that there is a common understanding that the exact opposite is true about modern science. That modern science does *not* deal with God or with the universe as God's creation, is so basic to our understanding of modern science that it only ever needs to be mentioned if someone transgresses this understanding, as for instance, a religious fundamentalist might; but scientists among themselves never need to discuss it, nor does it appear in science textbooks, nor in philosophical discussions today about 'what is science?' Yet I think we would all agree about it. This assumption is one of the most basic things that the members of the modern scientific community hold in common.

The same was the case with natural philosophy. Natural philosophers often spoke and wrote explicitly and at length about God in the course of pursuing their natural philosophy. But no-one needed to mention that the discipline was *by its very nature God-oriented*, unless they were afraid that natural philosophy was about to be taken over by the irreligious. Indeed it seems that the primary (possibly the only) circumstances under which anyone attempted to revise the prevailing form of natural philosophy or to produce a new version of it, was *in order to fight perceived atheism*. It is, I think, a fair conclusion to draw from Father Buckley's recent book *At the origins of modern atheism* that this was the case with Descartes and with Newton — and also with Berkeley and with Leibniz and others.[15]

In making this claim about the point of natural philosophy for its practitioners — that it was centrally about God and God's universe — it is of course essential for me to be able to show that it was the case with the most famous natural philosopher of all, Newton, who is also the natural philosopher whose natural philosophical work is most readily and insensibly turned into modern science by us. Fortunately for me, Newton actually makes some of the rare explicit public statements to the effect that the point of natural philosophy was about God and His universe.

As this material from Newton is now very well-known I will simply mention just a few points. In the first place, what initially motivated Newton to create his new natural philosophical system was the atheism he perceived in the natural philosophy of Descartes; as Newton's friend John Craig wrote in 1727, "the reason of his [Newton's] shewing the errors of Cartes' philosophy was, because he thought it was made on purpose to be the foundation of infidelity...".[16] And as Newton himself wrote to Richard Bentley in 1692, "When I wrote my treatise about our Systeme, I had an eye upon such Principles as might work wth considering men for the beleife of a Deity & nothing can rejoyce me more then to find it usefull for that purpose".[17] Then of course there is the General Scholium in the second edition of the *Principia*, which is explicitly all about God, His powers, His characteristics, and the way He runs His universe. It also contains that now most celebrated sentence, "And thus much concerning God; to discourse of whom from the appearances of things, does certainly belong to Natural Philosophy". That is how that sentence is rendered in Motte's translation of 1729, but in fact Newton actually wrote it in Latin: *Et haec de Deo: de quo utique ex phaenomenis disserere, ad Philosophiam experimentalem pertinet.* We would be within our rights to translate this passage more strongly than Motte does, perhaps even as "it is the *role* of experimental (natural) philosophy to discourse of God from the phenomena".[18]

Given what I said earlier, we should expect to find that Newton was driven to make public these exceptionally explicit remarks about the role and point of natural philosophy — to say, that is, the sort of thing about natural philosophy which was so taken-for-granted that it did not usually need to be said — as part of a *defensive* strategy. And this indeed was the case. For Newton's natural philosophy, as presented in the first edition of the *Principia*, although itself conceived to fight the atheism Newton detected in Descartes's natural philosophy, had in its turn been condemned as atheistical (materialist) by Leibniz.[19] And what is Newton's way of responding to this charge in the General Scholium of the second edition of the *Principia*? He responds by specifying *the nature of God* — and thus showing that the universe as he, Newton, had described it was the appropriate universe for this — the true — God to have created and to be running.

Newton's own particular brand of Christianity was Unitarian or Arian and this entailed a particular view of the powerfulness of God: that His power was unlimited. Newton felt it would have been impolitic for him to have admitted publicly that he was Unitarian.[20] But if we hear in mind that his particular Christian commitment came *first in time* for Newton, then his Arianism could help us to understand more precisely why he would have seen Descartes's natural philosophy as atheistical, and hence why he thought the world needed him to work out a better because truer account of natural philosophy, and why his own natural philosophy then had the characteristics that it did, for it of course had to describe a universe appropriate for a Unitarian's God to have made and to be running.[21] This Unitarian/Arian commitment and view of God on Newton's part might therefore explain, for instance, why he gave such a large role to *force*, operating across empty space without contact, and why his account of the universe presented it as being almost totally empty of matter. Again, recognizing that Newton's commitment to his Unitarian religion came prior to his undertaking natural philosophy, might help us understand why Newton, in particular, would want to introduce an *experimental* and mathematical form of natural philosophy, and one depending so strongly on personal experience of the phenomena.[22] An indication of why he might have wanted to do so is given in a (Latin) draft of the final query to the 1706 *Opticks*, where Newton wrote:

Metaphysical arguments are intricate and understood by few. The argument which all men are capable of understanding and by which the belief of a Deity has hitherto subsisted in the world is taken from Phaenomena. We see the effects of a Deity in the creation and thence gather the Cause[,] and therefore the proof of a Deity and what are his Properties belong to experimental philosophy.[23]

What I am trying to suggest here is that religion was not 'opposed to' nor 'influential on' natural philosophy as natural philosophy was practised and as it was repeatedly re-created in slightly differing forms over the centuries. Rather, particular religious attitudes went into *constituting* each and every variety of natural philosophy, because natural philosophy was itself about God and His universe. The religious commitments of certain people led them to engage actively in natural philosophy in the first place (beyond what they were taught in the universities), and the *particular* forms of their religious commitments led them to pursue it in particular ways.

Now, usually, and this brings us back to the "Just So" stories, natural philosophy does not strike most of us as alien today when we study Newton or other people who would have called themselves 'natural philosophers'. In our desire not to be whiggish, we often take care to *call* Newton's enterprise 'natural philosophy' because that is what Newton called it, even though the expression is much more awkward to type and much longer than the word 'science'. But as soon as we have done that, we take one of two options about its identity, both of which come to the same thing, and both of which ignore the actual identity of natural philosophy. In the first option we simply treat natural philosophy as *being* an early version of modern science. In this option, we spontaneously call Newton and other natural philosophers 'scientists'. Thus, according to us, Descartes was (supposedly) a 'scientist' and Newton was a 'scientist'; even Aristotle was (supposedly) a 'scientist'. In our more contorted sentences, we can sometimes then end up with 'scientists' engaged in 'natural philosophy' and *consequently* making (modern) 'scientific discoveries'! (I could quote thousands of instances of this, but it is so common that it would be invidious to pick out any one. We all do it.)

The other option we sometimes take is to speak as if there were *two parallel* enterprises for the investigation of nature in, say, the seventeenth century: natural philosophy *and* science. We might mean by this that someone's 'natural philosophy' was his general cosmological or ideological outlook, whereas his 'science' was the facts about nature that he found within such a natural philosophy; this seems to be the attitude of John Schuster, for instance.[24] But the difference between them never otherwise seems to be spelled out by us (at least I have never come across an instance of anyone else doing so),[25] and in practice we again actually treat natural philosophy as an awkward early name for science but one we are obliged to employ sometimes in order to respect actors' categories. The point is that the issue about the identity of natural philosophy as an enterprise — and therefore the way or ways in which its identity was *different* from that of modern science — simply disappears under the language we use; and of course we use that language because we usually take the two enterprises to be the same, because we assume they have the same identity. But what we need to notice here, is that this translation of identity, this translation of one thing into another — the translation of natural philosophy into (modern) science — is done by us, the historians. The confusion of the two identities has been made by us, not by the practitioners of natural philosophy.

If we look at natural philosophy at the hands of Newton or any other individual in the seventeenth century, we are looking at just one moment in the long history of natural philosophy. That natural philosophy was about God and His universe had been established long before Newton's day. The origins of natural philosophy lie in the thirteenth century, and it would appear that it was created as a discipline in the first place *in order to be* a weapon in a fight about power and religious orthodoxy.[26] This was why natural philosophy was initially given as its subject-matter everything dealt with in a particular set of books by Aristotle, which stretched from physics to the soul. As a consequence natural philosophy, through its whole career, would always be about God, and about His creation, and always about the Christian God, however conceived at a given time.

III

If we were to start to take natural philosophy seriously, as the God-centred study of nature that it was for the people who conducted it (rather than as some study of nature which was struggling to be objective and to free itself from the fetters of religion, or as some odd amalgam of 'science' and religion or of 'science' and theology), then certain consequences would follow for us, and for the conduct of our historical research.

In the first place, if their natural philosophy was, by its very nature, God-oriented in the way I have been claiming for these people of the past, then the question arises of what was the difference that *they* recognized between natural philosophy and religion? Centrally it lay in the fact that religion deals with *revelation*, whereas (natural) philosophy does not. Under the heading "Statements on Religion", Newton himself wrote "that religion and Philosophy are to be preserved distinct. We are not to introduce divine revelations into Philosophy nor philosophical opinions into religion."[27]

Secondly, we would have to find a more precise understanding of *natural theology*, for as historians we have hitherto given to natural theology the role of relating nature to God its creator, and of fighting perceived atheism — that is, the roles of natural philosophy. Natural theology might date from Plato (if we follow Clement Webb) or from Raymond of Sebond and his interpreter Montaigne (if we follow Buckley) or from the Boyle lectures (if we follow Gascoigne).[28] In the form in which we usually study it, at the hands of Boyle, Derham, Ray and others, natural theology was a new study in the seventeenth century.[29] It argued *from* the findings of natural philosophy *to* the existence and attributes of God (and sometimes to men's role in society).

Thirdly, if what I have said about the identity and role of natural philosophy is true — viz., that the point of natural philosophy was to discuss God via His creation, and to put forward accounts of God as glimpsed through His creation and (reciprocally) of His creation as evidence of His traits — a characterization which is certainly not true of modern science, then it raises the question of *when and why people stopped looking for God in nature*. That·is, when did people stop doing natural philosophy, and start doing science? For the last forty years or so we have been claiming this change-over as having taken place in the seventeenth century, and celebrating it under the title of 'the Scientific Revolution of the Seventeenth Century'. But if what I have said above about Newton is true — that Newton was still practising natural philosophy into the early eighteenth century (and not practising modern science), and that his greatest work, the *Principia*, is a perfect instance of natural philosophy — then that can no longer be considered to be the case.[30]

Fourthly, we need a better understanding of the seventeenth (and other) century meaning of '*scientia*' and 'scientific knowledge'. Traditionally '*scientia*' had meant *demonstrative knowledge* — that is, of effects through their causes — and especially using the syllogism. Mathematics was seen as giving an inferior form of certainty to that given by syllogistic reasoning.[31] In the course of the seventeenth century (that is, the period from Galileo to Newton) certain people tried to give mathematics *high* status and make it *replace* the syllogism as the arbiter of what counts as true in natural philosophy. So while there is in the course of the seventeenth century a revolution (or perhaps an attempted revolution) in how *scientia* is got, nevertheless the transformation happened *within* natural philosophy — as the title of Newton's book reveals: *The mathematical principles of **natural philosophy**!* It is a transformation *within* natural philosophy, not a transformation *of* natural philosophy (into modern science). Perhaps we should abandon altogether the concept of the 'Scientific Revolution of the Seventeenth Century' which we historians created, and introduce instead the concept of a '(Natural) Philosophical Revolution of the Seventeenth Century', which matches more closely the experiences and practices of the people of the seventeenth century itself?

Finally let me try and distinguish my position from that of other historians working today, lest anyone think I have been living in a cave for the last three decades and not heard the news, or that I am trying to reinvent the wheel. The difference between my position and that of other historians may be said to revolve around the meaning of an expression I have been using: that natural philosophy was 'about God'. For this can be taken in two senses, a weak and a strong one. Other historians may be said to take it in the weak sense. They have in recent years taken great pains to look at the theological content, implications, and religious commitments of certain natural philosophers, especially in the sixteenth and seventeenth centuries, and seen how this affected and related to the work on nature undertaken by these people. Some historians, such as Funkenstein, have explored at great length the relation of *theology* to the 'scientific' work of natural philosophers in the seventeenth century. That is to say, it is now a commonplace in our discipline that the men of the sixteenth, seventeenth and other earlier centuries whose work on nature we study, shows very often a great concern with God, with the world as God's creation, with the theological implications of particular views of nature and with the religious purposes and significance of the study of creation. The work of Robert Boyle, Isaac Newton and others has been studied in this light. Such work certainly shows that natural philosophy was 'about God', in the weak sense that certain natural philosophers were highly concerned with God, religion and theology. Yet such work nevertheless, in my view, still misses the central point, which is the strong sense of saying that natural philosophy was 'about God'. The point is not that the 'scientific' work of particular natural philosophers was theologically or religiously concerned or informed (which is the weak sense).[32] The point is that *natural philosophy* as such was a discipline and subject-area whose *role and point* was the study of God's creation and God's attributes. Thus, no-one ever undertook the practice of natural philosophy without having God in mind, and knowing that the study of God and God's creation — in a way different from that pursued by theology — was the point of the whole exercise. *All* natural philosophy was *always* like this; when people stopped having this understanding of their goal in their considerations of nature then they *necessarily* stopped doing natural philosophy, and started engaging in a discipline or enquiry which was, in this most fundamental of ways, different in its identity from natural philosophy. This strong sense of the expression indicates that natural philosophy was not just 'about God' and His creation at those moments when natural philosophers were explicitly talking or writing about God in their natural philosophical works or activities. It was, by contrast, 'about God' and His creation the whole time. We have so far, as historians, come to notice the first phenomenon and begun to take it seriously. It is now time to notice that the whole enterprise of natural philosophy had God and God's creation as its whole subject-matter and point. We will thereby be able to appreciate that each and every natural philosopher would bring his particular religious views to his pursuit of natural philosophy and would wish to see and describe *his* view of God in His creation. Hence each and every kind of Catholic, and also Lutherans, Calvinists, Latitudinarian members of the

Church of England, Puritans, eirenicist Protestants, Arians, each and every sect and sub-division of Christian and even deists, would all of necessity, by the very fact of their religious commitments, have somewhat differing concepts of God, His nature and attributes, and therefore about how His nature can be seen or uncovered in the world He created. This is the basic reason why there were varieties of, and developments in, natural philosophy over the centuries: because different people brought different understandings of God to this God-centred study, this study whose whole point was to talk of God through His creation, whose role, as Newton put it, was "to discourse of God" from the created universe.

The point I have been trying to make is very simple, but experience has shown me that it is very hard to convey and grasp because the categories that we historians of science bring to our work keep getting in the way. We are forever thinking and talking about the past in terms of 'science' and 'scientists', and in terms of the relations of 'science' and theology to each other, or of 'science' and religion to each other. But 'science' in the sense in which we mean the term, is simply an inappropriate category for us to use if we are to understand the nature and point of that now-defunct discipline, natural philosophy. Similarly, the distinction between 'science' and theology is a modern-day distinction which cannot legitimately be applied to the practice of natural philosophy in the seventeenth and other centuries.[33] And although, as our practice shows, we can if we choose indeed take instances of the practice of natural philosophy in the past and divide each of them up into bits which *to us* look like science and *to us* look like theology, we should not fool ourselves that by making this artificial and anachronistic dissection of natural philosophy we have thereby ascertained what natural philosophy was about and what its identity was. Natural philosophy *for its practitioners* most definitely did not consist of science and theology in some odd (or even easy) amalgamation, but was a discipline and enterprise *sui generis*.

REFERENCES

1. *The correspondence of Isaac Newton*, ii, 1676–87 (Cambridge, 1960), ed. by H. W. Turnbull, 434, letter Halley to Newton, 7 June 1686.
2. *Ibid.*, 437, letter Newton to Halley, 20 June 1686.
3. Alan Gabbey, "Newton and natural philosophy", in R. C. Olby, G. N. Cantor, J. R. R. Christie and M. J. S. Hodge (eds), *Companion to the history of modern science* (London, 1990), 243–63, p. 245.
4. *Ibid.*, 251, 260.
5. For Newton's anatomical interests, see for instance Plate II in A. Rupert Hall and Marie Boas Hall (eds), *Unpublished scientific papers of Isaac Newton: A selection from the Portsmouth Collection in the University Library, Cambridge* (Cambridge, 1962).
6. William A. Wallace, "Newton's early writings: Beginnings of a new direction", in G. V.

Coyne, s.j., M. Heller and J. Życiński (eds), *Newton and the new direction in science* (Vatican City, 1988), 23–44.

7. Betty Jo Teeter Dobbs, *The foundations of Newton's alchemy or "The hunting of the greene lyon"* (Cambridge, 1975).

8. *The correspondence of Isaac Newton*, v, 1709–13, ed. by A. R. Hall and Laura Tilling (Cambridge, 1975), 391, 398–9.

9. J. E. McGuire and P. M. Rattansi, "Newton and the 'pipes of Pan' ", *Notes and records of the Royal Society*, xxi (1966), 108–43, p. 138.

10. The term 'scientist' dates only from 1833. See Sidney Ross, "Scientist: The story of a word", *Annals of science*, xviii (1962), 65–85.

11. John A. Schuster, "The scientific revolution", in *Companion to the history of modern science* (ref. 3), 217–42, pp. 224–5.

12. Amos Funkenstein, *Theology and the scientific imagination from the Middle Ages to the seventeenth century* (Princeton, 1986); David C. Lindberg and Ronald L. Numbers (eds), *God and nature: Historical essays on the encounter between Christianity and science* (Berkeley, 1986); *Science in context*, iii (1989), no. 1, " 'After Merton': Protestant and Catholic science in seventeenth-century Europe".

13. Biot claimed that after his mental breakdown in 1692 Newton "n'a plus donné du travail nouveau sur aucune partie des sciences", and went on to write on the prophesies of Daniel and the Apocalypse of St John. Biot was mystified as to why such a precise mind as Newton's could allow himself to be convinced by such uncertain conjectures, but concluded that religious freedom and political freedom were of a piece in seventeenth-century England, hence "les savants anglais de cette époque prennaient plaisir à mêler aux recherches des sciences les discussions théologiques". It was Sir David Brewster who thought Biot was casting a slur on Newton — but because Biot claimed Newton went mad, not because Newton "mixed research in the sciences with theological discussions" as Biot claimed. See *Biographie universelle*, 2nd ed. (xxx, 1854), 366–404, esp. pp. 390 and 401; Sir David Brewster, *Memoirs of the life, writings, and discoveries of Sir Isaac Newton* (2 vols, Edinburgh, 1855), ii, chapter xvii.

14. See ref. 9 above, and J. E. McGuire, "Force, active principles, and Newton's invisible realm", *Ambix*, xv (1968), 154–208; J. E. McGuire, "Newton's 'Principles of philosophy': An intended preface for the 1704 *Opticks* and a related draft fragment", *The British journal for the history of science*, v (1970), 178–86; J. E. McGuire, "Newton on place, time, and God: An unpublished source", *The British journal for the history of science*, xi (1978), 114–29.

15. Michael J. Buckley, s.j., *At the origins of modern atheism* (New Haven, 1987).

16. As quoted in Brewster, *Memoirs* (ref. 13), ii, 315.

17. *The correspondence of Isaac Newton*, iii, 1688–94, ed. by H. W. Turnbull (Cambridge, 1961), 233, letter Newton to Bentley, 10 December 1692.

18. For Motte's translation, see *The mathematical principles of natural philosophy by Sir Isaac Newton, translated into English by Andrew Motte 1729*, Introduction by I. Bernard Cohen (2 vols, London, 1968), ii, 391–2. For the original Latin, see *The correspondence of Isaac Newton*, v, 1709–13 (ref. 8), 397, letter Newton to Cotes, 28 March 1713, "At the end of the last Paragraph but two now ready to be printed off I desire you to add after the words [nihil aliud est quam Fatum et Natura.] these words: [Et haec de Deo: de quo utique ex phaenomenis disserere, ad Philosophiam experimentalem pertinet.]".

19. The course of this dispute has been recently unravelled by Rupert Hall and by Steven Shapin: A. Rupert Hall, *Philosophers at war* (Cambridge, 1980); Steven Shapin, "Of Gods and kings: Natural philosophy and politics in the Leibniz-Clarke disputes", *Isis*, lxxii (1981), 187–215.

20. Herbert McLachlan, *The religious opinions of Milton, Locke and Newton* (Manchester, 1941), shows beyond doubt that Newton was a Unitarian. See also more recently James E. Force and Richard H. Popkin, *Essays on the context, nature and influence of Isaac Newton's theology* (Dordrecht, 1990).

21. For example, Newton wrote: "One principle in Philosophy is ye being of a God or Spirit infinite eternal omniscient omnipotent" — which are among the special characteristics of a Unitarian's God — "& the best argument for such a being is the frame of nature & chiefly the contrivance of ye bodies of living creatures.... And to lay aside this argumt is (very) unphilosophical", Cambridge University Library Add. Ms. 3970.3 folios 479r–v, as cited in McGuire, "Newton's 'Principles of philosophy'" (ref. 14), 183. See also the essay by Newton, "De gravitatione et aequipondio fluidorum", printed in Hall and Hall, *Unpublished papers* (ref. 5), 89–156.

22. This is to deliberately reverse our customary way of approaching Newton and the Newtonian universe, in which we take Newton's account to be, in essentials, true, and therefore ask how Newton 'discovered' it, and instead it opens up questions about how Newton *constructed* that particular view of the universe. On such "construction", see Bruno Latour, *Science in action: How to follow scientists and engineers through society* (Milton Keynes, 1987).

23. As quoted from Cambridge University Add. Ms 3970.9 f.619 in Simon Schaffer, "Natural philosophy and public spectacle in the eighteenth century", *History of science*, xxi (1983), 1–43, p. 4.

24. Schuster writes (*Companion* (ref. 11), 225): "Each system of natural philosophy rested on four structural elements whose respective contents and systematic relations went a considerable way towards defining the content of that system: (1) a theory of substance (material and immaterial), concerning what the cosmos consists of and what kinds of bodies or entities it contains; (2) a cosmology, an account of the macroscopic organisation of those bodies; (3) a theory of causation, an account of how and why change and motion occur; (4) an epistemology and doctrine of method which purports to show how the discourses under (1), (2) and (3) were arrived at and/or how they can be justified, and how they constitute a system." Throughout this article Schuster constantly speaks of Newton's and other people's "scientific and natural philosophical work" as if they are different enterprises, see for instance p. 217.

25. It is quite striking how, even when the term 'natural philosophy' is in the title of their books or articles, modern historians of science feel no need at all to explain what natural philosophy is or was — because they are assuming it was the same as modern science, or that they were parallel enterprises. To give just a few instances of this phenomenon, none of the following works defines natural philosophy or distinguishes it from science, despite having the term in their titles, and despite the term being one which is not in common use today: I. Bernard Cohen, *Isaac Newton's papers and letters on natural philosophy, and related documents* (Cambridge, Mass. 1958, repr. 1978); Edward Grant and John E. Murdoch (eds), *Mathematics and its application to science and natural philosophy in the Middle Ages* (Cambridge, 1987); P. Heimann, "Newtonian natural philosophy and the Scientific Revolution", *History of science*, xi (1973), 1–7; the list could be continued to great length.

26. This story will be detailed in Roger French and Andrew Cunningham, *Before science: The invention of natural philosophy* (forthcoming).

27. Quoted by William H. Austin, in "Isaac Newton on science and religion", *Journal of the history of ideas*, xxxi (1970), 521–42, p. 522.

28. See Clement C. J. Webb, *Studies in the history of natural theology* (Oxford, 1915); Buckley, *op. cit.* (ref. 15), 74 for Raymond of Sebond and Montaigne; John Gascoigne, "From

Bentley to the Victorians: The rise and fall of British Newtonian natural theology", *Science in context*, ii (1988), 219–56.

29. Neal C. Gillespie, "Natural history, natural theology, and social order: John Ray and the 'Newtonian ideology' ", *Journal of the history of biology*, xx (1987), 1–49, sees natural theology as an ancient practice revived in a particular way in Restoration England; I am not happy with the categories he uses to argue this. I am grateful to an anonymous referee for drawing this piece to my attention.

30. Evidently God was taken out of Newton's natural philosophy at some later date; if we need to locate this act with one individual then there is probably no better candidate than Laplace — colleague of Biot (see ref. 13 above). For some of the transformations of Newtonian natural philosophy in the eighteenth century see Simon Schaffer: "Newtonianism" in *Companion* (ref. 3), 610–26; "Natural philosophy", in George Rousseau and Roy Porter (eds), *The ferment of knowledge* (Cambridge, 1980), 57–91; "Scientific discoveries and the end of natural philosophy", *Social studies of science*, xvi (1986), 387–420.

31. See Nicholas Jardine, "Demonstration, dialectic and rhetoric in Galileo's *Dialogue*", in Donald R. Kelley and Richard H. Popkin (eds), *Shapes of knowledge in early modern Europe* (forthcoming).

32. Funkenstein (ref. 12) for instance pursues this line, applying the categories of 'science' and theology to the seventeenth century, claiming that "to many seventeenth-century thinkers, theology and science merged into one idiom, part of a veritable secular theology such as never existed before or after", p. ix; but 'science' in the sense in which Funkenstein means it had *never* existed before this time, and did not come into existence for a long time after this period. A similar position is taken by John Hedley Brooke in *Science and religion: Some historical perspectives* (Cambridge, 1991), which appeared while the present article was in proof. As far as I can see, Brooke's whole argument is posited on the possibility of perceiving 'science' and 'religion' in the past as opposed or at least contrasted categories, despite his historiographical considerations in his Introduction, and despite his comment (p. 7), "if we prejudge what we mean by science and religion, we might be in no position to appreciate the distinctiveness of Newton's vision. There would be a degree of artificiality in asking how Newton reconciled his 'science' and his 'religion', if he saw himself pursuing a form of 'natural philosophy', in which the two interests were integrated." Amen to that.

33. See my article "Getting the game right: Some plain words on the identity and invention of science", *Studies in history and philosophy of science*, xix (1988), 365–89.

III

De-centring the 'big picture': *The Origins of Modern Science* and the modern origins of science

ANDREW CUNNINGHAM and PERRY WILLIAMS

> What had happened to him was that the ways in which it could be said had become more interesting than the idea that it could not.
> A. S. Byatt, *Possession*

Like it or not, a big picture of the history of science is something which we cannot avoid. Big pictures are, of course, thoroughly out of fashion at the moment; those committed to specialist research find them simplistic and insufficiently complex and nuanced, while postmodernists regard them as simply impossible. But however specialist we may be in our research, however scornful of the immaturity of grand narratives, it is not so easy to escape from dependence – acknowledged or not – on a big picture. When we define our research as part of the history of science, we implicitly invoke a big picture of that history to give identity and meaning to our specialism. When we teach the history of science, even if we do not present a big picture explicitly, our students already have a big picture of that history which they bring to our classes and into which they fit whatever we say, no matter how many complications and refinements and contradictions we put before them – unless we offer them an alternative big picture.

This paper is based on the principle that big pictures are both necessary and desirable: that if our subject is to provide not merely accumulated information or discourse without meaning, but vision, growth, understanding and liberation – as our students have a right to expect of us and as we have a right to demand of ourselves – then we need to think explicitly about the overall picture of the history of science which we present and within which we work. On this principle, the problem which we now face is not the existence of big pictures in general but the continued existence of the particular big picture on which our discipline was founded, having been established in the early years of its professionalization and embodied in textbooks such as Herbert Butterfield's *The Origins of Modern Science*. The power of this old big picture, and the difficulty with which our

We are very grateful to Jim Secord for his invitation and encouragement to give an earlier version of this paper to the British Society for the History of Science conference 'Getting the Big Picture' in May 1991; to the respondents at that meeting, who inspired us to explain certain points more fully; to Nick Jardine and Harmke Kamminga for their expert and critical advice on several matters; and to John Christie, Ole Grell, Jonathan Hodge, Jim Secord, and an anonymous referee for various helpful suggestions. And finally we are grateful to our students, who have been our original audience and our original critics during the ten years that we have been developing this argument.

discipline is moving away from it, is revealed by the fact that *The Origins of Modern Science* is still in print, in paperback, over forty years after it first appeared, and that students continue to have their first and most formative encounter with the subject either through this book or through others that rely upon variously modified versions of the same big picture.[1]

That big picture is one which in principle covers the whole of human history in a single grand sweep; science is taken to be as old as humanity itself, so that the history of science can in principle run continuously from prehistoric megaliths and Bronze Age metallurgy to the human genome project. In practice, certain periods are selected for attention: for example, classical antiquity, the Middle Ages, and the early modern period. The 'scientific revolution' of the seventeenth century is regarded as a key event; Butterfield's title proclaimed that it represented 'the origins of modern science',[2] that is to say 'modern' as distinct from 'ancient' or 'medieval' science. It was supposed to mark the true beginning of the modern world and the abandonment of the ancient and medieval world, and for that reason Butterfield called it 'the supremely important field for the ordinary purposes of education', the one piece of the history of science which students of both the Arts and the Sciences should know. Butterfield and his followers were indeed successful in making the 'scientific revolution' 'supremely important'; today, it is almost unheard of for a History of Science course not to cover it in some way or another, and it is about the only piece of our discipline which has any currency in the general intellectual world.

This is now rather unfortunate, since over the past ten years the 'scientific revolution' concept has become increasingly difficult to sustain. In the first section of this paper, we survey the problems which have arisen with this concept since Butterfield's time. We realize that these will already be familiar to many readers. Our reason for rehearsing them here is to try to construct an account which explains why we are in our present position: an account which does not claim that Butterfield and his followers were idiots or worked with a defective historiography (as can so easily be done when attacking a former generation of historians), that is to say judging them by our lights and making them out to be failures, but one which acknowledges that they were brilliantly successful in terms of *their own aims*. Those aims we reconstruct as being the formulation and establishment of a concept encapsulating the particular big picture of the history of science which they wanted to promote: a big picture which was itself based on certain assumptions about the nature of

1 Herbert Butterfield, *The Origins of Modern Science 1300–1800*, 2nd edn, London, 1957. The chief rival as an introductory textbook in English is probably *The Scientific Revolution 1500–1800: The Formation of The Modern Scientific Attitude*, New York, 1954, by Butterfield's pupil, A. Rupert Hall, now in its second edition as *The Revolution in Science 1500–1750*, Harlow, 1983. See also Thomas S. Kuhn, *The Copernican Revolution: Planetary Astronomy in the Development of Western Thought*, Cambridge, Mass., 1957; C. C. Gillespie, *The Edge of Objectivity: An Essay in the History of Scientific Ideas*, Princeton, 1960; E. J. Dijksterhuis, *The Mechanization of the World Picture*, Oxford, 1961; Richard S. Westfall, *The Construction of Modern Science: Mechanism and Mechanics*, New York, 1971.

2 The title was also used by Koyré for a series of lectures at Johns Hopkins University in 1951. See his *From the Closed World to the Infinite Universe*, Baltimore, 1957, p. ix. Compare also the sub-title of Hall, 1954, op. cit. (1). 'The origins of modern science' was also the title of Chapter 1 of Alfred North Whitehead's *Science and the Modern World*, New York, 1925, the printed version of his Lowell lectures given the same year; he too was referring to the seventeenth century.

science which were common amongst science-supporting intellectuals in the years just before and just after the Second World War. What we are striving for here is a symmetrical account, which distances us from the legacy of a previous generation of historians, not by saying 'we're right and they're wrong', but by saying 'circumstances have changed'. In particular, what we are arguing is that the reason why the concept of the 'scientific revolution' is in trouble is, not that more or better research has been done, but that we have doubts about those beliefs about the nature of science and the big picture of the history of science which that concept encapsulated and promoted. The implication of this argument is that the concept of the 'scientific revolution' cannot be revived by tweaking, modification, or the addition of a few more 'social factors', or even a heavy dose of sociology of knowledge. And if the old big picture is going, then the 'scientific revolution' must go too. Indeed, trying to hold on to the concept may be damaging, for it will hinder us from developing a new big picture.

In so far as a new big picture is now being developed, inside and outside the profession of the History of Science, it seems to be pluralist in nature;[3] that is to say, it is based on the principle that there are many possible ways of knowing and studying the world, and that science is just the particular way-of-knowing currently dominant in our culture. Hitherto, in the old big picture, all ways-of-knowing-the-world which seemed sensible to the historian have been appropriated to a single, unitary 'history of science'. In the new big picture, a general history of ways-of-knowing-the-world across the whole of human history would have to be the history of many different things, rather than of one single thing at different stages of development. Working within this new big picture, it becomes important to understand the origin of our own culture's dominant way-of-knowing, though this is now to be seen not as the origin of modern science, by contrast with ancient or medieval science, that is the transition from one stage of science to another, but as the origin of science itself – or, as we personally prefer to call it (for reasons to be explained), the *invention* of science. A number of historians in recent years have attempted to discuss the origin of science, in the new sense, but almost without exception they have located it in the period of the old 'scientific revolution', on the assumption that these canonical events, suitably reinterpreted, correspond to the changes they are trying to identify. In principle, of course, there is no reason why science should not have originated at that time; but we believe that this happens not to be the case, and in the second section of this paper

3 Others, including other authors in this issue of the *BJHS*, would probably prefer to express this in the language of postmodernism. We are deliberately not using this language, not on the grounds of disagreement with postmodernism intellectually, but because we believe in using the minimum possible theory in an exposition, and in this case we think that all the necessary ideas can be adequately expressed in plain speech. The moral and political principle behind this stance is analogous to that behind the movement for 'intermediate technology', which seeks to develop solutions to the material problems of developing countries using only tools and materials which can be obtained locally – that is to say, without the importing of high technology which would tend to increase economic and political subordination to industrialized countries. The aim is to provide useful assistance in a form which does not lead to exploitative political relationships.

However, for those readers who want to explore how the ideas in this paper relate to postmodernism, we recommend not only the usually-cited Lyotard *The Postmodern Condition* but also Zuzana Parusnikova, 'Is a postmodern philosophy of science possible?', *Studies in History and Philosophy of Science* (1992), **23**, 21–37, which unusually for a work on postmodernism is not itself written in postmodern language.

III

we will argue that the period 1760–1848 is a much more convincing place to locate the invention of science.

We offer this sketch of the invention of science as a heuristic for teaching and research, and a contribution to the development of a new big picture, one which does not privilege one particular kind of knowledge. Its main significance, we think, is that it helps to demarcate more clearly the place in the big picture which our culture occupies. In the third section of this paper, we turn to the question of how we should deal with the remainder of the picture – that is to say, almost all of it. In the old big picture, science was taken to be a human universal, and so could act as a neutral framework on which to organize ways-of-knowing across the whole span of human history and human cultures. If we no longer assume that science is neutral and universal, a new big picture will require our vision to be jolted out of our culture and made aware of its contingency. We mention three forms of such 'de-centring', as we call it, the first of which is relatively close to existing practice, the second of which is more difficult to imagine but still, we think, quite feasible, and the third of which points towards a subject completely different from the History of Science as we know it now and which we ourselves can barely imagine, although some of the readers of this, starting from a more advanced baseline, may be able to do better – perhaps even eventually create it.

THE ORIGINS OF MODERN SCIENCE

Wise is the child that knows its father.
    English proverb

It is now well established that our present concept of the 'scientific revolution' of the seventeenth century, whatever precedent it may have had in earlier writing, was forged by a number of scholars during the 1940s, chiefly Alexandre Koyré, whose work was then beginning to be taken up in the USA, and Herbert Butterfield, who promoted the concept in Britain, for example with his 1948 Cambridge lectures which were the basis for *The Origins of Modern Science*.[4] This first generation of those working professionally on the History of Science were generalists, not only from necessity, because there were too few of them to specialize, but because they had a big picture of the history of science which they were eager to communicate. That big picture was itself based on their view of what science was, for at this period the main reason for working on the history of science was to speak out on behalf of science itself and to explain its nature and importance.[5] To understand

4 Roy Porter, 'The scientific revolution: a spoke in the wheel?', in *Revolution in History* (ed. Roy Porter and Mikuláš Teich), Cambridge, 1986, 290–316, see 295. Precedents for the concept can be traced back to the eighteenth century; see John R. R. Christie, 'The development of the historiography of science', in *Companion to the History of Modern Science* (ed. R. C. Olby, G. N. Cantor, J. R. R. Christie and M. J. S. Hodge), London, 1990, 5–22, especially 7–9; also I. Bernard Cohen, *Revolution in Science*, Cambridge, Mass., 1985, 51–101.

5 For very clear evidence of this in the American context, see Arnold Thackray, 'The pre-history of an academic discipline: the study of the history of science in the United States, 1891–1941', in *Transformation and Tradition in the Sciences: Essays in Honor of I. Bernard Cohen* (ed. Everett Mendelsohn), Cambridge, 1984, 395–420, especially 402–5.

fully the concept of the 'scientific revolution', and the problems which we are now having with it, we need to understand the view of science that through it they were trying to promote.

Amongst the various kinds of people who were interested in promoting science at that time, we can distinguish at least three ways of characterizing its nature, which they used in various combinations. The first was philosophical, defining science as a particular method of enquiry, producing knowledge in the form of general causal laws, preferably mathematical, as in the physical sciences, or which could be reduced to this form. This characterization of science was a legacy of nineteenth-century positivism, and it was very strong in the 'logical positivist' position articulated in the 1930s by the Vienna Circle; but even those philosophers who were anti-positivist in their stance, such as, in their different ways, Emile Meyerson and Karl Popper, continued to accept it. Few doubted that the scientific method existed, or disputed the centrality of issues such as laws, explanation and prediction to the business of defining philosophically what science was.[6]

A second, less academic, way of characterizing science was to see it in essentially moral terms, as the embodiment of basic values of freedom and rationality, truth and goodness, and the motor of social and material progress. Those who characterized science in this way saw it as the disinterested pursuit of truth, undeflected by the sway of emotions, and free from personal, political or economic interest. They generally believed that rational, scientific thought, if more generally distributed amongst the population, would put an end to misunderstanding, prejudice and social conflict, and even to fascism and totalitarianism; a classic expression of this belief was Robert Thouless's *Straight and Crooked Thinking*, first published in 1930, which had the double aim of setting out the principles of 'straight thinking', exemplified by science, and of exposing dishonest rhetorical tricks and so freeing people from manipulation by politicians.[7] The promoters of science also believed that practically applied science could bring an end to suffering and want; futuristic utopias from this period tended to show happy contented people wearing plastic clothing, free from drudgery in a world of automated factories and domestic robots, their only problem being finding things to do with their time.[8]

A third way of characterizing science was as a universal human enterprise. By this, we mean that science was seen as the expression of an innate human curiosity, a general and universal desire to understand the world, that was a fundamental part of human nature and

6 Emile Meyerson, *Identity and Reality*, London, 1930; original French edition, 1908. K. R. Popper, *The Logic of Scientific Discovery*, London, 1959; original German edition, 1935.

7 Robert H. Thouless, *Straight and Crooked Thinking*, 1st edn, London, 1930. For example: 'A really educated democracy, distrustful of emotional phraseology and all the rest of the stock-in-trade of the exploiters of crooked thinking, devoid of reverence for ancient institutions and ancient ways of thinking, could take conscious control of our social development and could destroy those plagues of our civilisation – war, poverty, and crime' (244–5). A later example of the association of science and freedom (here construed as opposition to Soviet totalitarianism) is *Science and Freedom*, London, 1955, being the proceedings of a conference convened by the Congress for Cultural Freedom and held in Hamburg 23–26 July 1953.

8 See, for example, the vision of the far distant future ('Everytown 2036') in the 1936 film *Things to Come*, loosely based on H. G. Wells's *The Shape of Things to Come: The Ultimate Revolution*, London, 1933; the 1934 film *Plenty of Time for Play* (excerpted in the Open University television programme *The All-electric Home* written and presented by Gerrylynn Roberts, from the course A282 'Science, technology and everyday life 1870–1950'); the BBC television film *Time on our Hands*, first transmitted in 1963.

human thought throughout time and space. Many propagandists for science represented it in this way, deliberately challenging the older arts-based concept of humanity and humanism which prevailed in the university curriculum and in intellectual life generally;[9] thus George Sarton spoke of science as 'the *new* humanism', and Julian Huxley coined the phrase '*scientific* humanism' to represent his beliefs.[10] Others use the phrase 'science and civilization' in order to express their claim that science was central to human civilization properly understood – or according to some, that science *was* human civilization;[11] the best-known example of this phrase today is probably Joseph Needham's grand project on *Science and Civilisation in China*, which was originally conceived in the late 1930s.[12]

These were the three most usual characterizations of science amongst those seeking to promote it in the 1940s, and historians of science naturally incorporated elements of all of them as they developed the concept of 'the scientific revolution'. In the first place, the philosophical characterization of science led them to conceive 'the scientific revolution' as the seventeenth-century transformation of human thought into a form close to that of 'scientific knowledge' as defined by twentieth-century philosophers of science. Its defining events were taken to be those in which knowledge approximating to the ideal of general mathematical causal laws seemed to be achieved; since this ideal derived from the physical sciences, naturally enough the physical sciences supplied almost all the defining events – the Copernican revolution in astronomy, the Galilean revolution in mechanics, and the Newtonian synthesis, which were conceived in terms of mathematization and the 'mechanization of the world picture'.

At the same time, the moral characterization of science meant that the historians who developed the concept of 'the scientific revolution' attributed to these defining events liberal values such as freedom and independence of thought against superstition and Church dogma. They regarded the main theme of 'the scientific revolution' as being the elevation of experience above tradition and authority, or the rise of research and experiment as against the study of ancient texts; and this was supposedly exemplified in Copernicus's rejection of Ptolemy, Galileo's rejection of Aristotle and his challenge to the Catholic Church, and – a sole example from the 'life' sciences – Harvey's rejection of Galen. Because these historians conceived 'the scientific revolution' in terms of the

9 Thackray, op. cit. (5), 411.

10 George Sarton, *The History of Science and the New Humanism*, Bloomington, 1962, original edn 1930; Julian Huxley, 'Scientific humanism', in his *What Dare I Think? The Challenge of Modern Science to Human Action and Belief*, London, 1931, 149–77. C. P. Snow's *The Two Cultures and the Scientific Revolution* (Cambridge, 1959) was also an expression of this view; it is often forgotten that his reason for pointing out the cultural divide between the arts and the sciences was to complain that arts people did not sufficiently understand and respect the sciences. (Snow's 'scientific revolution', incidentally, was the change resulting from the application of science to industry, which he dated not earlier than 1920.)

11 Thackray, op. cit. (5), 401, 408. The identification of science and human thought was also common; for example, 'it was as though *science or human thought* had been held up by a barrier until this moment [i.e. until 'the scientific revolution']' (Butterfield, op. cit. (1), 7, our emphasis).

12 Cambridge, 1954– , i, 11. Jacob Bronowski's *The Ascent of Man* was commissioned by the BBC as a counterpart to a similarly epic television series on the history of art by Sir Kenneth Clark; J. Bronowski, *The Ascent of Man*, London, 1973, 13. One may surmise that it was extremely irritating to Bronowski that the title 'Civilisation' had already been commandeered by Clark for his subject; but in the event, his own title made an even stronger claim: that the history of science was the history of human *evolution* (ibid., 19–20).

advancement of free thought, they generally saw politics, religion and economic circumstances only as factors impeding its progress. But at the same time, the way in which they associated science with material advance and social harmony led them to emphasize such events as Bacon's prophecy of power over nature, and the formation of the Royal Society and the Académie Royale as instances of scientific co-operation.

Finally, the fact that these historians characterized science as a universal human enterprise meant that they conceived 'the scientific revolution' as only a revolution *within* science. Since they supposed the scientific enterprise to be a fundamental part of human nature, the history of science, for them, was in principle as long as the history of the human race itself; they counted all respectable knowledge about the natural world, wherever and whenever it was found, and however it had been produced, as some form of science. Thus despite the revolutionary nature of what they believed had happened in the seventeenth-century period, they supposed that there was still an essential continuity with the past; the ancients had engaged in essentially the same activity as the seventeenth-century heroes; they had tried to answer the same questions (such as the problem of motion), only they had not done so as well or to such an advanced level. The 'scientific revolution', despite its tremendous significance, thus consisted only of 'picking up the opposite end of the stick' or putting on 'a different kind of thinking-cap';[13] it was only a change of approach to the same, supposedly eternal, problems.

All these defining characteristics of 'the scientific revolution' now seem rather dubious. The most obvious explanation for this change is internalist (that is to say, in terms of things internal to the History of Science discipline): more recent detailed, specialized studies have undermined the general big picture of 'the scientific revolution' as it was first conceived.[14] For example, with respect to the philosophical characterization of 'the scientific revolution', it has proved very difficult to fit developments in the 'life' sciences into the mould of mathematization and mechanism; Harvey's discovery of the circulation of the blood, once seen as the exemplary case of the physiological application of mechanical ideas (e.g. the heart as a pump) is now generally accepted to have been the result of an essentially Aristotelian investigation into the 'final cause' of the heart's motion and structure, into the ways in which this organ served the purposes of the soul.[15] Even in the physical sciences, seventeenth-century mechanical philosophy has been found to have been much less close to the twentieth-century ideal than was once imagined; seventeenth-century natural

---

13 Butterfield, op. cit. (1), 7, 5.

14 One problem raised by more recent research, which we are not attempting to discuss here, is that the length of 'the scientific revolution' has expanded enormously, as everyone has tried to climb on the bandwagon. Now that it has been extended to the end of the eighteenth century in order to include Lavoisier (Butterfield wrote of 'The postponed scientific revolution in chemistry'; op. cit. (1), ch. 11), and back to the high medieval period to trace the origins of Galilean mechanics (as in the work of Alistair Crombie, following Pierre Duhem), we are faced with a scientific revolution which spanned maybe five centuries. As Roy Porter has nicely put it, compared with *Ten Days that Shook the World*, this is an extraordinarily leisurely revolution (op. cit. (4), 293).

15 See, for example, C. Webster, 'William Harvey's conception of the heart as a pump', *Bulletin of the History of Medicine* (1965), **39**, 508–17; Andrew Cunningham, 'Fabricius and the "Aristotle project" in anatomical teaching and research at Padua', in *The Medical Renaissance of the Sixteenth Century* (ed. A. Wear, R. K. French and I. M. Lonie), Cambridge, 1985, 195–222; Andrew Cunningham, 'William Harvey: the discovery of the circulation of the blood', in *Man Masters Nature: 25 Centuries of Science* (ed. Roy Porter), London, 1987, 65–76.

philosophers have been revealed to have described the universe in terms of not only matter and motion but also spirits and powers, the whole being dependent on a metaphysics anchored in a precise and complex theology.[16]

The moral characterization of 'the scientific revolution' too has been weakened by more recent research, as new historiographies laying emphasis on the role of 'context' or 'external factors' have made it more difficult to maintain the prime role of free, independent thought. For example, the work of Robert K. Merton started a new tradition of seeing the cardinal intellectual changes of 'the scientific revolution', in England at least, as a consequence of Puritanism (or, more generally, of Reformation and Counter-Reformation theologico-politics), and hence as a consequence of the development of capitalism.[17] At the same time, studies of the rise of the mechanical arts have revealed that practical technology played an important role in those intellectual changes, and was not simply a consequence of them. In general, it is now widely recognized that religion, politics and economics to a large extent facilitated, instead of impeding, those changes which supposedly defined 'the scientific revolution': some would say they *produced* or *were constitutive of* them.

Finally, difficulties have arisen with the characterization of 'the scientific revolution' as a stage in an enterprise as universal as (supposedly) human nature itself. A greater respect for the categories, values and enterprises of historical actors has led to the appreciation that, say, when the ancient Greek philosophers made reference to the soul or the Divine in their writings on the natural world they were not making a poor effort at modern science but were succeeding brilliantly at ancient Greek philosophy, the whole point of which was the cultivation of the soul for this life and the next. In the same way, it has been realized that when the supposed heroes of 'the scientific revolution' such as Newton used theology, mysticism, alchemy and biblical chronology in their study of the natural world this was neither insanity[18] nor a failure to be properly 'scientific' but part of a coherent attempt to reach a deeper understanding of the Christian God by studying His creation. It is now being recognized that even the post-'scientific revolution' way of studying the natural world was very different from what we now call science – so different, indeed, that it has taken a great deal of effort to recover it; and there is an increased awareness of the very deep differences between the ways the natural world has been conceived and studied in ancient, medieval, Renaissance, early modern and modern times.

16 A key work here was J. E. McGuire and P. M. Rattansi, 'Newton and the "pipes of Pan"', *Notes and Records of the Royal Society* (1966), **21**, 168–43. See also Betty Jo Teeter Dobbs, *The Foundations of Newton's Alchemy: Or 'The Hunting of the Greene Lyon'*, Cambridge, 1975. For current thinking on this subject, see Simon Schaffer, 'Occultism and reason', in *Philosophy, its History and Historiography* (ed. A. J. Holland), Dordrecht, 1985, 117–43; and Simon Schaffer, 'Godly men and mechanical philosophers: souls and spirits in Restoration natural philosophy', *Science in Context* (1987), **1**, 55–85.

17 The new interest in Merton is indicated by the reprint in 1970 of *Science, Technology and Society in Seventeenth Century England*, New York, originally published in *Osiris* in 1938; and by *Puritanism and the Rise of Modern Science: The Merton Thesis* (ed. I. Bernard Cohen), New Brunswick, 1990. See also Steven Shapin, 'Discipline and bounding: the history and sociology of science as seen through the externalism–internalism debate', *History of Science* (1992), **30**, 333–69. A leading example of how the 'Merton thesis' could be reapplied is Charles Webster, *The Great Instauration: Science, Medicine, and Reform, 1626–1660*, London, 1975.

18 This was what Biot claimed in his entry on Newton in the *Biographie universelle*, 2nd edn, Paris, 1854, xxx, 366–404, especially 390 and 401.

New historiography and new research on the history of science, then, provides one kind of explanation for why the original defining characteristics of 'the scientific revolution' now seem dubious.[19] But a more profound explanation, we believe, can be made in terms of things *external* to the discipline, in particular the weakening of those assumptions about the nature of science which were incorporated into the concept of 'the scientific revolution' when it was first developed. First, the philosophical characterization of science has become progressively weaker since the 1960s, when doubts began to grow about the existence of a single logically-defined scientific method. The most direct challenge came from Paul Feyerabend, whose *Against Method* argued that the scientific method as defined by Popper and others simply did not work; he claimed that what *had* worked historically, in the sense of producing the scientific knowledge which we now had, was a whole variety of methods, which could not be reduced to a single logical procedure. Thomas Kuhn's *Structure of Scientific Revolutions* was already providing a framework for thinking about a multiplicity of methods, by arguing that all aspects of methodology, including theory evaluation, were relative to a particular scientific community's set of shared beliefs, values, techniques, and exemplary problems – their paradigm, for short.[20] There was also heavy criticism of the view that all scientific knowledge could be reduced to a single unified science in the form of general causal laws; some philosophers argued that this could not be achieved for the whole range of the physical sciences, let alone the historical sciences, such as geology, or the natural historical sciences, such as botany.[21]

In the next place, the moral characterization of science as an embodiment of the highest standards of intellectual probity and as the motor of social and material progress was also challenged by the radical, feminist and environmental movements of the 1960s, 1970s and 1980s. From a radical perspective, scientific expertise began to be seen as a form of power and control, mistrust having been prompted especially by scientists who used their authority to suppress opposition to the nuclear industry, and it began to be argued that no kind of knowledge is ever truly free of value-judgement or economic and political interest. Feminist analysis called into question the very desirability of objectivity and control over

19 As David C. Lindberg and Robert S. Westman put it in the introduction to their edited collection *Reappraisals of the Scientific Revolution*, Cambridge, 1990, the last twenty years have seen 'highly focused studies [which] took root and began subtly to undermine the wall on which Humpty Dumpty sat' (p. xviii).

20 Paul Feyerabend, *Against Method: Outline of an Anarchistic Theory of Knowledge*, London, 1975; Thomas S. Kuhn, *The Structure of Scientific Revolutions*, 2nd edn, Chicago, 1970. For an interesting exegesis of Feyerabend's much-misunderstood philosophy, see José R. Maia Neto, 'Feyerabend's scepticism', *Studies in History and Philosophy of Science* (1991), **22**, 543–55.

21 Jerry Fodor, 'Special sciences, or the disunity of science as a working hypothesis', *Synthese* (1974), **28**, 77–115; Alan Garfinkel, *Forms of Explanation: Rethinking the Questions in Social Theory*, New Haven, 1981, ch. 2. We should admit that many scientists, physicists especially, still believe in the possibility and desirability of reduction and unification. For instance Phil Allport, 'Still searching for the Holy Grail', *New Scientist*, 5 October 1991, 55–6; Allport works in the High Energy Physics Group at the Cavendish Laboratory in Cambridge. In this article, he reasserts his commitment to the philosophy of Karl Popper, by contrast with that of Nancy Cartwright, and quotes with approval Gerald Holton, a Harvard professor of physics and historian of science, speaking in 1981 of theoretical physics being engaged in a quest for a Holy Grail: nothing less than 'the mastery of the whole world of experience, by subsuming it ultimately under one unified theoretical structure'. Interestingly, in view of the religious analogy which we draw in the final section of the present paper, Allport suggests that this 'monotheorist' ambition could never have become established without a long tradition of *monotheist* belief.

III

nature; when seen as a masculine attempt at emotional detachment from the world, these basic scientific ideals could seem positively pathological. And as the repercussions of the attempt to control nature became better known, and new terms such as pollution, acid rain and greenhouse effect entered the language, science and technology have ceased to be regarded as forces only for good; it became clear that they could create material problems which they were unable to solve.[22]

Finally, the characterization of science as the universal and trans-cultural knowledge-producing enterprise was weakened when this general disenchantment with modern science led to a search for ways of knowing the natural world that seemed not to share in its objectionable features. Whether a better way was sought through oriental religions, feminist epistemology, or the traditional culture of native Americans, the appreciation of the worth of alternative knowledge forms has led to a growing sense that the pursuit of science is not a fundamental, universal feature of human nature or human civilization, but simply one among many actual or possible ways of knowing the world.

These changes in the assumptions about the nature of science over the last twenty-five years have led to various attempts to rewrite the history of the 'scientific revolution'. One set of attempts in the late 1970s, springing particularly from the radical, feminist and environmental critiques of science, followed the traditional characterization of the 'scientific revolution', but re-evaluated it as a bad thing rather than a good thing: the origin of modern forms of repression and exploitation, rather than progress and liberation.[23] Other attempts, freed by the dwindling authority of philosophy of science over historiography, sought to add qualifications and provisos to the 'scientific revolution' concept, first by the introduction of 'external factors' such as religion, politics and economics, and then insights derived from the methods of anthropology and sociology – most recently the sociology of knowledge.[24] But as more and more elements have been added on to the 'scientific revolution' concept in an effort to make it convincing, the problem has arisen that the concept is no longer obviously coherent – a problem now acknowledged quite generally, and poignantly expressed by David Lindberg and Robert Westman in their 1990 *Reappraisals of the Scientific Revolution*.[25]

22 For the radical challenge, see for example Jerome R. Ravetz, *Scientific Knowledge and its Social Problems*, Oxford, 1971; the *Radical Science Journal*, which started publication in January 1974. For the feminist challenge, see for example Hilary Rose, 'Hand, brain, and heart: a feminist epistemology for the natural sciences', *Signs: A Journal of Women in Culture and Society* (1983–4), 9, 73–90; Sandra Harding, *The Science Question in Feminism*, Milton Keynes, 1986. For the environmentalist challenge, see for example *The Limits to Growth* (ed. Club of Rome), London, 1972; *Only One Earth* (ed. Barbara Ward and René Dubos), London, 1972.

23 Carolyn Merchant, *The Death of Nature: Women, Ecology, and the Scientific Revolution*, San Francisco, 1980. Brian Easlea, *Liberation and the Aims of Science: An Essay on the Obstacles to the Building of a Beautiful World*, London, 1973; *Witch Hunting, Magic and the New Philosophy: An Introduction to Debates of the Scientific Revolution 1450–1750*, Brighton, 1980; *Science and Sexual Oppression: Patriarchy's Confrontation with Woman and Nature*, London, 1981.

24 For example, Steven Shapin and Simon Schaffer, *Leviathan and the Air-pump: Hobbes, Boyle, and the Experimental Life*, Princeton, 1985. Even more recently, there has appeared *The Scientific Revolution in National Context* (ed. Roy Porter and Mikuláš Teich), Cambridge, 1992, which unfortunately does not explore the question of what the scientific revolution was or whether it existed at all.

25 Op. cit. (19). The deliberate aim of the collection, and the conference which preceded it, was 'to offer at least a partial remedy' for the 'distressing situation' of the complete absence of 'a [general] picture fully consistent with recent developments [in scholarship]' (pp. xix–xx). 'Does any unity emerge' from the articles?

We do not want to discuss here the last twenty years or so of (to use Lindberg and Westman's image) attempts to put Humpty Dumpty together again. Our argument here is that such attempts are doomed to failure, because the 'scientific revolution' concept was specifically created to encapsulate a particular big picture and a particular view of the nature of science which, while extremely convincing in the 1940s, now seem increasingly implausible. The problem with the 'scientific revolution', we maintain, is not that insufficient research has been done, or that an up-to-date historiography is needed, or more 'external factors', or more sociology of knowledge, or discourse analysis, or whatever the current intellectual fashion happens to be. The problem is that historians are ceasing to believe in a single scientific method which makes all knowledge like the physical sciences, or that science is synonymous with free intellectual enquiry and material prosperity, or that science is what all humans throughout time and space have been doing as competently as they were able whenever they looked at or discussed nature; so small wonder that a concept developed specifically to instantiate these assumptions ceases to satisfy. If this position is correct, then (to use Nicholas Jardine's happy phrase) the 'scientific revolution' needs to be not so much rewritten as written off.[26]

This is not to say that those historians of science who have been studying the seventeenth century have been wasting their time; clearly something of great importance happened then with respect to the investigation of nature. We are not going to make any statement here about what that something might have been. We will, however, put forward our recommendation that whatever-it-is should not be referred to as 'the scientific revolution'. People at the time spoke, either proudly or contemptuously, of 'the new philosophy', and perhaps this phrase would be a better basis for conceiving and naming what happened then; or if it is felt necessary to include the word 'revolution', then we might speak of 'the mathematical revolution', since a great change in the scope and use of mathematics was at least one of its components. But we will leave *The Origins of Modern Science* there, since we now want to turn to the question of what kind of big picture we might construct to replace the old one.

THE MODERN ORIGINS OF SCIENCE

It was like having a nightmare about a man who had got it into his head that τριήρης was the Greek for 'steamer', and when it was pointed out to him that descriptions of triremes in Greek writers were at any rate not very good descriptions of steamers, replied triumphantly, 'That is just what I say. These Greek philosophers... were terribly muddle-headed, and their theory of steamers is all wrong'. If you tried to explain that τριήρης does not mean steamer at all but something different, he would reply, 'Then what does it mean?' and in ten minutes he would show you that you didn't know; you couldn't draw a trireme, or make a model of one, or even describe exactly how it worked. And having annihilated you, he would go on for the rest of his life translating τριήρης 'steamer'.

    R. G. Collingwood, *Autobiography*

they ask rhetorically, and conclude that 'the reader will have to decide' (p. xx). If two scholars of such calibre, after several years of effort, can find no unity to put forward, then there is little hope of the rest of us faring any better.

  26 Nicholas Jardine, 'Writing off the scientific revolution' (a review of Lindberg and Westman's *Reappraisals*), *Journal of the History of Astronomy* (1991), **22**, 311–18.

A new big picture must take account of the changed view of the nature of science. The old big picture was based on the conception of science in the time of Butterfield and Koyré – rooted in transcendent timeless logic and embodying absolute moral values of freedom, rationality and progress: a universal human enterprise. But a new big picture must be based on the emerging re-conception of science as historically contingent and embodying the values, aims and norms of a particular social group: one amongst a plurality of ways of knowing the world. In a new big picture, what we refer to as 'science' can no longer be used as a general defining framework; it must be seen as limited, bounded in time and space and culture.

The best way of establishing such a new big picture – both of fixing it in our own minds and of teaching it to a new generation of students – is surely to focus on those boundaries: to identify the origins of science and to explain how science came into being. To describe this project as identifying the origins of science unfortunately makes it sound like the project of Butterfield and others of investigating 'the origins of modern science', so it is necessary for us to reiterate that what they claimed to have identified was the appearance of the modern form of something transcendent: the definitive and most complete realization of something which had existed in potential throughout all human history; hence their frequent use of organic metaphors such as 'birth' and 'emergence', which implied an embryonic pre-existence and the unfolding of a pre-ordained plan. By contrast, identifying the origins of science, in the revised sense, would mean finding the first appearance, the first practice, of something which is distinct and specific to our own region of time and space, rooted in the particular circumstances of our culture.

In this section we will be proposing that the origins of science in this revised sense can be located in Western Europe in the period sometimes known as the Age of Revolutions – approximately 1760–1848.[27] This proposal may be somewhat unexpected, since, as far as we are aware, all those authors to date who have discussed the origins of science in the revised sense have assumed that these coincide with 'the origins of modern science' in the old big picture; in other words, that they are to be found in the so-called 'scientific revolution'.[28] Of course the major changes in the seventeenth century, in which one may continue to believe even while abandoning the old big picture and the 'scientific revolution' with it, certainly make it, at first sight at least, a plausible location for the origins of science, in the revised sense; but the extent to which the old big picture and its underlying view of

27 This takes its cue from the usage of E. J. Hobsbawm, *The Age of Revolution, 1789–1848*, London, 1962.

28 For example, Merchant, op. cit. (23), and Easlea, op. cit. (23), aimed to rewrite the history of the 'scientific revolution' as the origin of the present politically-oppressive way of knowing the world. The recent wave of anti-scientistic writings consistently ascribe to the 'scientific revolution' the origin of the scientistic outlook which they criticize; see for example Bryan Appleyard, *Understanding the Present: Science and the Soul of Modern Man*, London, Picador, 1992, especially ch. 2, 'The birth of science'. Mary Midgley's sophisticated and accessible *Science as Salvation: A Modern Myth and its Meaning*, London, 1992, is a partial exception. While she repeats the common view that it was in the seventeenth century that 'modern science first arose' (p. 1) and that matter began to be regarded as inert and passive, all interesting properties being attributed to God alone, she also relates present-day scientism to the expulsion of God from the investigation of nature in the nineteenth century. 'It is surely extraordinary that nineteenth- and twentieth-century thinkers have supposed that they could take over this attitude to matter unaltered, while eliminating the omnipotent Creator who gave sense to it, as well as the immortal soul which took its status from him' (p. 76).

science have permeated our discipline and its key concepts, such as the 'scientific revolution', makes it necessary to treat such an automatic assumption with caution. We believe that if historians of science were to come to this question of the origins of science afresh, without reference to the lines laid down by the founders of the discipline, they would locate them elsewhere. To explain why we are locating the origins of science in the Age of Revolutions, we will begin by making explicit the historiographic principles on which we have been working – principles which are, we believe, now relatively consensual within the profession.

The first of these principles, and the one that underlies and gives force to the others, is that the basic values and norms of science (its ideology, that is) are things which need explanation, rather than things which are to be taken for granted, as being above explanation, or to be used (without question) to explain other things. To put it another way, for example, the main question we used to ask about the 'objectivity' of science used to be: 'in what does the "objectivity" of science consist?'. The new question, by contrast, might be said to be: 'how did science come to have "objectivity" ascribed to it?' This principle follows from the values and norms of science no longer being seen as defined by absolute moral criteria or timeless logic, and the pursuit of science no longer being seen as a universal characteristic of human nature; if these things do not derive from some transcendent realm, then they are susceptible to, and require, historical explanation in the usual way.

The second principle is about the relation of knowledge to society, the position being that the knowledge in any society is an integral product of that society, and embodies within it the values and social relations of that society. This assumption has been used to good effect in 'New Left' work from the 1960s, and also in the more recent approach of the sociologists of science (many of whom are not politically of the Left at all).[29]

The third principle of historical enquiry concerns actors' categories. The work of Skinner and Dunn and others[30] has led us all (at least in principle) to try to respect and seek to understand the terms and categories in which the past people we study actually thought of and performed their work – in order to appreciate its authentic meaning and identity. In the recent formulation by Martin Rudwick: 'A non-retrospective narrative of any episode in the history of science should be couched in terms that the historical actors themselves could have recognised and appreciated with only minor cultural translation [to help the modern reader understand it]'[31] – and this respecting of actors' categories should not be limited to the moment when the historian is making his or her historical exposition,

29 Robert M. Young, 'The historiographic and ideological context of the nineteenth-century debate on man's place in nature', in *Changing Perspectives in the History of Science: Essays in Honour of Joseph Needham* (ed. Mikuláš Teich and Robert M. Young), London, 1973, 344–438. Steven Shapin, 'History of science and its sociological reconstructions', *History of Science* (1982), **20**, 158–211.

30 Q. R. D. Skinner, 'Meaning and understanding in the history of ideas', *History and Theory* (1969), **8**, 3–53. John M. Dunn, 'The identity of the history of ideas', *Philosophy* (1968), **43**, 85–104. Adrian Wilson and T. G. Ashplant, 'Whig history and present-centred history', and T. G. Ashplant and Adrian Wilson, 'Present-centred history and the problem of historical knowledge', *The Historical Journal* (1988), **31**, 1–16 and 253–73 respectively.

31 Martin J. S. Rudwick, *The Great Devonian Controversy: The Shaping of Scientific Knowledge Among Gentlemanly Specialists*, University of Chicago Press, 1985, 14.

III

but should apply also to the categories within which the historian conducts his or her researches as well.

Finally the fourth principle is to do with projects of enquiry: that it is necessary to identify the particular and specific 'projects of enquiry' in which people in the past were engaged in their investigations of nature. This is a particular form of the more general principle of 'Question and Answer' (as it was originally termed by Collingwood):[32] when we read texts from the past, we need to ask ourselves, 'to what *question* – both what immediate question, and what project of enquiry – in the life and world of the person who wrote it, was this text the *answer* for its author?'. For without knowing the project that a particular historical actor was engaged on, the results arrived at by that historical actor are meaningless to us; the answer is meaningless without the question. By assigning, consciously or unconsciously, the wrong project or the wrong question to the historical actors we are investigating, the results or answers that they arrived at are given the wrong meaning by us (usually a modern-day meaning). These four historiographic principles are, we believe, now relatively consensual within the profession; for example 'Whig' or 'present-centred' history would violate all four criteria, and few would now defend it.[33]

But as well as forbidding certain things, these principles can be of positive assistance in approaching the problem of the origins of science. For instance, taking the final two together, Principle 4, about projects of enquiry, suggests that we should direct our attention, not simply to statements about the natural world in past texts, but to the precise enterprises of which these thoughts and statements were part and which gave them their identity and meaning. Principle 3, actors' categories, suggests that we should take past people's own accounts of their enterprises, including their own names for them, very seriously indeed.

Following these two principles, we can start by noticing and taking seriously the fact that it was not until the beginning of the nineteenth century that the term 'science' was used for the enterprise of investigating the natural world in the way that it is used today.[34] The word 'science', deriving as it does from the Latin word *scientia*, existed prior to the nineteenth century of course. But it was not restricted to the investigation of nature, for it was used for all disciplines which dealt in terms of theory (or for the theoretical side of

32 R. G. Collingwood, *An Autobiography*, first published 1939, reprinted 1978, Oxford University Press, ch. 5, 'Question and answer'. A recent development of this philosophical argument is Nicholas Jardine, *The Scenes of Inquiry: On the Reality of Questions in the Sciences*, Oxford, 1991.

33 The term 'Whig' history derives from Herbert Butterfield, *The Whig Interpretation of History*, London, 1931. Unfortunately many people use the term as an all-purpose smear for historiography of which they disapprove, without having read *The Whig Interpretation*, and occasionally without even being aware of its existence. On this issue we recommend Wilson and Ashplant, op. cit. (30). (They propose the term 'present-centred' as a more precise and more general substitute for 'Whig'.)

34 Sadly, there have been virtually no critical discussions of the changing meaning of the word 'science'. It is probably significant that the most accessible account we know of was *not* the work of a historian of science: Raymond Williams, *Keywords: A Vocabulary of Culture and Society*, London, 1976, s.v. 'science'. There has been one good study of the word 'scientist': Sydney Ross, '"Scientist": the story of a word', *Annals of Science* (1962), **18**, 65–86. This, however, is a good deal less threatening to the discipline; the invention of the word 'scientist' can be interpreted in terms of 'professionalization' – that is, merely a change in the organization of essentially the same activity.

all disciplines), theory which was based on firm principles, and for the knowledge generated within such disciplines. Thus most of the disciplines concerned with the natural world were 'sciences', but grammar was also a science, and so was rhetoric, and so was theology. Indeed theology was often regarded as 'the Queen of the Sciences' right up until the end of the eighteenth century, whereas today it would not qualify as a science at all under our modern meaning of the term. What this difference in the application of the term 'science' should lead us to see is that for hundreds of years before this time, when western people studied the natural world they did so not as 'science' but within other disciplines, the disciplines of either 'natural history', or 'mixed mathematics' or especially 'natural philosophy'.

Yet the meaning of the most important of these, 'natural philosophy', to the historical actors themselves who actually practised it, has scarcely been investigated by historians of science. Instead it has been treated as if it was transparent: almost without exception historians of science have simply translated it as 'science' or treated it as if it meant 'science' in the modern sense, as though the two terms were interchangeable in the past;[35] or historians have thought of the term 'natural philosophy' as meaning a sort of 'general world-view', a cosmology within which the detailed, empirical investigations of science itself were performed. It is only recently that a few historians of science have regularly begun to leave the term 'natural philosophy' as it is. This has opened up the possibility of recognizing that while natural philosophy was itself indeed an investigation of the natural world which was sometimes empirical and sometimes even experimental, yet it was nevertheless one which was radically different from 'science' in the modern sense.

For the whole point of natural philosophy was to look at nature and the world *as created by God*, and as thus capable of being understood as embodying God's powers and purposes and of being used to say something about them. This is what Newton, the most famous practitioner of natural philosophy, was saying when he commented in 1692 'Et haec de Deo: de quo utique ex phaenomenis disserere, ad Philosophiam experimentalem pertinet', it is the role of experimental natural philosophy to discourse of God from the phenomena. Newton shared this view of the role and purpose of natural philosophy with all other natural philosophers, as he did his belief that the study of natural philosophy, properly conceived and pursued, was a bulwark against atheism.[36] Natural philosophy scrutinized, described, and held up to admiration the universe as the true God had created it and kept it running. To the modern ear, accustomed to the distinction between 'science' and 'religion', and to a clear-cut distinction between the 'sacred' and the 'secular', this may sound as though natural philosophy was merely an aspect of theology (and particularly that it was 'natural theology'). But this was not the case: natural philosophy was an autonomous study separate from theology and from natural theology, but whose practitioners had at the forefront of their minds, as Creator of the universe they were studying, the same God whose attributes the theologians studied from other points of view.

35 One spectacular instance we have come across is of the title of Newton's 1687 work being rendered into English as 'Mathematical Principles of Natural *Science*'.
36 See Andrew Cunningham, 'How the *Principia* got its name; or, taking natural philosophy seriously', *History of Science* (1991), **29**, 377–92.

We will probably catch the appropriate attitude of reverence toward the Creation that a natural philosopher necessarily held, if we regard natural philosophy not as a sacred study, but as a godly or pious one, which could be conducted by men both in and out of holy orders.[37] To confuse natural philosophy with science is to repeat Collingwood's nightmare about the man who had got it into his head that trireme was the Greek for 'steamer'.[38] The historical study of the trireme of natural philosophy is still at an early stage, but that does not mean that we can afford to interpret the term 'natural philosophy' as meaning the 'steamer' of science.

Thus the principles of 'projects of enquiry' and 'actors' categories' suggest that a major change in nomenclature applied by the historical actors themselves might well mark a change in the identity of the discipline under which they investigated nature. On this basis, a strong candidate for the origins of science, in the revised sense, should be the period when people stopped using the term 'natural philosophy' to refer to the identity of their project of investigating nature, and started, for the first time, to speak of 'science' or 'the sciences' referring *only* to the sciences of nature. And this period was around the beginning of the nineteenth century.[39]

Now this period is one which has already been identified as highly significant in the history of science. For one thing, it was the period when many new disciplines for the investigation of nature were created. Through Lamarck in France and Treviranus in Germany the discipline of 'biology' was created in the years around 1800 as a new discipline to replace the old study of the 'animal economy' or 'animated nature', which had dealt with the nature of those things endowed with 'soul' (*anima*, in Latin): biology now covered the same area, but now defined in terms of 'life' (*bios*, in Greek). The 'soul', the very thing which had united animals and plants as an area of study, was to be dismissed from this new discipline, and even to be regarded as something 'unscientific'. Geology, and its subsidiary sciences, was another new discipline created at this period. In the hands of Cuvier and Lyell and others, an old discipline which had been described as a '*sacred* history', and in which one had studied the earth as created and modified by God, was replaced by a *secular* history to which questions about God, Creation and Providence were deemed irrelevant and inappropriate. Radical transformations took place also in the meaning and content of other old disciplines for the investigation of nature: there was for instance a new version of physics (as Robert Fox and Maurice Crosland have shown for Laplace and the Society of Arceuil in France, and Susan Faye Cannon has shown for England). Similarly, there was a new version created of that old discipline, physiology.[40]

37 The difficulty of finding a term which conveys the sense of Christian belief typical of early modern people and which informed the natural philosopher's view of the discipline and its subject matter (nature as created by God), but which does *not* oppose such a position to notions of objectivity, secularity, and science, is itself indicative of the distance of the identity of natural philosophy from that of science.

38 Collingwood, op. cit. (32), 64.

39 The argument of this paragraph is put in more detail in Andrew Cunningham, 'Getting the game right: some plain words on the identity and invention of science', *Studies in History and Philosophy of Science* (1988), 19, 365–89.

40 M. J. S. Hodge, 'Lamarck's science of living bodies', *BJHS* (1971), 5, 323–52. Dorinda Outram, *Georges Cuvier: Vocation, Science and Authority in Post-Revolutionary France*, Manchester, 1984. Timothy Lenoir, *The Strategy of Life: Teleology and Mechanics in Nineteenth Century German Biology*, Dordrecht, 1982. Roy Porter,

This period has also been identified as the one in which it first became possible for people interested in investigating the natural world to do so as a career. Professional organizations, too, can be found at this period. Although Britain lagged far behind France and the German states in these matters, one can find at this time even in England the beginning of the professional career and the first professional organizations of science, such as the British Association for the Advancement of Science (copied from a German model) or the reformed Royal Society: 'professional' in the sense that the gentleman amateur was beginning to be replaced by the professional (salaried) man as the model type of person who pursued the knowledge of nature. In producing this new professional of science, a great part was played by the universities and other institutions of higher learning, especially the French ones which had been reformed as a result of the French Revolution, and the new secular University of Berlin, which was the model for other universities in Prussia and the other German states.[41]

It is now also recognized that in this period a new kind of site dedicated to the production of knowledge about nature first became common and came to be seen as basic to research in the sciences of nature: the laboratory. In chemistry the term 'laboratory' had long been used for the workplace, but now laboratories began to be created in physics, physiology, and later other new sciences such as bacteriology. Professors of the nature-sciences in the German universities were provided by their respective states with their own laboratories and were expected to use them to find out new things about nature. This was the basis of the laboratory research careers of Müller and Liebig, of Helmholz, Virchow and Koch, who were expected to (and did) establish research schools based on teaching laboratories. This pattern of state-provided laboratories was to be envied and emulated by French men of science, by the British and by Americans. In the laboratory research methods were taught in practice, and new generations of experimental researchers into nature were reared. It is now clear that between them they established the laboratory as the main locus

The Making of Geology: Earth Science in Britain 1660–1815, Cambridge, 1977. Susan Faye Cannon, *Science in Culture: The Early Victorian Period*, New York, 1978, especially the chapter on 'The invention of physics'. Maurice Crosland, *The Society of Arcueil: A View of French Science at the Time of Napoleon I*, London, 1967. Robert Fox, 'The rise and fall of Laplacean physics', *Historical Studies in the Physical Sciences* (1974), 4, 89–136. *The Invention of Physical Science: Intersections of Mathematics, Theology and Natural Philosophy since the Seventeenth Century* (ed. Mary Jo Nye, Joan Richards and Roger Stuewer), Dordrecht, 1992. John Lesch, *Science and Medicine in France: The Emergence of Experimental Physiology, 1790–1855*, Cambridge, Mass., 1984. W. R. Albury, 'Experiment and explanation in the physiology of Bichat and Magendie', *Studies in the History of Biology* (1977), 1, 47–131.

41 For Britain see Jack Morell and Arnold Thackray, *Gentleman of Science: Early Years of the British Association for the Advancement of Science*, Oxford, 1981. Marie Boas Hall, *All Scientists Now: The Royal Society in the Nineteenth Century*, Cambridge, 1984. On France see for instance Dorinda Outram, 'Politics and vocation: French science 1793–1830', *BJHS* (1980), 13, 27–43; Robert Fox, 'Science, the university, and the state in nineteenth century France', in *Professions and the French State 1700–1900* (ed. Gerald Geison), Philadelphia, 1984, 66–145; Robert Fox, 'Scientific enterprise and the patronage of research in France 1800–70', *Minerva* (1973), 11, 442–73. On Germany see R. Steven Turner, 'The growth of professorial research in Prussia, 1818 to 1848 – causes and context', *Historical Studies in the Physical Sciences* (1971), 3, 137–82; on Berlin, see Elinor S. Shaffer, 'Romantic philosophy and the organization of the disciplines: the founding of the Humboldt University of Berlin', in *Romanticism and the Sciences* (ed. Andrew Cunningham and Nicholas Jardine), Cambridge, 1990, 38–54.

III

424   *Andrew Cunningham and Perry Williams*

of the creation and assessment of most natural knowledge. The laboratory was made the final arbiter of truth about nature.[42]

More controversially, the growing body of work on this period enables it to be characterized as a time when the investigation of nature was changed from a 'godly' to a secular activity. This does not mean that there was a decline in religious belief (although there probably was); the important thing is that religious beliefs became private. It is possible today for scientists to have religious beliefs, but these are supposed to be irrelevant to their science; their religion is supposed to be a matter *only* of private belief. As sociologists use the term, 'secularization' means religious institutions giving way to new social institutions in matters of politics, education, social policy and morality.[43] And this kind of change was just what happened in the Age of Revolutions as new political, legal and educational institutions were established across Europe, inspired by the *philosophes*, as with the educational systems of Prussia and Hannover, or imposed by the administrators of Napoleon's Empire. Despite the political changes following from the Bourbon Restoration, most of the new legal, educational and administrative institutions were preserved intact, and even in political terms no country returned to the *ancien régime*. What we are pointing to here is that, paralleling this creation of new secular institutions, there was the creation of new secular disciplines – or the desacralizing of old ones.[44] Epitomizing which is that wonderful if perhaps apocryphal moment when Laplace said to Napoleon – when the world's top physicist said to the world's most powerful man – that he had 'no need of the hypothesis' of God in his account of the Heavenly Mechanism, having effectively, in his *Traité de mécanique céleste* (1799–1825), taken God out of Newton's universe.[45] Coupling this with the observation about the abandoning of the term 'natural philosophy' previously used for the study of the natural world as created by the Christian God, our claim can be stated most simply: that 'science' was the new collective name of the new secular disciplines for studying the natural world as a secular object, for the discovery of abstract regularities in nature and for the exploitation of natural resources, for acquiring knowledge in a secular sense and for material and social improvement.

This is a sketch of these changes in broad brush strokes. Of course, when we come to look in detail, we find that there were local contests for the meaning of the important terms, and the term 'science' in particular was used by many who did not agree that the investigation of the natural world should be a secular activity. This was especially the case in England, which frankly is an exception to this pattern. But we believe that it is the exception which proves the rule, because it was an exception at every level. Although in

42 See the articles in *The Development of the Laboratory: Essays on the Place of Experiment in Industrial Civilization* (ed. Frank A. J. L. James), Basingstoke, 1989; in *The Investigative Enterprise: Experimental Physiology in Nineteenth-Century Medicine* (ed. William Coleman and Frederic L. Holmes), Berkeley, Calif., 1988; and in *The Laboratory Revolution in Medicine* (ed. Andrew Cunningham and Perry Williams), Cambridge, 1992.

43 See, for example, Tony Bilton *et al.*, *Introductory Sociology*, London, 1981, which defines secularization as 'the process through which religious thinking, practice, and institutions lose their social significance' (p. 531). The crucial word here is 'social'.

44 Owen Chadwick, *The Secularization of the European Mind in the Nineteenth Century*, Cambridge, 1975.

45 On the Laplace story see Roger Hahn, 'Laplace and the mechanistic universe', in *God and Nature: Historical Essays on the Encounter Between Christianity and Science* (ed. David C. Lindberg and Ronald L. Numbers), Berkeley, California, 1986, 256–76.

the eighteenth century England had been the source of inspiration for French and German intellectuals, the defensive reaction of the British to the French Revolution meant that England was slow to follow the transformations in the study of nature taking place on the Continent. In England the Church and the aristocracy stayed largely in control, and Oxford and Cambridge, the main institutions of learning, were kept avowedly Christian. Exceptional institutions of learning such as the University of London (later University College London), founded in 1826 to represent the radical, God-less and Utilitarian view, were deeply controversial. Even the attempt to found the British Association for the Advancement of Science on the German model of the Gesellschaft Deutscher Naturforscher und Ärzte was almost immediately hijacked by Cambridge dons such as William Whewell and devoted to the pursuit of a Christianized contemplative knowledge of nature.[46] Until mid-century, the most forward-looking amongst young British men went to France and the German states to study. Not until around 1860 did corresponding institutional and intellectual changes happen in England and enable it to catch up with the Continent; hence although we can find individual attempts to model English investigation of the natural world on the Continental pattern, yet still most such investigation by the English remained God-centred – that is, it remained Natural Philosophy – until then.[47]

We are proposing that the origins of science can be located as one aspect of the Age of Revolutions (with England as a partial exception, in that changes there took place later and more gradually than in Continental Europe). These revolutions, as conventionally characterized, are: (1) the French Revolution, beginning in 1789, which was a political revolution, concerned with radically transforming the political organization of society; (2) the industrial revolution, beginning in Britain in the 1770s, a revolution in the means of production, exchange and ownership of the wealth or resources of society; and (3) the post-Kantian intellectual revolution, centred on the German states, a revolution in what one should think and in who should be the intellectual masters of the future. As a result of these simultaneous and linked revolutions, a new middle class was consolidated, wielding the political power, the industrial power, and the intellectual power.

To locate the origins of science in these events nicely conforms to the second historiographic principle outlined above, that the knowledge of any society is an integral product of that society. For it is to be expected that the invention of a new form of intellectual activity should be the product of a major social change, such as the consolidation of a new social class with new power bases.

Locating the origins of science in the Age of Revolutions is also supported by the first

---

46 J. B. Morrell, 'Brewster and the early British Association for the Advancement of Science', in *'Martyr of Science': Sir David Brewster 1781–1868* (ed. A. D. Morrison-Low and J. R. R. Christie), Edinburgh, 1984, 25–9; Morrell and Thackray, op. cit. (41), especially 63–76, 165–75. For an argument that Whewell's entire philosophy was part of a programme to bring the investigation of the natural world back into the service of the Christian God and the established social order, see Perry Williams, 'Passing on the torch: Whewell's philosophy and the principles of English university education', in *William Whewell: A Composite Portrait* (ed. Menachem Fisch and Simon Schaffer), Oxford, 1991, 117–47.

47 Frank M. Turner, 'The Victorian conflict between science and religion: a professional dimension', *Isis* (1978), **69**, 356–76. W. H. Brock and R. M. Macleod, 'The scientists' declaration: reflexions on science and belief in the wake of *Essays and Reviews* 1864–5', *BJHS* (1976), **9**, 39–66. Ruth Barton, 'The X Club: Science, Religion, and Social Change in Victorian England', Ph.D. thesis, University of Pennsylvania, 1976.

historiographic principle, that the norms and values of science require explanation rather than being truths derived from some transcendent realm. For the emergent middle class, or to be more precise, the emergent *professional* middle class, drew its authority from all three revolutions, the intellectual, the political and the industrial. First they drew authority from the intellectual revolution, following which primacy was given to the autonomy of ideas and to the attendant concepts of 'originality' and 'genius', with intellectual achievement being considered to be above the market and something to be judged only by one's peers. Secondly the professional middle class drew authority from the political revolution, as developed into the ideology of liberalism: a philosophy not only of free trade but also of free enquiry (especially criticism of the old powers, Church and aristocracy); a philosophy arguing for the establishment of a new aristocracy of intellectual talent – a 'meritocracy', as we would say today. Thirdly the professional middle class also drew authority from the industrial revolution, for although this benefited the commercial-industrial middle class more directly, the two classes at this time saw their interests as essentially coincident and worked closely together, for example in promoting a vision of progress and prosperity, both social and material, in which the professional middle class's secular knowledge, disseminated by new secular education, would be vital. In these three revolutions can be seen the origin of values and aims – for example, genius, free enquiry, free exchange of ideas, objectivity, disinterestedness – which have been made an integral part of the identity of science.[48]

Finally, the location of the origins of science in the Age of Revolutions is supported by the observation that this period also saw the start of many of the particular stories about the history of science which were handed down to Butterfield's generation, and hence down to our own time. In the same way that long, distinguished national traditions were constructed to support the existence of the new nation states,[49] so a long, distinguished intellectual tradition was now also constructed to support the existence of the new enterprise, science.[50] Building on the twin traditions created in the eighteenth century by

48 On intellectual transformation and liberalism, see Raymond Williams, *Culture and Society, 1780–1950*, London, 1958, especially Introduction and chs. 1–2; M. H. Abrams, *The Mirror and the Lamp: Romantic Theory and the Critical Tradition*, London, 1971; Marilyn Butler, *Romantics, Rebels and Reactionaries: English Literature and its Background 1760–1830*, Oxford, 1981; R. Steven Turner, 'The growth of professorial research in Prussia, 1818 to 1848 – causes and context', *Historical Studies in the Physical Sciences* (1971), 3, 137–82; T. W. Heyck, *The Transformation of Intellectual Life in Victorian England*, London, 1982; Christopher Harvie, *The Lights of Liberalism: University Liberals and the Challenge of Democracy 1860–86*, London, 1976; Irene Collins, 'Liberalism and the newspaper press during the French Restoration, 1814–1830', *History* (1961), 46, 17–32; R. Hinton Thomas, *Liberalism, Nationalism and the German Intellectuals 1822–1847*, Cambridge, 1951.

49 See for example Eric Hobsbawm and Terence Ranger (eds.), *The Invention of Tradition*, Cambridge, 1984.

50 Apparently some people find this claim difficult to countenance; but an argument of this kind was being made as long ago as 1970, by Thomas Kuhn in *The Structure of Scientific Revolutions*, in the chapter on 'The invisibility of revolutions':

> Characteristically, textbooks of science contain just a bit of history, either in an introductory chapter or, more often, in scattered references to the great heroes of an earlier age. From such references both students and professionals come to feel like participants in a long-standing historical tradition. *Yet the textbook-derived tradition in which scientists come to sense their participation is one that, in fact, never existed.* For reasons that are both obvious and highly functional, science textbooks (and too many of the older histories of science) refer only to that part of the work of past scientists that can easily be viewed as contributions to the statement and solution of the texts' paradigm problems [op. cit. (20), 138, our emphasis.]

the French *philosophes* and the German Romantics, early nineteenth-century men of science located science's origin at one or the other of the periods in which these two groups had located the origin of their own forms of philosophy, namely the seventeenth century and ancient Greece respectively.[51] As they co-opted the people of these two periods to serve their own political and social ambitions, the new men of science naturally recast them in a contemporary mould: the ancient Greeks were represented as having established free enquiry into the natural world, appealing only to natural principles and not to gods, while Galileo, Bacon, Descartes, and Newton were represented as having led the way back to this Greek way of thinking and then gone beyond it by establishing a physicalist, mechanist science, free from the constraints and superstitions of religion.[52] This period, in other words, saw not only the beginning of the history of science, but also the origin of the specific 'Whiggish' or 'present-centred' traditions of the history of science which our own generation of historians have inherited, and from which we have been trying to move away for some twenty-five years. It was also this period which saw the origin of that tradition of political history which Herbert Butterfield first called 'the Whig interpretation'; this was created by Whigs or liberals writing constitutional histories, in which all the causes for which they were currently fighting were presented as having been anticipated long ago and fought for for centuries, so that their values seemed transcendent and eternal, their own victory historically-ordained.[53] Here, surely, is the reason why it has been so natural to adopt critically for the history of science a term originally coined in the context of political history: the triumphalist progressivist 'Whig' traditions in the history of politics and the history of science are twin traditions: they were constructed at the same time, by some of the same people, in the service of the same political interests.[54]

To sum up, historical scholarships over the last twenty years enables us to identify the Age of Revolutions as the period which saw the origin of pretty well every feature which is regarded as essential and definitional of the enterprise of science: its name, its aim (secular as distinct from godly knowledge of the natural world), its values (the 'liberal' values of free enquiry, meritocratic expert government and material progress), and its history.

For some concrete examples of how revolutionary changes have been made invisible by newly-constructed histories, we recommend *Functions and Uses of Disciplinary Histories* (ed. Loren Graham, Wolf Lepenies and Peter Weingart), Dordrecht, 1983.

51 On the *philosophes* see Christie, op. cit. (4), 7–8; I. Bernard Cohen, 'The eighteenth-century origins of the concept of scientific revolution', *Journal of the History of Ideas* (1976), **37**, 257–88. For the classic expression of the Romantic view of the Greeks, and of the supposed 'rediscovery' and emulation of the Greek approach in the Renaissance, see Jacob Burckhardt, *History of Greek Culture*, originally published as *Griechische Kultur*, 1898–1902; and his *Civilization of the Renaissance in Italy*, originally published as *Kultur der Renaissance*, 1860. On the historical roots of this image of the Greeks see Martin Bernal, *Black Athena: The Afroasiatic Roots of Classical Civilisation*, Vol. 1, *The Fabrication of Ancient Greece 1785–1985*, London, Free Association Press, 1987; Wallace K. Ferguson, *The Renaissance in Historical Thought*, Cambridge, Mass., 1948.

52 For an example of 'Whig' history of science being deployed for political ends, see John Tyndall, 'The Belfast address', in *Fragments of Science*, 6th edn, 2 vols., London, 1879, ii, 137–203.

53 See Butterfield, op. cit. (33). See also Peter J. Bowler, *The Invention of Progress: The Victorians and the Past*, Oxford, 1989.

54 For a perfect example of a Whig politician and historian creating a 'Whig' view of the history of the investigation of the natural world, see Lord Macaulay's famous essay on Francis Bacon.

It is on the strength of this scholarship that we propose that the origins of science should be regarded as being in this period, rather than in the seventeenth century. We must emphasize that we are not arguing for Butterfield's 'origins of modern science' or 'the scientific revolution of the seventeenth century' to be simply moved forward 150 years. Most historians who claim to have identified a fundamental change around 1800 have seen it as akin to the old 'scientific revolution'; indeed, some of them have even called it a 'second scientific revolution'.[55] But to see the change in this way is to stay within the old big picture, based on the old assumptions about the nature of science; thus the change is presented as just a further development in the supposedly universal and eternal human enterprise of investigating nature, just a matter of the same thing ('science') being suddenly done much better, with greater intensity, and being better organized. By contrast, what we are proposing in this paper is something more fundamental: that this period saw the origins of science, in the revised sense: that it saw the *creation* of science's particular and definitive aims, values and practice, not by derivation from some transcendent realm, but as a result of particular human activity in response to the local conditions of material life: an event not of *emergence* but more of *invention*. This term 'invention', which is our preferred term, helps to fix the revised view of science as a contingent, time-specific and culture-specific activity, as only one amongst the many ways-of-knowing which have existed, currently exist, or might exist; and for this reason the phrase which we propose for the fundamental changes which took place in this period is 'the invention of science'.[56] And we can of course now drop the qualifier 'modern', since the term 'science' can only be properly applied in our own time, the modern era.[57] What we are speaking of is therefore not the origins of modern science, but the modern origins of science.

This radical way of interpreting these changes may, we hope, be found useful for teaching and as a heuristic for future research. For the purposes of this paper it enables us to identify the boundaries of science in time and space and culture: to map out the

55 As far as Cohen (op. cit. (4), 97) could find, the term was first used by Thomas S. Kuhn ('The function of measurement in modern physical science', *Isis* (1961), **52**, 161–93, 190). However, the term is now also used in other ways; for example, Stephen G. Brush, *The History of Modern Science: A Guide to the Second Scientific Revolution, 1800–1950*, Ames, Iowa, 1988, uses it to refer to a *late*-nineteenth, early-twentieth century revolution, associated with the breakdown of classical physics. One author who as long ago as 1974 saw the change around 1800 as having been something like as fundamental as we are proposing was Arnold Thackray. He pointed out that in the 1750s,

> not only was the very name and function of the scientist not yet invented, but *science in the sense we know and use the term was unfamiliar to the English-speaking world of the mid-eighteenth century*. Natural knowledge certainly existed, and…the period's philosophers or men 'deep in knowledge' certainly included many…'well-versed in natural philosophy'. But the professional norms, occupational structures, values, goals and rewards associated with the scientist were as unknown as the word. [Our emphasis. 'The industrial revolution and the image of science' in *Science and Values* (ed. Arnold Thackray and Everett Mendelsohn), New York, 1974, 3.]

56 There is further discussion of the term 'invention' in Cunningham, op. cit. (39).

57 We think it necessary to make this point explicitly, because the report (in the *Newsletter* of the British Society for the History of Science) of our original conference paper, on which this present paper is based, suggested that we were attempting 'to seek transcendent criteria of "modernity"'. We were not. As we hope is now clear, we are not attempting to seek transcendent criteria of *anything*. In fact, that is exactly what we are arguing against.

boundaries of that part of the big picture which we occupy. And as the advocates of the 'scientific revolution' believed their story conformed to the historical evidence and accurately represented what had happened in the past, so we believe 'the invention of science' tells the true story about the origins of science: that those origins are modern.

DE-CENTRING THE 'BIG PICTURE'

History may be servitude,
History may be freedom. See, now they vanish,
The faces and places, with the self which, as it could, loved them,
To become renewed, transfigured, in another pattern.
    T. S. Eliot, *Little Gidding*[58]

On this view, the history of science becomes a relatively short and local matter: extending back less than 250 years, and largely confined to western Europe and America. What are we to say of the rest, of analogous knowledge before this time or in other cultures? In the old big picture, other forms of knowledge appeared as early and more primitive versions of science; thus big picture histories which had chapters on Indian and Chinese 'science' tended to place them after the chapters on tribal societies but before the chapters on ancient Greece. This evolutionary view (in accordance, that is, with evolution as popularly understood rather than as in neo-Darwinian theory),[59] showing growth and progress taking place along a single line, was possible only on the assumption that the pursuit of science was a fundamental part of human nature, a universal enterprise transcendentally derived. For a new big picture, in which science is just one amongst a plurality of ways of knowing the world, other forms of knowledge must be allowed to appear on their own terms, instead of being measured against a scientific framework. Science will appear only as the native knowledge-form of our own culture, not in a central or special place. What is required, we might say, is a kind of 'de-centring'.

'De-centring' is a psychologists' term, deriving from the theories of Mead and Piaget, according to which very young children's understanding of the world is at first entirely 'centred' on themselves; 'de-centring' thus describes the process by which children come to realize that external objects have permanence, that other people can have different visual perceptions of the same scene, and that other people can have different knowledge, interests, feelings and so on.[60] In a more general sense, de-centring is something which we continue to do repeatedly throughout our adult lives, as we identify yet another aspect of our own egotism, and realize that something which we thought was universal is actually

---

58 Excerpt from 'Little Gidding' in *Four Quartets*, copyright 1943 by T. S. Eliot and renewed 1971 by Esme Valerie Eliot, reprinted by permission of Harcourt Brace & Company, and Faber and Faber Ltd.

59 Strict neo-Darwinian theory, based on natural selection, implies that evolutionary lines branch like the twigs of a bush: a non-linear and non-hierarchical view of evolution (see, for example, Stephen Jay Gould, *Wonderful Life: The Burgess Shale and the Nature of History*, London, 1989, 27–45). Nevertheless, the popular view of evolution is one of a linear, progressive ascent. Peter Bowler has pointed out that it was *this* view of evolution which prevailed following the publication of the *Origin of Species*, not the non-teleological view which Darwin's theory of natural selection implied (*The Non-Darwinian Revolution: Reinterpreting a Historical Myth*, Baltimore and London, 1988).

60 See, for example, George Herbert Mead, *Mind, Self, and Society* (ed. Charles W. Morris), Chicago, 1934; Jean Piaget, *The Child's Conception of Reality* (tr. Margaret Cook), London, 1955.

peculiar to ourselves, or our group, our class, our nation, or our culture. To see science as a contingent and recently-invented activity is to make such a de-centring, and to acknowledge that things about our primary way-of-knowing which we once thought were universal are actually specific to our modern capitalist, industrial world.

This kind of de-centring is already beginning to be made in the history of religion, which until recently, in western European countries, was firmly centred on Christianity. This was quite explicitly and unapologetically the case in traditional providential Church History, in which other religions figured only in relation to Christianity; most spectacularly, Judaism was relegated to the status of a precursor. But even in the more liberal viewpoint, in which different faiths are thought of as moving towards the same end, or as being different parts of the same truth, the Christian-centredness remains, because that element of sameness is conceived in Christian terms; features of Christianity, such as monotheism or belief in the existence of an immortal soul, are taken to be constitutive of religion in general.[61] But during the course of the last two decades, a few writers, mainly working in comparative religion or (in Britain) the teaching of Religious Education, have begun to build a view of religion and its history that is not centred on Christianity but which tries to treat at the very least all the major world faiths symmetrically.[62]

The problem which we face in the History of Science is essentially the same. We too have the legacy of a big picture in which historical events have been interpreted as leading towards our own culture, providentially guided from a transcendent realm – in this case the transcendent element being objective truth, goodness or human nature, rather than the Christian God. There are parallels between the way in which the first practitioners of Christianity attempted to erase the separate identity of Judaism and the way in which the first practitioners of science attempted to erase the separate identity of natural philosophy; in both cases, the older texts were taken over by the new practice and reinterpreted as marking early stages in its development: a reinterpretation that was validated only by assuming the transcendent, eternal and universal nature of Christianity, in the one case, and of objective scientific knowledge, in the other. Just as we need a big picture of religion which does justice to the separate identity of Judaism and other non-Christian faiths, so we need a big picture of the history of science which does justice to the separate identity of natural philosophy and other past and present ways of knowing the natural world.

It is not too difficult to imagine a new big picture which is de-centred from our own position along the axis of time, which grants a separate identity to, for example, ancient

61 This Christian-centredness is revealed, for example, in the common statement that 'different religions are worshipping God in different ways'.

62 Ninian Smart, *The Religious Experience of Mankind*, New York, 1969, London, 1971; Don Cupitt, *Taking Leave of God*, London, 1980; Wilfred Cantwell Smith, *Towards a World Theology: Faith and the Comparative History of Religion*, London, 1981; Keith Ward, *A Vision to Pursue: Beyond the Crisis in Christianity*, London, 1991; Jean Holm, *The Study of Religions*, London, 1977. For both Christian priests and teachers of Religious Education, the practical political problems of living in a liberal multi-cultural (hence multi-faith) society have been an important stimulus to the development of this view. Interesting possibilities for an account of the invention of religion, paralleling our account of the invention of science, are raised by Wilfred Cantwell Smith, *The Meaning and End of Religion*, New York, 1962, London, 1978; as John Hick says in his foreword to the 1978 edition, 'he shows with full historical evidence that the concept of religions, as contraposed ideological communities, is a modern invention which the West, during the last two hundred years or so, has exported to the rest of the world' (p. xi).

Greek philosophy, medieval natural philosophy, and the modified forms of natural philosophy developed in the early modern period; historiography for these periods already exists which points clearly in this direction. De-centring from our own position in space, however, is a more difficult step to take; we are much less aware of the existence, let alone the separate identity, of ways of knowing the natural world outside our own culture, or even of those which we claim as ancestors. One way of seeing the necessity for such a de-centring is to look at a map of the world, preferably in the Peters projection,[63] and consider that almost all the material with which the History of Science discipline has been concerned comes from a tiny geographical area, about the same size as Zaire or the Sudan, and considerably smaller than Brazil. The only thing that is unusual about the countries in this area, apart from the fact that they are where we live, is that it was these countries which rose to world-domination during the nineteenth century, through the formation of overseas empires. It was only this historical accident that has meant that what began as their own native culture – by that time including that recent invention, science – has now become world-culture.[64] A spatially or rather geographically de-centred big picture would treat all native knowledge-forms with perfect symmetry.

But even de-centred temporally and spatially in this way, our big picture might still retain its present almost exclusive focus on cognitive knowledge, following our culture's peculiar elevation of theoretical knowledge to a higher status than practical knowledge. To de-centre further, then, we would need a big picture which dealt not only with cognitive knowledge, the knowledge of fact, but also with practical knowledge, the knowledge of skill: not only with *know-of*, but also with *know-how*. To some extent, this de-centring is already being made, particularly with the work on skills now being done in the Sociology of Scientific Knowledge, and with the growth of the History of Technology as a discipline. But even this new work retains a very strong centre, being concerned with subjects such as measurement and calculation, power sources and industrial processes: those things that have importance principally in industrial societies, and principally for men within industrial societies. Feminist and environmentalist analysis points to some further de-centring which is needed: away from the rather specialized technical achievements, such as travelling very fast, and building tall buildings, and killing people in large numbers; and towards more basic and more generally-appreciated technical achievements, such as making sure that people have enough to eat, and clothing to keep them warm, and a place

63 The Peters projection is claimed to provide a more accurate representation of the relative size of the Earth's major land areas than the more-familiar Mercator projection, which exaggerates the area of countries further away from the Equator (e.g. Europe). See Arno Peters, *Der Europa-zentrische Charakter unseres geographischen Weltbildes und seine Überwindung*, Dortmund, 1976.

64 Much of the literature on the exporting of Western knowledge is based on the assumption that the process has been more-or-less successful; for example, Peter Buck, *American Science and Modern China, 1876–1936*, Cambridge, 1980; James R. Bartholomew, *The Formation of Science in Japan*, New Haven and London, 1989. But interestingly, some recent works have emphasized the *difference* of the non-Western traditions, hence questioning how completely Western science retained its identity when transplanted. See for example Arnab Rai Choudhuri, 'Practising Western science outside the West: personal observations on the Indian scene', *Social Studies of Science* (1985), **15**, 475–505; *Science, Hegemony and Violence: A Requiem for Modernity* (ed. Ashis Nandy), Oxford, 1988; Masao Watanabe, *The Japanese and Western Science* (tr. Otto Theodor Benfey), Philadelphia, 1991, original Japanese edn 1976.

to sleep, and comfort and healing when they get sick. To incorporate such know-how within a big picture should surely be a feasible goal.[65]

De-centring could be taken still further. Cognitive knowledge – the knowledge of fact, and technical knowledge – the knowledge of skill, are only two aspects of the knowledge which each one of us possesses. There is also what we might call relational knowledge: the knowledge of acquaintance, the knowledge by which each of us relates to another, person to person, and in small groups, and in larger social and political units. There is also moral knowledge: the knowledge of value, by which each of us judges what is right and wrong. Would it be possible to write a history of human knowledge which, instead of following the positivist legacy of elevating cognitive knowledge (and only one kind of cognitive knowledge, at that) above all the rest, would treat fact, skill, human relations and morals with perfect symmetry?[66]

We do not apologize for raising such questions, even if we cannot yet see how they might be followed out. A big picture, as we understand it, is not just a summary of research results or a general theory of history, but a vision of the world and our place in it; and if such a vision is to remain alive, to grow and to change, then it is necessary for its limits to be constantly pushed beyond what is currently imaginable. Of course, all of us will probably continue to remain specialists, more or less; even when teaching general courses, for some considerable time to come we are likely to continue to specialize in the dominant Western traditions of knowledge – philosophy, natural philosophy, and science. But since these specialisms derive their meaning and identity from the big picture in which, explicitly or implicitly, they are placed, we should do our best to ensure that the big picture we use is one in which we believe and which is appropriate to our time and place and culture. We owe this to our students, to our public, and to ourselves.

---

65 This paragraph was inspired by Ursula K. Le Guin, 'The carrier bag theory of fiction', in her *Dancing at the Edge of the World: Thoughts on Words, Women, Places*, London, 1989, 165–70. See also Joan Rothschild (ed.), *Machina ex Dea: Feminist Perspectives on Technology*, New York, 1983. An interesting recent attempt at a 'big picture' history from an ecological perspective – i.e. of humanity's changing relationship to its environment, through the development of agriculture and industry – is Clive Ponting, *A Green History of the World*, London, 1991.

66 This paragraph was inspired by Hilary Rose, op. cit. (22).

IV

Science and Religion in the Thirteenth Century Revisited: The Making of St Francis the Proto-Ecologist Part 1: Creature not Nature

1.   Saints and Scientists

Discussion of the relation of science and the Christian religion in the past, especially in the medieval and early modern periods, is receiving renewed attention currently, at least among Anglophone scholars (Brooke, 1991; Brooke and Cantor, 1998; Moore, 1992). Not only are scholarly books on the theme being written which open new lines of approach, but also lectureships are being founded, and new postgraduate courses being established.[1] Our customary textbook accounts of the history of science, which are taken as authoritative statements of the relation of science and religion in the past by those outside the profession, present a picture of men of religion in the medieval period happily functioning also as men of science. Look at any book describing itself as a history of medieval science and you will find it full of men of religion, such as bishops and friars, being portrayed as early scientists. Among the most notable of these are Robert Grosseteste (c. 1175–1253), Bishop of Lincoln; Roger Bacon, Franciscan friar (c.1214–68); Albertus Magnus, Dominican friar (1193–1280); John Pecham, Franciscan friar and Archbishop of Canterbury (c.1230–92). Beyond the thirteenth century there is Nicholas of Cusa, Bishop of Brixen and Cardinal (1401–64), and others on into the seventeenth century. Recently an early saintly woman, Hildegard of Bingen (1098–1179), has been added to the roster of saint-scientists. The historian of science Ernan McMullin, while perhaps exceptional in his directness, encapsulated

---

[1] A lectureship has recently been established in Cambridge University on donated funds. A new postgraduate course has been established at the University of Leeds.

This article was published in *Studies in History and Philosophy of Science* 31.4 (2000), pp. 613–643. © 2000 by Elsevier Science Ltd.

the consensual view when he wrote: 'In the early growth of this science [i.e. physi-cal science], bishops like Grosseteste, Albert of Saxony and Oresme, as well as priests like Albertus Magnus, Bradwardine and Buridan had played a decisive part. In Galileo's own day, the contributions to science of priests like Copernicus and Mersenne were known to all...'. [2] So historians of science seem to have no problem with the view that science and religion could co-exist in the medieval period as practises and ways of thinking, and that they could also be practised at one and the same time and by one and the same person:[3] the professional religious, either as theologian or as a regular of one or other religious order, or as a priest. This is a view that I intend to question. The larger significance of raising this issue is that it leads on to questioning the very possibility of medieval science. That is to say, through the present case-study I shall be exploring the minimum criteria which would need to be satisfied if we are to be able to claim with assurance and evidence that any people in the medieval period pursued science (in our customary meanings of that term).

It is not possible here to look at all of the medieval men and women of religion to whom the practice of science has been ascribed, so I shall take a single example here, but stressing that the same kind of exercise could be made for each and all of the others too. My chosen example is a newcomer to the ranks of those medieval religious persons with an apparent stake in modern science: St Francis, in his role as proto-ecologist. He has been given this role by several historians in the last few decades, who have claimed that Francis had some universal 'love of Nature' which, while it was linked to religious belief and practice, was also separable from any particular practice, and which could and did make him a Saint of Nature. As Nature is what ecological scientists study and ecological campaigners seek to protect, this 'love of Nature' on the part of St Francis, and especially his concern for animals and plants, is implicitly taken by such historians to mean that Francis had a scien-tific (or at the very least, a proto-scientific) attitude to Nature. Edward Armstrong

---

[2] McMullin (1968), p. 32. See, for other examples among recent works by eminent scholars, Lindberg (1992) and Grant (1996); for an earlier version of this see, for example, Crombie (1952).

[3] Most historians of medieval science could be quoted to this effect. Edward Grant, for instance, recently wrote: 'the Christian accommodation with Greek science and philosophy, [was an] instrumental condition that facilitated the widespread, intensive study of natural philosophy during the late Middle Ages' (Grant, 1996, pp. 8–9). Again, 'With theology and natural philosophy related so intimately during the Middle Ages, and with arts masters forbidden to apply their knowledge to theology, it remained for the theologians to interrelate these two disciplines, that is, to apply science to theology and theology to science ... Theologians had remarkable intellectual freedom and rarely permitted theology to hinder their inquiries into the physical world. If there was any temptation to produce a "Christian science", they successfully resisted it ... The positive attitude of medieval theologians toward natural philosophy, and their belief that it was also a useful tool for the elucidation of theology, must be viewed as the product of an attitude that was developed and nurtured during the first four or five centuries of Christian-ity ... The amazing lack of strife between theology and science is attributable to the emergence [in the thirteenth century] of theologian-natural philosophers who were trained in natural philosophy and the-ology and were therefore able to interrelate these disciplines with relative ease' (pp. 84–85). He repeats these sentiments on pages 174–6. Although Grant sometimes uses the term 'natural philosophy' to refer just to physics in the medieval period, here and elsewhere in his book he is using it rather as a synonym for 'science'.

in his book, *St. Francis: Nature Mystic. The Derivation and Significance of the Nature Stories in the Franciscan Legend* (1973), looks towards what he terms 'the nature stories' in the Franciscan legend, and tries to account for the 'ephemeral efflorescence of *regard for nature* in medieval Italy' which he believes occurred around St Francis. In Armstrong's view, St Francis was 'an outstanding representative of that long line of Christians who, throughout the centuries, *delighted in and cared deeply for nature* in all its exquisite manifestations'. This 'Christian *compassion for nature*' was one of Francis's outstanding characteristics, according to Armstrong, who believes that by scrutinizing the Franciscan legends concerning Francis and the birds, the beasts and the plants, 'we may gain further insight into the mind of the saint, the intellectual climate of his time, and the outlook of his biographers' (Armstrong, 1973, pp. 2–3, 41, emphasis added). The course of Armstrong's book shows that what he means by this is that the time of St Francis was a time of great interest in Nature, that Francis himself had a great interest in Nature, and that Francis personally served as a focus and channel of such interest. Roger Sorrell, in his book *St Francis of Assisi and Nature. Tradition and Innovation in Western Christian Attitudes toward the Environment* (1988) seeks instances within the ascetic tradition of Christianity for the 'general feeling of affinity with the rest of creation', which he believes typical of St Francis (Sorrell, 1988, pp. 7–8, 31).[4] The thrust of Sorrell's book is that Francis must have been—and was—aware of the beauty and usefulness of creation; and while Sorrell believes this was a thorough-goingly religious attitude on the part of Francis, yet he also believes that it was similar to the modern environmental or ecological sentiment. These ascriptions to St Francis of a 'love of Nature' and of being a proto-ecologist have been taken up by scientists. For instance, a famous modern 'ecological' bacteriologist, René Dubos, has claimed that 'It is not unlikely that the Franciscan worship of nature, in its various philosophical, scientific, and religious forms, has played some part in the emergence of the doctrine of conservation in the countries of Western civilization and its rapid spread during the last century' (Dubos, 1974, p. 124).

Two things are being claimed by such writers. One is that St Francis himself had a love of and concern for Nature, which was the manifestation of an ecological or environmental awareness and conscience on his part (whatever else it might also have represented to him), and that this concern for Nature on his part was inherently scientific, both passionate and dispassionate, as it were. The second claim being made here is that his awareness of Nature fits St Francis to be seen as a proto-ecologist, and thus to be seen as an originator or precursor of the modern ecological movement and of the science of ecology—precursor, that is, of what

---

[4]It should be noted that both Armstrong and Sorrell take precautions to avoid anachronism with their vocabulary and concepts, and both are very aware of the Christian and spiritual dimension of what they regard as Francis's love for nature. However, as will appear below, I do not think they take the right precautions.

Dubos calls 'the doctrine of conservation ... during the last century', which covers both the technical science and the campaigning movement.

In addition to these accolades from historians and scientists, in November 1979 the pope also declared St Francis to be the patron saint of practitioners of ecology. In the words of the papal *Act*, Francis was declared patron saint of the cultivators of ecology (*oecologiae cultores*), an expression which, in its context, was probably intended to embrace both the practitioners of the modern scientific discipline of ecology and also the ecological campaigners who see the environment as a fragile resource which needs to be responsibly managed (*Acta Apostolicae Sedis*, 1983, pp. 1509–10).

Although at one level the attribution of the roles of proto-ecologist and patron saint of ecologists to a medieval saint is merely sentimental and symbolic, and should not be taken too seriously, at another level it is quite clear that, for those making these attributions, they are meant as perfectly serious accounts of St Francis and the early Franciscans, as any reader of the writings of Armstrong or Sorrell will discover. Such attributions have significant consequences, which cannot be dismissed lightly. They perform work for certain interested parties. On the one hand, from the history of science side, they contribute to shaping in a particular way the history of science stories that professional historians of science produce: that is, shaping them to buttress the claim that medieval men of religion were often also men of science. This, in its turn, supports a particular view of the nature of science in the past and the present—a view of what kind of practice it was and is—and of the nature of the relations between science and religion in the past and the present. And on the other hand, from the religious side, such attributions can be borrowed by advocates of particular religious positions today, and in particular by members of the Roman Catholic Church, to buttress the mirror-image claim: that medieval men of science were often also men of religion. In its turn this too supports a particular view of the nature of religion (especially Catholicism) in the past and the present—a view of what kind of practice it was and is—and of the nature of the relations of religion and science in the past and the present. So, however innocently or indulgently historians may make such claims, whether about St Francis or Grosseteste or Roger Bacon, or any other medieval man or woman of religion, they need to remember that there are interested parties ready to make serious use of light-hearted attributions.

This recent appearance of St Francis, a thirteenth-century saint, as proto-ecologist and as the patron saint of practitioners of ecology, a nineteenth- and twentieth-century science, raises several questions about the criteria for a legitimate history of 'medieval science'. First, what was St Francis's own attitude to Nature, if indeed he had one? The first of these two articles is dedicated to exploring this question. As will be seen, I shall not follow Armstrong and Sorrell in gathering together the apparent evidence for St Francis as an early sympathizer with Nature, and weighing up its validity toward making the case for Francis as an early ecologist. For this approach starts from the wrong premise—viz., that Nature is an appropriate category to use in our investigation here. While I shall certainly be investigating the thirteenth- and fourteenth-century legends of St Francis, yet I shall not be trawling them for anything which might look like modern environmentalism. My conclusion to this first part will be that St Francis did not have—and indeed could not have had—an attitude to Nature (in the sense of 'Nature' meant by Armstrong or Sorrell or by historians of science and of 'medieval science' in general), let alone the particular attitudes to Nature they ascribe to him. Thus my next questions will be: how and why did St Francis get *given* his great reputation as the Saint of Nature, and thence his reputations as proto-ecologist and appropriate patron saint of ecology? The second of these two articles will be devoted to answering these questions. As will become evident, for me these are questions centrally about the history of the Catholic Church in the nineteenth and twentieth centuries, and about the creation of a modern ecological conscience—not questions about St Francis or his putative attitude to Nature.

The study of anything to do with St Francis, including the present topic, is usually bedeviled by two problems, which we must dispense with before we can go further. One is that we come over-prepared to the investigation, for in the West we are all brought up with visual images of St Francis and animals, especially preaching to the birds (Fig. 1), and we have almost all of us also heard some of the stories of St Francis and other animals, such as the wolf of Gubbio, whom St Francis chastised for terrorizing the town-folk of Gubbio, and who thereafter became tame and friendly. These residual mental images mean that when we hear that St Francis has been proposed as a proto-ecologist or as patron saint of ecologists, then such roles for him seem to us natural and self-evident. There doesn't seem to us to be an issue to investigate, since this image of St Francis as Nature-lover seems timeless. However, as will be seen below, such popular images of St Francis are themselves not at all ancient and are in fact part of a modern construction of his characteristics and attributes, and in themselves prove nothing about the historic St Francis. The second bedevilment is that the study of St Francis in modern times has almost all been conducted by people with religious interests, and usually with a Catholic commitment to the sainthood and perfection of St Francis. Such people customarily believe that St Francis casts over all posterity since his time, including themselves, a transcendent charm, even a sort of spell (if Christian saints can indeed legitimately be said to cast spells). In the case, for instance, of the books of Edward Armstrong and Roger Sorrell, one finds their authors both assuming that St Francis is somehow mysteriously 'reaching out' to us or 'speaking' to us across the centuries, and that he *already* has a particular 'message' for us about Nature and our proper relation to the environment, and it behoves us to work out what this message is and to take heed of it. Armstrong, for instance, expresses hope that a survey of what he calls 'Franciscan natural history' will allow us to 'enlarge our vision and sympathies and at the same time come closer to Saint

Fig. 1. St Francis preaching to the birds. Engraving by J. Pelota after the thirteenth-century fresco by Giotto, held in the Louvre. Picture: Mary Evans Picture Library.

Francis' (Armstrong, 1973, p. 3), while Sorrell even dedicates his 1988 book 'To St Francis of Assisi, that most gentle and creative man, whose legacy all humanity lovingly passes down through the ages'. But St Francis is dead. Although for believers it may be feasible to think that we can hear dead saints sending messages from the past to the present, I really think we have to put aside such fetishism if we are going to produce dispassionate history about medieval people of religion and their involvement (or otherwise) in 'medieval science', since such an approach begs all the important questions. Unquestioned prior commitment to any image of St Francis—whether that of the 'Povorello', the Saint of Nature, the Saint of Personal Liberation, or the saint who reconciled science and religion, or whatever— is very likely to impede one in finding out the historical source and meaning of such images, which is what I am seeking to do here. This view of St Francis as having a 'message' for posterity is again part of the construction of his image in the nineteenth and twentieth centuries.

2. St Francis and Nature: A False Question

The modern historian who has had the most impact in encouraging people to think of St Francis as a proto-ecologist is Lynn White, Jr., an academic historian of medieval technology. In 1967 White published a piece in the American general science journal *Science*, under the title 'The Historical Roots of our Ecologic Crisis'. It has been reprinted many times, as ethicists have extended their concerns to environmental issues in recent years. It is regarded as one of the first pieces to have raised modern ethical issues about man's relation to his environment. White's argument is that the exploitation, and more particularly the *over*-exploitation, of Nature is uniquely Western, and dates from the High Middle Ages, having been made yet more of a goal during the Scientific Revolution of the seventeenth century, and having continued with increasing rapaciousness up to and including modern times. But not only is this over-exploitation of Nature Western, it is also, White claims, uniquely Christian. White describes Christianity as the most anthropocentric of all religions. Believing God had given him dominion over all other creatures, Christian mankind took God at His word. Thus, more than any other religion Christianity licenses mankind to treat, to maltreat and to mistreat Nature as he wishes. The effect of this Western and Christian arrogance towards Nature, White claims, is the ecological crisis evident in the modern world. But, says White, there is a rebel in the Western and Christian tradition to whom we could look back for guidance to find a way out of this crisis: St Francis. For St Francis (according to White) had a proper, ecologically sound and non-exploitative attitude to Nature. This is how White concludes his article:

> The greatest spiritual revolutionary in Western history, Saint Francis, proposed what he thought was an alternative Christian view of nature and man's relation to it: he tried to substitute the idea of the equality of all creatures, including man, for the idea of man's limitless rule of creation. He failed. Both our present science and our present technology are so tinctured with orthodox Christian arrogance toward nature that no solution for our ecologic crisis can be expected from them alone. Since the roots of our troubles are so largely religious, the remedy must also be essentially religious, whether we call it that or not. We must rethink and refeel our nature and destiny. The profoundly religious, but heretical, sense of the primitive Franciscans for the spiritual autonomy of all parts of nature may point a direction. I propose Francis as a patron saint for ecologists. (White, 1967, pp. 93–4.)

White's thesis was interpreted in some circles as being anti-Christian and anti-Catholic, and he received his due share of hate-mail. But the article promoted much scholarly interest. It lies at the basis of all modern views of St Francis as an early ecologist, and it was one of the inspirations for Armstrong and Sorrell in writing their recent books. This paper also led to the pope declaring Francis the patron saint of ecologists. So the claims put forward by White here were very influential thirty years ago, and have continued to be so. But we shall see in due course that, although White was a professional medievalist, every word he wrote about St Francis in this paragraph is erroneous.

A few words first about St Francis. He was born in Assisi in Umbria in Italy in 1181 or 1182, the son of a merchant. According to his biographers and hagiographers, he received the call when he was about 26 years old to embrace Lady Poverty, and he renounced his former life of pleasure and self-indulgence. He requested permission from the pope in 1210 to form a brotherhood to preach repentance, and wrote a Rule by which the brothers were to live. He called the Order the 'Friars Minor' to emphasise their humbleness. Many people 'inflamed by desire of the perfection of Christ, holding in contempt every worldly vanity, followed the footsteps of Francis' (Bonaventura, *Major Life*, 1973, 4: 6). Within just a few years the Order was so large that Francis organized it in a series of provinces, with the brothers going in twos or threes across the world, to draw the souls of the people away from the vanities of the world. In 1224 Francis, long obsessed by and weeping for the Passion of Christ, while reportedly in a state of ecstatic contemplation saw a seraph in a vision, who impressed the flesh of Francis with the stigmata: the same wounds in hands, feet and side that Christ had received on the cross. Francis died two years later, and was canonized by pope Gregory IX in 1228.

Shortly after Francis's death, stories of his life were assembled. These were called 'legends' in the medieval sense of the term: texts which were to be regularly read out loud (*legenda*) to the members of the Order. There were two such lives assembled by Thomas of Celano, the *First Life of St Francis* (*Vita Prima*, 1228/9) and the *Second Life of St Francis* (*Vita Secunda*, 1247), and then two by Bonaventura (destined to have his own sainthood recognized later), his *Major Life of St Francis* (*Legenda Major*, 1261) and his *Minor Life of St Francis* (*Legenda Minor*, of around the same date). A copy of Bonaventura's *Major Life* was ordered to be sent to every Franciscan house in Europe and to be regularly read out loud to the Brothers. In this way Bonaventura's account of the life and miracles of the saintly founder was meant to be a constant example to and resource for the Brothers, and to promote a particular view of the aims, attitudes and work of St Francis.

With respect to Francis's reputation for love of Nature, these early legends seem readily to offer much material to the modern historian. It is certainly the case that animals and plants, from wolves to worms, from flowers to forests, and even the elements fire and water, feature in the legends of Francis to an extent that they do not in the legends of any other saint. Many strange stories were told about him in relation to animals, plants and elements, unusual both in their quantity and in their content. Francis is reported as continually talking to them, reacting to them, speaking about them and showing consideration for them. He calls them all 'brother' or 'sister'. So the case for Francis being a 'Nature Saint', the primary candidate as the patron saint of Nature, and even as an early ecologist, would seem to be pretty much cut and dried when we first look at the legends. For all of them have a distinct section on Francis and his love for creatures, and some of them also have a chapter on the fact that this love was reciprocated. For instance, the *First Life* by Thomas Celano has a chapter (no. 29) entitled 'Of the love Francis bore all creatures on account of their Creator' ('De amore quem propter Creatorem in omnibus creaturis habebat'). Celano's *Second Life* has two chapters (nos 124–5) headed 'The love of the saint toward sensible and insensible creatures' ('Sancti amor ad sensibiles et insensibiles creaturas'), and 'How the creatures returned his love' ('Quomodo ipsae creaturae sibi amoris rependebant vicem'). Bonaventura's *Major Life* of Francis has a chapter (no. 8) 'On the affect of his piety and how irrational creatures seemed attracted to him',[5] while his *Minor Life* has a chapter (no. 5) entitled 'Creatures obey St Francis; God's condescension towards him' ('De obedientia creaturarum et condescensione divina'). Finally, we may cite the *Speculum Perfectionis*, an account of Francis written or assembled in the early fourteenth century; this has a chapter 'On his love for creatures, and of creatures for him' ('De amore ipsius ad creaturas et creaturarum ad ipsum').

But what do we actually know about St Francis and Nature (whether that term is capitalized or not)? It might be thought that we could discover this from the various accounts of his life and teachings written by his biographers and hagiographers, looking for appearances of the term *natura* in Latin sources, and its equivalent in other relevant languages, such as the Umbrian dialect, and for statements and acts by St Francis and his immediate followers relating to animals, plants and what we would today term the 'environment'. This is the route which has been taken by Armstrong and Sorrell and others.

However, the materials that we have available to reconstruct the life and doctrines of St Francis cannot properly be used in this direct and immediate way. There are two reasons for this.

The first reason is that all of the thirteenth- and fourteenth-century accounts of the life and teachings of St Francis are products of *contests* amongst his followers, and his was a legacy fought over extremely fiercely, almost as fiercely as the legacy of Christ Himself. The legends attribute much direct speech to Francis, but in fact Francis's own views on most matters (if indeed he had any) are now lost to us, beyond a small amount written in his own hand or at his dictation such as the Rules of the Order and his will (Cunningham, 1976), and no amount of work on the sources will reveal them to us: such work will only reveal what views he was *represented* as having.

As the issues facing the early Franciscans were immediate and of great importance, his followers sought to resolve them by appealing to particular images of

---

[5]Chapter 8 of Bonaventura's *Major Life* in the English translation is entitled 'Francis' Loving Compassion and the Love which Creatures had for him'; printed in *Omnibus*. The Latin text is: 'De pietatis affectu et quomodo ratione carentia videbantur ad ipsum affici' (Patres Collegii S Bonaventurae, 1927, p. 592).

St Francis, and in so doing of course they created just those images. What had Francis taught about how the true Minorite should live? Should the followers of St Francis be poor as Christ was poor? Should the Order own property? Should the Franciscans devote themselves to the active life, or the contemplative life, or to both? Should they preach or should they pray? How should they pursue their mystic path of religious devotion? Different, and in some ways contradictory, St Francises were created to resolve such issues. From the moment of Francis's death the Franciscan Order was plagued by rival interpretations of St Francis's intentions. For the first century after his death extreme friars, known as Zelanti and Spirituals, claimed that Francis had wanted the Order to be dedicated to total poverty as the only true route to spiritual purity. A more moderate group, the Community, opposed them and used appeals to the papacy to fight the separatist and prophetic tendencies of the Spirituals.[6] So inflammatory among the Franciscans was the issue of poverty, and its relation to spirituality, that a papal ruling in 1323 had to insist that it was heretical to maintain that Christ and the disciples had been poor!

In the early centuries the lives and legends of St Francis that were written all tried to find a Francis who supported one particular position in this squabble. On the one hand, the early legends all support the Community position. The two early *Lives* of Francis by Thomas Celano were written at the behest of anti-Spiritual Minister-Generals. The *Major Life*, written by Bonaventura in 1261 as Minister-General himself, was intended to replace all others and be the authoritative account of Francis and his intentions for the Order (Burkitt, 1932; Brooke, 1982). Together these three legends are often referred to today as the 'official' lives. Some time ago the historian Harold Goad showed that all these early lives put forward an image of what he called 'St Francis the Apostle': someone looking *outwards*, and urging his followers to look outward to the world, to preach, to convert, to be an example to others in holiness by themselves living in the world (Goad, 1926). In other words they present a Francis whose dedication to poverty is strong, but which is tempered by a realization that the mission of the Franciscans to the rest of the world can only be accomplished if it is not taken too literally.

It was only in the fourteenth and fifteenth centuries, in a second round of this strife, that legends of St Francis were written down which showed that he had supported the view that the Franciscan friar should be zealously dedicated to total poverty and to the inner spiritual life. This was the period when the stricter Observants went off to live in hermitages a more spiritual life than they believed their brothers the Conventuals were observing in their houses in towns. They produced what Goad has called an image of 'St Francis the Hermit', someone passionately

---

[6]The situation has been well described by Harold Goad: 'After [Francis's] death his followers divided into two or three distinctive sections, each striving jealously to impose on the whole body their own particular reading of the Common Rule. The more practical and apostolic men lived in large communities on the outskirts of great cities and ministered to the bodily and spiritual needs of the new suburbs … On the other hand, the mystic visionaries … retired into lonely hermitages upon the mountains …' (Goad, 1929, pp. 16–17).

devoted to poverty and abstinence and the personal spiritual life. These legends of Francis present the saint as looking *inwards* to the members of the Order (rather than outwards to the task of spiritualizing the lay world), looking to the individual friar and his spiritual state in his retreat from the world and in his commitment to the spiritual life. The stories of Francis that the compilers of these legends favour are ones whose principal topics are 'mystical experiences, visions and dreams, ecstasies, raptures and temptations, the possession of the soul by God, or by the Devil' (Goad, 1926, p. 161). The most important such works were the *Speculum Perfectionis* ('The Mirror of Perfection'), and the *Fioretti* ('The Little Flowers of St Francis'), both written or assembled around the 1320s, and often referred to today as the 'unofficial' lives. These bitter controversies between Conventuals and Observants were resolved only when pope Leo X separated the two groups in 1517, into the Large and Strict Observance, respectively (Moorman, 1968).

It is important to notice that the lives and legends of St Francis produced by both competing wings of the Order contain stories about St Francis and creatures. However, the writings of the Spirituals or Observants contain significantly more. If we keep in mind that the stories of Francis's relation toward other creatures were themselves part of campaigns to create particular *post factum* images about him, we will come to see why the Spirituals or Observants devoted so much energy toward this theme.

In asking our questions about St Francis and Nature, there is a second and more important reason why we cannot simply read the early materials on the life of Francis in a direct and immediate way. It is that the supposed 'love of Nature' ascribed to St Francis in the twentieth century is something which he could not have had. For neither St Francis nor his followers had the appropriate concepts available to them. Thus the supposed universal 'love of Nature' which all modern writers on the topic take to underlie St Francis's concern with animals and plants is a figment.

To make this point about the availability (and non-availability) of concepts clear, it is first necessary to establish some of the terms and concepts that the early Franciscans actually did have available to them, and which they used with great frequency. I am referring to the trio of concepts: creator, creation, creature. Obviously, in the Christian tradition, God is the *creator*, who in a series of creative acts made His *creation*, which consists of His *creatures*, of which man is one. The referent of the term *creature* was to a relationship of superior and inferior, of total freedom in the case of one of the parties in the relationship, and its counterpart, total dependence, in the case of the other. It was a term with an even stronger force than the term 'servant' has. The term and status of servant implies its counterpart, that of master. Similarly, the term and status of creature constantly implied its counterpart, that of the creator, God. The term constantly drew attention to the fact that the creatures were the products of God's creative acts, and completely subject to Him. For the characteristic which most defines creatures is that they are not autonomous. They, their existence, their qualities, depend completely on the will of their creator. This is what Catholic theology teaches about the status of creatures:

> To be created is to be not of itself but from another. It is to be non-self-sufficient. This means that deep within itself it is in a condition of radical need, of total dependence. This is to be before God 'as though one were not', i.e. to stand before Him without autonomy... Human artifacts continue to exist after the work of fashioning them is completed—they survive their authors. The creature, however, which has its very existence from God, is in a daily total dependence that does not cease. (Ehr, 1967, p. 420)[7]

Not only is the creature in total dependence on the creator, but it is used by God, the creator, to fulfil His goals. The creature receives its fulfilment in being a channel for God's purposes, and in being available for His acts. In the thirteenth century it was accepted that the creatures are in a hierarchy, descending through the nine orders of angels to man, and thence to animals, plants and everything else down to the elements themselves (Lovejoy, 1936). Man's pivotal position in this hierarchy is what makes him unique. Made in the image of God, he participates both in the spirituality of the angels, and in the materiality of lesser creatures. This enables him to transcend the materiality of his body, while remaining a little lower than the angels. Man, the high-point of earthly creation, was given dominion by God over the lower parts of creation; he lost this dominion at the Fall, but it was renewed by Christ's birth as a man—a creature. Christ's role in the status of creatures was thus important. For, 'since man in existing is like the rocks, in having life is like the trees, in sensing is like the animals, and in thinking is like the spirits', and since all creatures participate in man in a reciprocal manner, so all creatures may be said to be exalted through the assumption of man to the deity, through Christ who was God made man (Bartholomew of Pisa, 1590, p. 30 verso, col. 1). This means that all creatures are sacred, not just by having been created by God, but by this exaltation that they have received through Christ's incarnation. But man is superior to all other earthly creatures: 'Every creature confesses and cries out: God has made me on account of thee, O Man!', St Francis is reported as having said.[8] None of this was controversial in Catholic dogma in the early thirteenth century.

It should be clear from all this that (1) when one spoke of creatures, one was speaking of them as products of God's creative acts; so in this sense, when one spoke of creatures one was speaking of Him who had made them: God. And (2) when one took creatures into one's serious consideration in the thirteenth century, it was *in order to think about God* via them. So it is essential for us to remember that 'creatures' was not (as it is primarily today) just a neutral shorthand term to

---

[7]This is in accord with Vatican II: see 'Creation' in Deretz and Nocent (1968). For the text of *Gaudium et Spes*, the Pastoral Constitution from Vatican II on which this draws, see Abbott (1966).

[8]Thus the *Mirror of Perfection*: 'Omnis enim creatura dicit et clamat: "Deus me fecit propter te, homo"' (Sabatier, 1928–31, vol. 1, ch. 11, para. 118, p. 333).

refer to all animals and insects, with no allusion express or implied to God or His creation. Rather, *the very use of the term* referred to God the creator and His creative acts.

The relation of creature to creator was so central to Francis and the early Franciscans that it was embodied in striking ways in the religious Order which Francis founded. In the first place the structure of the Order itself—the distinctive and novel way in which its members gained their daily bread—was built around this relationship. For Francis insisted that the brothers should beg from door to door for their daily bread, beg from what in his will he called 'the Lord's table', that is, from those who had it in their power to give, and who thus fulfill divine providence by giving. A modern-day Franciscan Capuchin has described this commitment to poverty as 'the special form at once of his [Francis's] faith in God and of his relationship with creatures. God, to him, was the Great Father, the Infinite Love encompassing the creation with watchful solicitude'. In his recognition of this supreme truth, Francis acquired the conviction that

> the creature's life is fulfilled in a trustful dependence upon the Divine solicitude: for so will the creature be brought into accord with the love of the Creator. As a religious principle therefore absolute poverty—the poverty of the beggar—is man's response to God's solicitude. But the creature whilst it is the recipient of the bounty of God, is also the channel of that bounty: the creation itself is God's providence in action, except where the Divine Law is frustrated by the will or sin of man. Hence as on the one hand the Divine solicitude demands our entire trust and dependence, on the other it calls for the fulfillment in ourselves of the act of providence so that in us the Divine bounty be not impeded. (Cuthbert, 1912, pp. 27–8.)

By insisting that his brothers begged for their bread, Francis was throwing himself and them completely on God's solicitude, recognizing their complete dependence on Him, thus bringing the creature into accord with the love of the creator, and providing no impediment to God in the fulfilment of His bounty towards His creatures. Francis is reported to have said that the words of the Psalmist, 'Man should eat the food of angels' are fulfilled in God's poor, the friars, 'because the bread of angels is that which has been begged for love of God and given at the inspiration of the angels, and gathered from door to door by holy poverty' (Bonaventura, *Major Life*, 7.7). In other words, begging made both the Franciscans and the donors, in their role as God's creatures, more spiritually fulfilled, and was itself an act of religious piety.

A second role that the relation of creatures to creator played in the Franciscan Order was as a central part of their own fundamental spiritual and devotional practice. Bonaventura, who is considered the second founder of the Order (Minister-General 1257–74) himself practised, and claimed that Francis had practised, *ecstatic contemplation* as a means of approaching God mystically. According to Bonaventura, the stigmata (the wounds in hands, feet and side that Christ received during the crucifixion) that the seraph gave to Francis on Mount Alverna in 1224 were received by Francis in just such a state of ecstatic contemplation. Bonaventura taught the Franciscans how to achieve such states: one should make a mental and spiritual journey, mounting as if it were by a ladder from the visible to the invisible. According to Bonaventura this is what Francis had done: 'of all creation he made a ladder by which he might mount up and embrace Him who is all-desirable'; he rose spiritually to the creator via His creatures (Bonaventura, *Major Life*, 9.1)[9] This mystical practice is one that the Franciscans learnt from the writings of the early Christian mystic Dionysius, known now as Dionysius the Pseudo-Areopagite. Strictly speaking, what one considers on this spiritual journey, if one is a Franciscan, are the *properties* of things, not the things themselves (French and Cunningham, 1996, ch. 9). The first four of the seven stages of spiritual illumination deploy the properties of things. In the first stage, for instance, one should contemplate properties such as the origin, magnitude, multitude, beauty, plenitude, operation and order of a particular created thing. Each of these properties of things is a testimony to a particular aspect of God. The *magnitude* of particular created things, for instance, indicates the immensity of the power, wisdom and goodness of the triune God. Similarly, the *operation* of created things 'by its very variety shows the immensity of that power, art and goodness which indeed are in all things the cause of their being'.

> He who is not illumined by such great splendor of created things [wrote Bonaventura] is blind; he who is not awakened by such a great clamor is deaf; he who does not praise God because of all these effects is dumb; he who does not note the First Principle from such great signs is foolish. Open your eyes therefore, prick up your spiritual ears, open your lips, and apply your heart, that you may see your God in all creatures, may hear Him, praise Him, love and adore Him, magnify and honor Him, lest the whole world rise against you. (Bonaventura, 1953, p. 13.)

Some early Franciscans produced large works which, while they may look like early encyclopedias to our untutored eyes, were actually written to help devotees contemplate the properties of created things, such as Bartholomew the Englishman's great volume of c.1230–50, *De Proprietatibus Rerum*, 'On the Properties of Things' (first printed in Cologne in 1472). It will be evident that the point of contemplating the properties of things on the first step of the ladder of ecstatic contemplation was not that these were parts of 'Nature', or 'natural' products or part of the 'natural environment', but that the properties of created things led the ecstatic Franciscan to God ('from the visible to the invisible') because God had made them: they were His creations, His creatures, and thus exemplified His characteristics, His properties. And following this route of ecstatic contemplation did not take the Franciscan nearer the creatures, but *further away* from them. For,

---

[9]Sorrell, in his determination to portray Francis as a 'nature-mystic', dismisses this in a footnote (Sorrell, 1988, p. 175, n. 82).

as a modern historian has rightly observed, '"The ascent of the ladder of created things" is, after all, only another name for a progressive *contemptus mundi*'.[10]

We find also that the creator and His relation to His creatures is precisely the theme of the famous 'Canticle', variously known as the 'Canticle of the Sun' (*Canticum Solis*) or the 'Canticle of the Creatures' (*Canticum Creaturarum*), which, according to the legends, Francis composed in an ecstasy.[11] The title (or *incipit*) of the poem in the manuscripts is, 'incipiunt Laudes Creaturarum Quas Fecit Beatus Franciscus ad Laudem et Honorem Dei', that is, 'Here begin the praises of the creatures which Blessed Francis composed to the praise and honor of God'. The first lines speak of praises, glory and honor and every blessing being due to You, most high God, whose name no man is worthy of speaking. Then:

> Be praised, my Lord, with all Your creatures,
> Especially brother Sun,
> Who gives the day and through whom You illuminate.
> He is beautiful and radiant with great splendor:
> He is the symbol of You, the Most High.
> Be praised, my Lord, by sister moon and the stars,
> In heaven You made them, bright, precious and beautiful.
> Be praised, my Lord, by brother wind,
> And the air, both cloudy and clear and any weather,
> Through which You support Your creatures.
> Be praised, my Lord, by sister water,
> Which is very useful, and humble and precious and chaste.
> Be praised, my Lord, by brother fire,
> Through which You light up the night,

---

[10]Lovejoy (1936), p. 93. writing about the seventeenth century treatise of the Jesuit Cardinal Bellarmine, *De Ascensione Mentis in Deum per Scalas Creaturarum*.

[11]The English version of the canticle, as offered in Sabatier (1894b), pp. 305–306, is by Matthew Arnold and not to be trusted, except as a piece of English poetry; the following Umbrian text is as given there on pp. 304–305, and also in Sabatier (1894a), pp. 349–350.

> Altissimu, onnipotente, bon signore,
> tue so le laude la gloria e l'onore et onne benedictione.
> Ad te solo, altissimo, se konfano
> et nullu homo ene dignu te mentovare.
>   Laudato sie, mi signore, cum tucte le tue creature
> spetialmente messor lo frate solo,
> lo quale jorna, et illumini per lui;
> Et ellu è bellu e radiante cum grande splendore;
> de te, altissimo, porta significatione.
>   Laudato si, mi signore, per sora luna e le stelle,
> in celu l'ài formate clarite et pretiose et belle.
>   Laudato si, mi signore, per frate vento
> et per aere et nubilo et sereno et onne tempo,
> per le quale a le tue creature dai sustentamento.
>   Laudato si, mi signore, per sor acqua,
> la quale è multo utile et humele et pretiosa et casta.
>   Laudato si, mi signore, per frate focu,
> per lo quale ennallumini la nocte,
> ed ello è bello et jucundo et rubustoso et forte.
>   Laudato si, mi signore, per sora nostra matre terra,
> la quale ne sustenta et governa
> et produce diversi fructi con coloriti flori et herba.

And it is beautiful and joyful and vigorous and strong.
Be praised, my Lord, by sister our mother earth,
Which sustains and rules us,
And produces various fruits with colorful flowers and grass.

This, then, is the theme of the poem: the *creatures* praise their *creator* on whom they depend, and these *creatures* are shown to be the *means through which* God their *creator* chooses to act in His *creation*; they are the channels of His bounty. Just as man is a creature, so the sun, moon and stars, the wind, water, fire and earth, are also creatures; so Francis calls them, in his customary way, 'brother' and 'sister'. The canticle repeats Francis's constant message to other creatures to praise their creator. So this is not some early ecological poem about the value of Nature, or the beauties of Nature, expressing Francis's love for all natural things in a generalized way, and recognizing that they have equal status and value with mankind, as it is usually taken to be nowadays. No: it is by contrast a very strong and religiously informed celebration of the relation between the creator and His creatures, specifying the total respect and praise that the dependent creatures should offer their creator, describing how He carries out His will through His creatures. It is an instance of thirteenth-century religiosity, embodying the most orthodox— not revolutionary—theological sentiments about the relation of the creator to His creation.[12]

The final significance of the relation of creator and creature in the Order to which I shall refer here was that the early legends reveal that St Francis himself had a unique relation to all other creatures. For within the Franciscan Order it was believed that Francis, alone of men, had been personally given dominion over all creatures. He was a second Adam. In his *Major Life of Francis*, Bonaventura wrote, 'We should have the greatest reverence therefore, for the piety of this blessed man, which had such wonderful charm and virtue that it could bring savage animals into subjection and tame the beasts of the forest, training those which were tame already and claiming the rebelled nature of the beasts against fallen mankind to obedience to him. This is truly that divine virtue which subjects all creatures to itself and is "all-availing, since it promises well both for this life and for the next"' [I Tim. 4,8].[13] An important late fourteenth-century Franciscan work put it like this:

---

[12] As far as I know I am the first modern historian to read this canticle in the light of the creator–creature relationship, and as an expression and celebration of that relationship. Other historians, not interpreting it in this way, have found the preposition '*per*' ambiguous or unclear. Sorrell (1988, ch. 6, 'The controversy over the *Canticle's* meaning') has argued that '*per*' is causal, meaning 'for' or 'because of'. Hence for him the overall meaning of the poem is 'Be praised, my Lord, because of all you have made'. He relies on 'non-official' early sources to back up this argument, especially the *Legend of Perugia* and the *Mirror of Perfection*. The desire on Sorrell's part to link Francis with modern environmentalism makes this argument over the *Canticle* read like special pleading. For some of the reasons why the *Canticle* has been read as it has over the last two centuries, see the second paper in this pair.

[13] 'Pie igitur sentiendum de pietate viri beati, quae tam mirae dulcedinis et virtutis fuit, ut domaret ferocia, domesticaret silvestria, mansueta doceret et brutorum naturam homini iam lapso rebellem ad sui obedientiam inclinaret. Vere haec est, quae cunctas sibi creaturas confoederans, valet ad omnia, promissionem habens vitae, quae nunc est, et futurae' (*Omnibus*, pp. 697–8, translation modified; Patres

> As every Creature was disobedient to *Adam*, for his disobedience to God Almighty, so every Creature was familiar, and obedient to *St. Francis*; who kept all Gods Commandments, *He hath put all things under his Feet, &c.* He hath set him over the Works of his Hands, and he might deservedly say what's said in the Gospel, *All things are deliver'd over unto me from my Father.* It hath pleased God that all things should obey, and comply with *St. Francis.*[14]

Having seen how central the concepts of creator, creation and creature were to medieval people, and their pervasive role in early Franciscan religious life, let us now compare this to some of the views put forward by modern writers on St Francis, Nature and the environment. For the sake of brevity, I will take Lynn White's opinions as representative. In the final paragraph of his 1967 article in *Science* (quoted above), where he called for Francis to be made patron saint of ecology, and put him forward as a pattern of the proper relationship of mankind to Nature, White commits a number of egregious errors. I will deal with four of them.

Firstly, White claims that Francis 'tried to substitute the idea of the *equality of all creatures*, including man, for the idea of man's limitless rule of creation. He failed.' This is an error because Francis did not claim this nor try to do this, and White presents no evidence to support his assertion that Francis did so. If Francis had claimed this, then he would probably have been committing a heresy. For all creatures are emphatically *not equal* in Catholic dogma, as we have seen, and it is not a point that Francis did or (given what we know from elsewhere about Francis's orthodoxy) would have wanted to challenge. Indeed the early Franciscans, and in all likelihood Francis himself too, had a profound sense of the hierarchy of angels above them, through whom one could, as man, ascend to God, as they regularly did in their practice of ecstatic contemplation. There is every reason to think that they had an equally strong sense of the hierarchy of animals, plants and elements continuing beneath man. Moreover, far from believing in the *equality* of creatures, Franciscans (as we saw just now) believed that Francis had been personally given *dominion over* all creatures by God! What Francis did celebrate about other creatures, according to Bonaventura, was that they have the same *origin* as man, not that they are equal to him: he loved them *not* for themselves but because they all reminded him of God.[15] So much for the equality of all creatures.

White makes his second error when he then speaks of the 'profoundly religious, but heretical sense of the primitive Franciscans for the spiritual autonomy of all

---

Collegii S Bonaventurae, 1927, p. 597). The passage from 1 Timothy 4, 8 reads (in part): 'godliness is profitable unto all things, having promise of the life that now is, and of that which is to come'.

[14]Folio 249 of the 1510 edition of the *Conformities*, as cited and translated (and attacked) in Alberus (1679), pp. 93–4.

[15]Bonaventura, *Major Life*, 8 : 6: 'Consideratione quoque primae originis omnium abundantiore pietate repletus, creaturas quantumlibet parvas fratris vel sororis appellabat nominibus, pro eo quod sciebat eas unum secum habere principium' (*Omnibus*, p. 692; Patres Collegii S Bonaventurae, 1927, p. 594). Bonaventura's conclusion to this chapter tells how these episodes about creatures are to be read: that is, that they are instances of Francis's capacity to tame and subject other creatures to his will.

parts of nature'. Had the primitive Franciscans actually held that view, then White would have been right: it would have been heretical. But they did not, and again he presents no evidence that they did. The reason that they did not is because, as we have seen, they had a profound sense of the concept of *creature*, something which we have now lost.

If White had been using the appropriate term, 'creatures', instead of the sloppy expression 'all parts of nature', the third of his errors here might have been more evident to him. For in this sentence he speaks of the Franciscans having a sense of 'the spiritual *autonomy* of all parts of nature'. Had he written of 'the spiritual autonomy of creatures' one hopes he would have paused, because of course creatures have no autonomy in the Catholic tradition, a tradition which Francis was not challenging. So much for the autonomy (spiritual or otherwise) of creatures.

The fourth and perhaps most important error White commits in this paragraph concerns a term we have not yet dealt with directly, *nature*, where he speaks of '*all parts of nature*'. Francis and his followers did not think either in terms of 'nature' nor of 'parts of nature', in the senses which White means. Here the problem is that White is using 'nature' in modern ways, where it has come to replace 'creation': we use it primarily as a neutral term to refer to the universe and all its contents, without any reference to a creator or a creation. We do this, of course, because for us the Big Bang theory and the theory of evolution have been substituted for older stories of the origin of the universe and its contents in a series of creative acts by a creator God, who continues to sustain His creation and His creatures.

While they did not—could not—speak of Nature in our modern sense, literate medieval people did of course use the Latin term *natura* and its vernacular equivalents. There were two primary senses (French and Cunningham, 1996, ch. 4). The first was the nature-of-a-thing, in an Aristotelian sense: the unfolding of the attributes that make a thing what it is. Thanks to the computerised concordance of the early legends of St Francis in the *Corpus des Sources Franciscaines*, we can say with complete certainty that it was primarily in this sense that the writers of the early legends used the term *natura* and its derivatives. For instance, Thomas of Celano employed the term in this sense in his *First Life* of Francis: vice became second nature to Francis as a youth; the birds rejoiced in a wonderful way according to their nature; Francis was serene by nature; Francis near death lost his natural warmth. Similarly, Bonaventura in his *Major Life* speaks of Francis gladly accepting the gift of a sheep 'in his love of innocence and simplicity, two virtues which the nature of a sheep recalls' (Godet, 1975, 8: 7; Mailleux, 1974).

The second main meaning of the term *natura* available in the thirteenth century for Francis and his contemporaries was nature-as-a-generative-principle. This can sound temptingly like our modern-day concept, and we need to take care not to conflate our modern meaning with this old one. This second meaning, which has roots in Platonic philosophy, treats Nature in an allegorical way, as if she (it is always she) acts to generate beings in the world on behalf of God. But this usage always means Nature-as-true-servant-of-the-Creator, His handmaid. While this meaning was available in the thirteenth century, nevertheless the computerised concordance of the early legends of St Francis reveals little if any usage of the term in this sense by those recording and celebrating his life and virtues. Perhaps Celano in his *Second Life* means *natura* in this sense when he writes of Francis that he 'will be found true and trustworthy unto whom nature, the law, and grace will be witnesses'.[16] This is also probably the sense in which Bonaventura uses the term in the prologue to his *Major Life*: the impression of the stigmata on the saint's body 'was not the work of *nature's powers* or of any human agent, it was accomplished by the miraculous power of the Spirit of the living God alone'.[17] But this is the nearest usage I can find of the term in this sense in the earliest legends. What is striking is that we do not find the creation as a whole described as Nature, nor the creatures referred to as parts of Nature. However, other Franciscans did at other times sometimes use the term in this way. One of them, for instance, could in the late fourteenth century write of 'natura, idest omnia creata' (Bartholomew of Pisa, 1590, f. 2 recto, a). But the phrase itself (which is taken from Augustine)—'nature, that is all created things'—makes clear that this *Nature = creation*. To think or speak of Nature as an autonomous creative entity or process, separate or distinct from God, as we do today when we use the term, would have been heretical in the thirteenth century. If we lose sight of the fact that for medieval people in general, and in this case for Franciscans in particular, Nature = creation, then we are substituting our thought patterns for theirs, and removing the creator God and His created world from their concepts, together with the relationship of creator to creature. So Lynn White's claim that Francis 'proposed what he thought was an alternative Christian view of nature and man's relation to it' is completely wrong.

Perhaps there is a legitimate way for us to use the modern term and concept 'Nature' when speaking about the historic St Francis? Some modern writers on St Francis speak, for instance, in terms of Francis's 'delight in Nature and [his] sense of Nature as a sacramental revelation of God', or of his 'joyous intimacy with Nature and [his] habit of seeing in the visible creation the Hand of the Creator'.[18] Certainly here the term 'Nature' in its modern sense is each time glossed as the

---

[16]'verus erit, dignusque fide, cui natura, lex et gratia testes erunt' (para. 203, emphasis added). The law and grace (but not nature) are allusions to John 1.17: 'For the law was given by Moses, but grace and truth came by Jesus Christ'.

[17]'signaculum...quod in corpore ipsius fuit impressum, non per naturae virtutem vel ingenium artis, sed potius per admirandum potentiam Spiritus Dei vivi' (*Omnibus*, p. 632, emphasis added; Patres Collegii S Bonaventurae, 1927, p. 558).

[18]Cuthbert (1912), pp. 21, 20. The discussion of these issues in Cuthbert (1915), otherwise so judicious, is vitiated by Cuthbert's constant use of the modern concept of Nature. Cuthbert is here seeking religious romanticism in the thought of Francis; as we shall see in due course in the second paper of this pair, it is the romantic view of mystical medievalism which is the source of Cuthbert's misplaced concern with Nature here.

thirteenth-century terms 'creation' and 'creator', or as 'sacramental revelation of God', and these glosses provide a legitimate way of talking about Francis's attitudes, because they reproduce or paraphrase thirteenth-century concepts. But even this usage does seem an unnecessary multiplication of entities, constantly introducing modern concepts only to have to gloss them back every time into thirteenth-century ones. Moreover, the repeated introduction of the term 'Nature' on the part of modern writers betrays a continued desire on their part to impose that modern concept on the world of St Francis. If we were to talk constantly about creator, creation and creatures, and to stop talking about Nature, we would save ourselves a lot of extra work as well as potential and actual ascription of concepts and opinions to medieval people which they could not possibly have had. Slipping back and forth between the terms creature, Nature, parts of Nature, creation, is, if I may use the expression, second nature to us. But it is a habit that it would be good for us to break when we speak of St Francis and other medieval people of religion.

So what was all that stuff about Francis and the creatures? What do those stories mean, especially the animal stories which have become so popular nowadays, and why might they have accrued around the figure of St Francis? This question needs to be answered in a number of ways which are inter-linked, and all of which flow from the view of Francis that the early Franciscans held of him: as the most perfect, pure and holy man to have lived since Christ. Stories about Francis and creatures appear in all the legends, from the earliest lives by Celano and Bonaventura. But the largest number of such stories were added to the original ones in the early fourteenth century by Franciscans of the Spiritual, or hermit, persuasion. Harold Goad has described these stories as 'the natural product of the hermit section of the Franciscan Order' (Goad, 1926, p. 161). The wolf of Gubbio is just one such example out of dozens of stories added by the hermit section. Similarly, the 'Canticle of the Creatures', which, as we have seen, rehearses the relation of creator and creature so succinctly, though mentioned in the early legends, was first recorded *in extenso* only in a Spiritual document of a century later, the *Speculum Perfectionis*, presumably because the Spirituals, the most other-worldly of the Franciscans, held its sentiments so dear. In other words, the more that Franciscans were concerned with promoting the spiritual, the inward-looking, the devout, image of Francis as the model for the Order, the more they gave credence to remarkable stories about Francis's relations with other creatures, especially animals, and added them to the legends. This is the opposite of what we might expect at first glance, where our modern disposition would be to expect Francis to be portrayed looking more at creatures—because we would have translated this into 'Nature' in our parlance—the more outward-looking, the more secular, the less spiritual, he was. But we would be wrong.

In discussing the roles of these stories about St Francis and creatures, I shall, for the sake of brevity, restrict my discussion almost entirely to the most 'official' and least Spiritual accounts: the legends of Francis produced by the Community friar and leader of the Order, St Bonaventura. These are the same themes that the Spiritual friars were later to use in amplified form in constructing their rival image of 'St Francis the Hermit'. First we need to deal with the love and the beauty. Yes indeed, Francis is described as loving (showing *amor* toward) other creatures, which is something extremely unusual, even for a saint. The grounds for this were to demonstrate Francis's God-fullness, his closeness to God, which made him love other creatures as God Himself loved them. For instance, when he was preaching to the birds, telling them about the benefits God bestows on His creatures and the praise they owe Him in return, then 'it was only right', Bonaventura wrote, 'that St Francis, who was *so full of God*, should have felt such tender affection for these creatures lacking reason', and they for him.[19] He also recognized the beauty (*pulcher, pulchritudo*) of creatures. But this is always in the sense of the beauty of creatures leading him to Him who is Beauty itself, that is, God and Christ:

> In everything beautiful, he saw Him who is Beauty itself, and he followed his Beloved everywhere by His likeness imprinted on creation; of all creation he made a ladder by which he might mount up and embrace Him who is all-desirable. By the power of his extraordinary faith he tasted the Goodness which is the source of each and every created thing, as in so many rivulets. He seemed to perceive a divine harmony in the interplay of powers and faculties given by God to His creatures and like the prophet David he exhorted them all to praise God.[20]

This is like a refrain in the legends: the beauty of creatures leads one to the beauty of God and Christ; and indeed Francis's love for and praising of God is the only context in which the beauty of creatures is ever mentioned. This is strikingly shown in one of the earliest legends where we find that God, in calling the young man Francis to His service, sent him a long illness. As a consequence of his illness, Francis began to think of 'things other than he was used to thinking upon'.

> He went outside one day and began to look at the surrounding country with great interest. But the beauty of the fields, the pleasantness of the vineyards, and whatever else was beautiful to look upon, could stir in him no delight. He wondered therefore at the sudden change that had come over him, and those who took delight in such things he considered very foolish. From that day on, therefore, he began to despise himself and to hold in some contempt the things he had admired and loved before.[21]

[19] Bonaventura, *Minor Life*, emphasis added. 'Iuste quidem *vir Deo plenus* [Gen. 41, 38] ad creaturas huiusmodi ratione carentes pio ferebatur humanitatis affectu' (*Omnibus*, p. 819; Patres Collegii S Bonaventurae, 1927, p. 671). In Genesis 41.38, Pharaoh described Joseph as full of the spirit of God.

[20] Bonaventura's *Major Life*. 'Contuebatur in pulchris Pulcherrimum et per impressa rebus vestigia prosequebatur ubique Dilectum, de omnibus sibi scalam faciens, per quam conscenderet ad apprehendendum eum qui desiderabilis totus. Inauditae namque devotionis affectu fontalem illam bonitatem in creaturis singulis tamquam in rivulis degustabat, et quasi caelestem concentum perciperet in consonantia virtutum et actuum eis datorum a Deo, ipsas ad laudem Domini more prophetae David dulciter hortabatur' (*Omnibus*, p. 698; Patres Collegii S Bonaventurae, 1927, pp. 597–8).

[21] Celano, *First Life*, ch. 2:'. . . die quadam foras exivit et circumadiacentem provinciam coepit curiosius intueri. Sed pulchritudo agrorum, vinearum amoenitas et quidquid visu pulchrum est, in nullo eum potuit delectare. Mirabatur propterea subitam sui mutationem, et praedictorum amatores stultissimos reputabat. Ab ea itaque die coepit seipsum vilescere sibi, et in contemptu quodam habere, quae prius in admiratione habuerat et amore' (*Omnibus*, pp. 231–2, with modifications; Patres Collegii S Bonaventurae, 1927, 63 pp. 7–8).

634

Thus, interestingly enough, it looks as though Francis as a young man, before he received his mission, did take delight in the beauty of fields and vineyards—that is, he had a sense of them being beautiful independent of any relation to their Creator—and then *lost* this delight when he received his mission! Thereafter, he loves creatures and the beauty of creatures, but he does so only because he loves their Creator, and because the creatures remind him of their Creator and of His beauty.

The second role that creatures play in the legends of St Francis is as *symbols*. Not only did every creature remind Francis of God because He was its Maker, but each had a particular symbolism which reminded Francis constantly of the life and passion of Christ. Francis seems to have been particularly resourceful in finding such parallels. And absolutely everything in creation reminded Francis of Christ. He has been described as living 'in the midst of a forest of symbols' (Gilson, 1938, p. 72). Water was symbolic of baptism, stones of Christ the keystone, trees of the Cross. 'He reserved his most tender compassion for those creatures which are a natural reflection of Christ's gentleness', wrote Bonaventura, 'he often rescued lambs, which were being led off to be slaughtered, in memory of the Lamb of God who willed to be put to death to save sinners'.[22] Even the worm reminded Francis of Christ, who had once likened Himself to a worm, and thus Francis moved the worm away from the danger of being trampled on. One certainly has to acknowledge that these extreme obsessions about the role of creatures as symbols of Christ and his passion make Francis unusual, even for a saint, but the religious sentiment behind them is conventional. Perhaps it is this superabundance, this excess of piety, that fits one for sainthood.

The third way in which Francis is portrayed by the legend writers as being in a special relation to other creatures, is that in him spiritual purity is rewarded with true power. Francis had achieved perfect purity, the only man to be like this since Christ; as he was pure, so he inherited the powers of Adam over all other creatures. The stories show the creatures recognizing and acknowledging this fact. For the creatures obey his commands. 'Such was his pure love of God', Bonaventura wrote in his *Major Life*, 'that Francis had arrived at a point where his body was in perfect harmony with his spirit, and his spirit with God. By divine ordinance it was established that all creation, which must spend itself in the service of its Maker, should be miraculously subject to his will and command. Not only did creation (*creatura*) serve this servant of God at his will, but everywhere the providence of the Creator condescended to his wishes', so that Francis could perform miracles such as making water flow from the rock, or cause the waves to have no effect

---

[22]'Illas tamen viscerosius complexabatur et dulcius, quae Christi mansuetudinem piam similitudine naturali praetendunt ... Redemit frequenter agnos, qui ducebantur ad mortem, illius memor Agni mitissimi, qui ad occisionem duci voluit pro peccatoribus redimendis' (*Omnibus*, p, 692; Patres Collegii S Bonaventurae, 1927, p. 594).

IV

Science and Religion in the Thirteenth Century Revisited: Part 1 635

on the boat from which he was preaching.[23] 'Fire lost its burn and water its taste at his wish; an angel came to cheer him by his light, showing that the whole of creation waited upon his material needs, so holy had he become'.[24]

Thus, however much today we might want to see Francis as having some special friendly, or ecologically aware, relation with other creatures 'as themselves', or as autonomous beings or as equals with man, we nevertheless have to conclude that according to the legends—which are our only resource for Francis's views and attitudes—his relationship to them was actually one of power. He, alone of all men, had dominion over all other creatures: 'every creature was familiar, and obedient to St Francis', (Alberus, 1679, p. 93–4, citing folio 249 of the 1510 edition of the *Conformities*), or in more modern language, 'the most timid or ferocious became fearless or tame at his side' (Cuthbert, 1912, p. 163). He was not in a position of equality with them or they with him. Nor did he respect them 'for themselves' as we might think today. The Papal Encyclical of Pius XI, issued in 1926, is quite correct on this: 'The herald of the great King [i.e. Francis] did not come to make men doting lovers of flowers, birds, lambs, fishes or hares; he came to fashion them after the Gospel pattern, and to make them lovers of the cross' (Pius XI, 1926, p. 26).

We have seen, I hope, the role of creatures in the world-view of these medieval men, St Francis and early Franciscans, and the inappropriateness of our using a term such as 'Nature' in its modern connotations to describe the relation that obtained between creator and creature, or creature and creature. But there is one final point which needs to be made while talking about Francis and 'Nature', and while it should really go without saying, the point nevertheless needs to be made explicitly. It is this: that St Francis would not have recognized our terms or concepts 'environment' and 'ecology', which are crucial to the existence of a domain of concern, or a science, or a discipline, of ecology. Unfortunately this very assumption has been implicit in the approaches of White, Armstrong and Sorrell. These concepts and this vocabulary would have been alien to Francis and his contemporaries. And it is not just that they *did not* think about the environment and ecology, but that they *could not* do so. These concepts were not to be fashioned for hundreds of years. In the meantime Francis and the Franciscans were living their own lives—which were full-time *religious* lives—perfectly equipped with their own concepts, which exquisitely matched their needs.

While Francis certainly had a reputation for having the power of a second Adam

[23]'Quia enima ad tantam pervenerat puritatem, ut caro spiritui et spiritus Deo harmonia mirabili concordarent, divina ordinatione fiebat, ut creatura Factori [Creatori] suo deserviens [Wisdom, 16, 24], voluntati et imperio eius mirabiliter subiaceret … Non solum creatura servo Dei serviebat ad nutum, sed et Creatoris ubique providentia condescendebat ad placitum' (*Omnibus*, p. 669; Patres Collegii S Bonaventurae, 1927, p. 581).
[24]'[…vir iste munditiae quantaeque virtutis] ad cuius nutum suum ignis ardorem contemperat, aqua saporem commutat, angelica praebet melodia solatium, et lux divina ducatum, ut sic sanctificatis viri sancti sensibus omnis probetur mundi machina deservire' (*Omnibus*, pp. 670–1; Patres Collegii S Bonaventurae, 1927, p. 582).

over other creatures, he did not (and could not) have a reputation for the 'love of Nature' and of the natural world, either in his lifetime nor for centuries after his death. In fact, Francis had other reputations which were far more important and significant than any reputation for being a saint of Nature could have been (even had it made any thirteenth-century sense), and which were far better grounded in the Bible, the basic text of the religious life, than any 'love of Nature' could have been. Indeed, Francis had some extraordinary reputations, which gave him and his Order a most important series of roles in the working-out of the high drama of Christian history. Looking after the claims of Nature or the environment, however, were not among them. Indeed, even the reputations of Francis's extraordinary relations with creatures, which we have just been exploring, came to be neglected in favor of his other roles.

3. The Historical Roles of St Francis and his Order

The first and most important of these roles of St Francis was as a Second Christ (Fig. 2). What reputation could be greater than this? The image of Francis's life as conforming to that of Christ comes initially from Celano and Bonaventura. This reputation had a well published career—literally. On no less than three occasions were large books assembled by members of the Order to show how the life and virtues of Francis conformed to those of Christ. The first and most important of these was by an Italian Franciscan friar of the Observant persuasion, Bartholomew of Pisa, whose *Golden Book: or the Conformities of the Life of the Blessed Francis to that of Lord Jesus Christ, our Redeemer* was completed in 1399 after fifteen

Fig. 2.   St Francis as Second Christ. From Bartholomew of Pisa, Liber aurens *(1590), title page.*

years' work.[25] After the invention of printing it was published several times in the sixteenth century. Bartholomew arranged his conformities in the shape of a tree with fruits. 'In this tree', he wrote, 'I set 20 branches, of which 10 are to the left and 10 to the right, on each branch I put four fruits, or two conformities about Christ and two about the blessed Francis, and in this way they add up to 80 fruits or conformities, of which 40 are about Christ and 40 about the blessed Francis.' Bartholomew says that Francis's life conformed to that of Christ 'as far as it is possible that a mortal man could conform to Jesus Christ' (p.1 *verso*, col. 1). At the Reformation the early Protestants, including Luther, attacked the *Conformities* as blasphemous, for claiming divine status for a mere man. In his book with the deliberately insulting title *The Koran of the Barefoot Friars*, one such German Protestant, Erasmus Alberus, treated this image of Francis as typical of Catholic lies and superstition.[26] A Franciscan, Henricus Sedulius, defended the work in 1607 in an *Apology* published in Antwerp (Sedulius, 1607). The second such work appeared in 1651 with a comprehensively self-explanatory title: *Prodigy of Nature, Portent of Grace; that is, the Acts of the Life of our Seraphic Father Francis regulated and co-opted to the Life and Death of Our Lord Christ. In the First Column are Described the Mysteries of the Redeemer of the World, Beginning from his Eternal Predestination to his Glorious Ascension into Heaven, and in the Second Column the Conformities, Similitudes and Parallels of the Seraphic Father, divided into forty-five Headings.* It was produced by a Spanish Franciscan friar, Alva y Astorga, who had been on a Franciscan mission to Peru (Alva y Astorga, 1651). The third appeared in 1656–60, written by the Vicar of the Convent of Bolland in Liège: a *Treatise on the Conformities of the Disciple with his Master, that is our Seraphic Father St Francis with our Lord Jesus Christ.* Its author, the Franciscan friar Brother Valentin Marée, devotes three volumes to portraying Francis as walking step by step with Jesus, 'and this not only with respect to the state of his most holy life, but also with respect to his most doleful and most loving Passion, his Death, his Resurrection, Ascension, glorification, exaltation: in brief, in respect to all his other mysteries, including the most memorable and most signal: like a little madman who goes running after his master, so this good man always found himself following his Lord JESUS' (Marée, 1656–60, Avant-propos, B4 recto).[27] This was intended as a real counter-blast to Protestant criticism of St Francis.

---

[25]I have translated this version of the title of the work as given in the incipit to the 1513 printed version: Bartholomew of Pisa (1590). For a modern scholarly edition of the text see Patres Collegii S Bonaventurae (1906–12).

[26]First published in German in 1542, and in Latin translation as Alberus (1542). For an English version of this, see Alberus (1550). The new English version produced in the wake of the Popish Plot, is fuller: Alberus (1679).

[27]'et cela non seulement en ce qui regarde l'estat de sa tres-saincte vie, ains aussi celuy de sa tres-douloureuse et tres-amoureuse Passion: celuy de sa Mort, de sa Resurrection, Ascension, glorification, exaltation; bref en ce qui regard l'estat de tous les autres siens mysteres, iusques aux plus memorables, et plus signalés: tellement qu'a la façon d'un petit sot, qui va courant apres son maistre, ce bon homme se trouvera tousiours rangé à la suitte de son Seigneur IESUS...'.

Understandably, the celebration of the role of Francis, the founder of their Order, as a second Christ seems to have been something of an obsession amongst the Franciscans. But in only one of these works can I find any significant celebration of Francis's relationship with other creatures, the 1607 defence of the *Conformities* by Sedulius, responding to some minor sarcasm of the Protestant attack on St Francis.[28] And even here what Sedulius celebrates is Francis's unique *piety*, shown (among other things) by his relation to other creatures.[29] So it would seem that by the late fourteenth century (the date of the *Golden Book*) Francis's multiple special relations to creatures were not particularly celebrated within the Order any longer; at the least, they had ceased to have any significant role in the representation of Francis's holiness. As a modern historian has said, in the sixteenth and seventeenth centuries 'St Francis was exalted as the thaumaturgist, rather than as the lover of the Creatures' (Goad, 1929, p. 22).

In a brilliant piece of iconographic and exegetical analysis, John V. Fleming has recently brought back to attention the fact that Francis had many other roles and reputations too, all of which were recognized by his contemporaries and followers, and all of which could be given visual representation emblematically in ways which, to persons steeped in the Bible, spoke clearly of Francis's highly important personal role in the working-out of Christian history.[30] Fleming has analyzed the late fifteenth-century picture by Giovanni Bellini, best known as 'St Francis in the desert'. In the course of his analysis he shows that from Bonaventura onwards, Franciscans represented Francis in an apocalyptic framework, especially as the Angel of the Sixth Seal who is the messenger of the next age,[31] thus making it

[28]These works on Francis as a second Christ are all long, extremely laborious works, and while I have perused them all, I have perforce had to use their own contents listings and indices as my main guides to whether they discuss Francis's relationship to other creatures at any length. So it is possible that I have missed something, though unlikely.

[29][Introduction to chapter 8, 'On his piety and love (*affectus*) for Christ the Saints, and irrational creatures':] 'Those who define *piety* correctly, write *legitimate and due honour and love toward God and one's parents*. We define it a little more fully, understanding also piety toward the Saints and even the beasts, but in the contemplation of God, benevolent love. For St Bonaventura, writing about the piety of St Francis, took for piety such love (*amor*) towards irrational creatures. But impious people savage piety, or they bite into it, their teeth never leaving an impression. [There are 11 paragraphs making separate points; only two concern irrational creatures: paragraphs 10 and 11:] 10. (The Alcoran of the Barefoot Friars says) He called all creatures by the name of brother or sister, even thieves, even wolves. [Sedulius:] The old tag is relevant here: He who loves me also loves my dog. The cause of this love was explained perfectly by St Bonaventura [= ch. 8.51: "From a consideration of the first origin of all, filled with a more abundant piety, he used to call the creatures, even the most insignificant, by the name of brother or sister: for this reason, that he knew they all have one origin." [Sedulius:] Therefore, what they take for insanity is wisdom.' etc., my translation. Paragraph 11 concerns the sheep which Francis told to kneel before the altar to the blessed Virgin.

[30]Fleming (1982). I love this book: it is a model of what scholarship in this area should be like, in goal, method and results.

[31]*Apocalypse* (*Revelations*) 7: 2–3, in the translation offered by Fleming (which better represents the Vulgate than the King James' version), reads: 'And I saw another Angel descending from the rising of the sun, having the sign of the living God; and he cried with a loud voice to the four Angels, to whom it was given to hurt the earth and the sea, saying: Hurt not the earth, nor the sea, nor the trees, till we sign the servants of God in their foreheads'. On Francis as Angel of the Sixth Seal see also McGinn (1979).

possible for others of a more prophetic bent to interpret Francis as someone having a major role in the Joachimist prophecies about the end of time. Fleming shows that in the Bellini picture Francis was also being portrayed as being, or as playing out the role of, a second Moses, with the Franciscan Order as a pilgrimage, an extension of the Exodus. He is similarly Aaron, Moses' brother. He is a Desert Father in the mode of St Anthony, 'the father of monks'. He is a second Elijah: Elijah was a prefiguration of Christ, and Francis a postfiguration. Franciscan theology of history, Fleming shows, gave the Franciscan Order 'an unparalleled role within the Church ... The Franciscans have a special vocation of leadership, renewal, and prophetic purification within the Church. They are, so to speak, the institutional memory within the New Israel of the deliverance of the Exodus' (Fleming, 1982, p. 160).

The reputations of St Francis—who he was, what he stood for, what was the role of the Order he had founded, what kind of religious life his followers were supposed to live—all these were again a matter of controversy in later centuries. In the course of the Counter-Reformation in the sixteenth century the Observants split as a number of yet more earnest reform groups were formed, including the Capuchins, each seeking to outdo the other in their commitment to the absolute poverty they believed Christ and Francis had practised and which they should imitate. The Discalced refused even to wear sandals in their practice of poverty and self-denial; Recollects and Reformed sought a more severe discipline in remote houses; the Capuchins observed the Rule 'to the letter' and were itinerant preachers. In general, recruitment to the Order, in all its many fractions, was quite strong, and the Franciscans remained the most popular and numerous of the mendicant Orders, though from the mid sixteenth century they had their traditional missionary, exegetical and preaching roles challenged by the Jesuits. I am not aware of St Francis's distinctive relations with other creatures, so important to the Franciscans of the thirteenth and early fourteenth centuries, being a subject of importance during these disputes.

Then the Franciscan movement declined, and the views of the Franciscans on any matters at all became less and less significant. And, even if they had still been glorying in St Francis's special relations with other creatures, they would not have been heard. For in the eighteenth century the Franciscans were, like the other Orders, the subject of attack by the growing number of people who were anticlerical in their views; and suppressions of the friaries in various countries, especially as a consequence of the spread of the French Revolution, meant that in the early nineteenth century the Order was only a shadow of what it had been in the eighteenth. Now marginalized within Catholicism, outside Catholicism the Order and its founding saint commanded no interest at all. It has been remarked that in 1826, on the sixth centenary of Francis's death, the Order and the cult of Francis were negligible: 'there were no world-wide celebrations, because the whole world outside the Roman Catholic communion thought of him, if it thought at all, as a dead Roman Catholic' (Seton, 1926, p. 247). Things were to be quite different a century later.

This list of the roles and reputations of St Francis and his Order between the thirteenth and the nineteenth centuries may not be complete, and none of this of course precludes St Francis also having had yet further roles and reputations. But I trust that it is now clear that he is precluded from having had the reputation in his own time, or for centuries after, of being a 'Nature lover' or having a generalized 'love of Nature' in any of the senses in which we would normally use, or want to use, those terms, reputations which might justify us today in claiming him as an early ecologist. For not only did Francis and his followers not have the necessary concepts available to them, but nor did they have the inclination. For when they were speaking of or referring to creatures, which it seems that Francis and his immediate followers did frequently, they were actually engaged in coherent and important activities, all of which had direct and vital *religious* functions for them in their lives as full-time professed *religious* men with particular mystical, Dionysian, spiritual practices. There was neither room nor occasion for them to indulge in some universal 'love of Nature', even if this had been a thinkable thought.

The new historical reputation of St Francis as having a supposed universal 'love of Nature' derives from moments in the great revival of interest in him which took place in the course of the nineteenth century. Indeed, Francis acquired a whole range of new reputations, of which this was only one. The legends and other historic material on the life of Francis proved rich enough to be used to fashion many new, and sometimes contradictory, images of the saint. Each of these new reputations was contested. Thus, just as the first lives of the Saint had been written in the course of thirteenth- and fourteenth-century struggles to determine how the future of the Franciscan Order ought to go, so most of the many new lives of Francis which were written in the nineteenth century—there were at least twenty published between 1825 and 1900[32]—were also written in the course of struggles, this time to determine how the future of the Catholic Church itself ought to go. These issues will be dealt with in the second article of this pair.

Are we in a position to come to any conclusions about religion and science in the thirteenth century and their relationship (or lack thereof), or about the involvement (or not) of men of religion in the practice of science in the thirteenth century, or about the validity of 'science' as a possible medieval practice? Obviously, a first conclusion is that we need to keep distinct the thirteenth-century concepts of creature and *natura*, and not mistake one for the other. With St Francis it looks as though the enthusiasm of modern historians to find proto-ecological attitudes in St Francis has meant that they have mistaken a thoroughly religious position about

---

[32]From a count primarily of works listed in the catalogue of the British Library. There have been dozens of new lives of Francis in the twentieth century; I stopped counting when I reached sixty.

creatures for a proto-scientific one about Nature, which, unfortunately, in effect means that they have mistaken religion for science. But while the category 'Nature' is now clearly not a suitable one for us to use in our accounts of St Francis, we cannot extrapolate from this and say that *natura* in any sense was not a topic of interest and a category of thinking for other men of religion in the period. We will only know from a case-by-case investigation what *natura* meant, and why it was deployed (if indeed it was) by particular men of God in the medieval centuries. A second conclusion, working from the St Francis case, is that in general it is *prima facie* unlikely that men of religion were practising science in the thirteenth century, or even proto-science, in the sense in which we understand those terms today, given the fact that being a professed man of religion was a full-time job and a life commitment which gave one a view of this world and the next radically different from the scientific world-view of today.

References

Acta (1983) = *Acta Apostolicae Sedis*, Vol. 71, 2 (Rome).

Abbott, Walter M. (ed.) (1966) *The Documents of Vatican II* (translated from Latin) (London and Dublin: Geoffrey Chapman).

Alberus, Erasmus (1542) *Alcoranus Franciscanorum. Id est, Blasphemiarum et nugarum Lerna, de stigmatisato idolo, quod Franciscum vocant, ex Libro Conformitatum* (Frankfurt: Petrus Bambacchius).

Alberus, Erasmus (1550) *The Alcoran of the Barefote Friers, that is to say, an heape or numbre of the blasphemous and trifling doctrines of the wounded Idole Saint Frances taken out of the boke of his rules, called in latin,* Liber conformationum, trans. n.n. (n.p.: R.G. excudebat).

Alberus, Erasmus (1679) *The Alcoran of the Franciscans, or a Sink of Lyes and Blasphemies, Collected out of a Blasphemous Book belonging to that Order, called The Book of Conformities. With the Epistles of Dr Martin Luther and Erasmus Alberus, detecting the same. Formerly printed in Latine, and now made English, for the Discovery of the Blasphemies of the Franciscans a Considerable Order of Regulars amongst the Papists,* trans. D. S. (London: L. Curtise).

Alva y Astorga, Petrus (1651) *Naturae Prodigium Gratiae Portentum. Hoc est. Seraphici P. N. Francisci Vitae Acta ad Christi D. N. Vitam et Mortem Regulata, et Coaptata, in Prima Columna describuntur Redemptoris Mundi Mysteria, incipiendo ab eius aeterna Praedestinatione, usque ad gloriosam ipsius ad Caelos Ascensionem, et in altera correspondente, Conformitates, Similitudines, ac Parallela Seraphici Patriarchae, in quadraginta quinque Titulos divisa* (Madrid: I. de Paredes).

Armstrong, Edward A. (1973) *St Francis: Nature Mystic. The Derivation and Significance of the Nature Stories in the Franciscan Legend* (Berkeley: University of California Press).

Bartholomew of Pisa (1590) *Liber Aureus, Inscriptus Liber Conformationum Vitae Beati, ac Seraphici Patris Francisci ad Vitam Iesu Christi Domini Nostri. Nunc denuo in lucem editus, atque infinitis propemodum mendis correctus a Reverendo, ac doctissimo P. F. Ieremia Bucchio Utinensi sodali Franciscano Doctore Theologo laboriosis, ornatissimisque lucubrationibus illustratus. Cui plane addita est perbrevis, & facilis historia omnium virorum; qui sanctitate, probitate, innocentia vitae, ac doctrina, ecclesiasticisque dignitatibus, in Franciscana Religione usque ad nostra haec tempora excelluerunt,* written 1385 (Bologna: Alexandrus Benatius).

Bonaventura, Saint (1953) *The Mind's Road to God. St Bonaventura,* trans. George Boas, The Library of Liberal Arts, vol. 32 (Indianapolis: The Bobs-Merrill Company Inc.).

Bonaventura, Saint, *Major Life,* See *Omnibus* (1973).

IV

Brooke, Rosalind B. (1982) 'Recent Work on St Francis of Assisi', *Analecta Bollandiana* **100,** 653–676.

Brooke, John H. (1991) *Science and Religion: Some Historical Perspectives* (Cambridge: Cambridge University Press).

Brooke, John H. and Cantor, Geoffrey (1998) *Reconstructing Nature. The Engagement of Science and Religion*, The Glasgow Gifford Lectures for 1995–6 (Edinburgh: T & T Clark).

Burkitt, F. C. (1932) 'St Francis and Some of his Biographers', in F. C. Burkitt, H. E. Goad and A. G. Little (eds), *Franciscan Essays II*, British Society of Franciscan Studies, Extra Series, vol. 3 (Manchester: The University Press), pp. 19–39.

Crombie, A. C. (1952) *Augustine to Galileo. The History of Science A.D. 400–1650* (London: Falcon).

Cunningham, Lawrence S. (1976) *Saint Francis of Assisi* (Boston: Twayne Publishers).

Cuthbert, Father, O.S.F.C. [Lawrence Anthony Hess] (1912) 'St Francis and Poverty', in *Franciscan Essays (I) by Paul Sabatier and Others*, ed. A. G. Little, British Society of Franciscan Studies, Extra Series, vol. 1 (Aberdeen: The University Press), pp. 18–30.

Cuthbert, Father, O.S.F.C. [Lawrence Anthony Hess] (1915) *The Romanticism of St Francis and Other Studies in the Genius of the Franciscans* (London: Longmans, Green and Co.).

Deretz, J. and Nocent, A., O.S.B. (eds) (1968) *Dictionary of the [Second Vatican] Council* (London: Geoffrey Chapman).

Dubos, René (1974) 'Franciscan Consecration versus Benedictine Stewardship', in *Ecology and Religion in History*, ed. D. and E. Spring, (New York, Evanston: Harper Torchbooks), pp. 114–36.

Ehr, D. J. (1967) 'Creation, Theology of', *New Catholic Encyclopedia*, vol. 4 (New York: McGraw-Hill), pp. 419–23.

Fleming, John V. (1982) *From Bonaventure to Bellini. An Essay in Franciscan Exegesis*, Princeton Essays on the Arts, no. 14 (Princeton, NJ: Princeton University Press).

French, Roger and Cunningham, Andrew (1996) *Before Science: The Invention of the Friars' Natural Philosophy* (Aldershot: Scolar).

Gilson, Etienne (1938) *The Philosophy of St Bonaventure*, trans. Dom I. Trethowand and F. J. Sheed (London: Sheed and Ward).

Goad, Harold E. (1926) 'The Dilemma of St Francis and the Two Traditions', in W. Seton (ed.), *St Francis of Assisi: 1226–1926: Essays in Commemoration. With a Preface by Professor Paul Sabatier* (London: University of London Press), pp. 127–62.

Goad, Harold E. (1929) *The Fame of St Francis of Assisi. The First Walter Seton Memorial Lecture* (London: University of London Press).

Godet, Jean-François (ed.) (1975) *Sancti Bonaventurae Legendae Major et Minor Francisci*, Corpus des Sources Franciscaines, vol. 2 (Louvain: Publications du CETEDOC, Université de Louvain).

Grant, Edward (1996) *The Foundations of Modern Science in the Middle Ages. Their Religious, Institutional and Intellectual Contexts* (Cambridge: Cambridge University Press).

Lindberg, David (1992) *The Beginnings of Western Science. The European Scientific Tradition in Philosophical, Religious, and Institutional Context. 600 B.C. to A.D. 1450* (Chicago: Chicago University Press).

Lovejoy, Arthur O. (1936) *The Great Chain of Being. A Study of the History of an Idea*, (Cambridge, MA: Harvard University Press; repr. 1960, New York, Harper Torchbooks).

Mailleux, Georges (ed.) (1974) *Thesaurus Celanensis*, Corpus des Sources Franciscaines, vol. 1 (Louvain: Publications du CETEDOC, Université de Louvain).

Marée, F. (1656–60) *Traicte des Conformitez du Disciple avec son Maistre: c'est a dire, du Seraphique Pere S. Francois avec nostre Seigneur Iesus Christ, contenant en soy tous les Mysteres de leur Passion, Mort, &c.*, 3 vols. (Liège, Flanders).

McGinn, Bernard (1979) *Visions of the End: Apocalyptic Traditions in the Middle Ages*, Records of Civilization, Sources and Studies, no. 96, ed. W. T. H. Jackson (New York: Columbia University Press).

McMullin, Ernan (1968) 'Science and the Catholic Tradition', in Ian G. Barbour (ed.), *Science and Religion. New Perspectives on the Dialogue* (New York: Harper and Row), pp. 30–42.

Moore, J. R. (1992) 'Speaking of "Science and Religion"—Then and Now [Review of J. H. Brooke, *Science and Religion*, 1991]', *History of Science* **30**, 311–23.

Moorman, John (1968) *A History of the Franciscan Order from its Origins to the Year 1517* (Oxford: Clarendon Press).

Omnibus (1973) = Habig, Marion A. (ed.) (1973) *St. Francis of Assisi, Writings and Early Biographies, English Omnibus of the Sources for the Life of St. Francis*, 3rd ed. (Chicago: Franciscan Herald Press).

Patres Collegii S Bonaventurae (1927) = *Legendae S. Francisci Assisiensis Saeculis XIII et XIV Conscriptae*, Analecta Franciscana sive Chronica Aliaque Varia Documenta ad Historiam Fratrum Minorum Spectantia, ed. Patres Collegii S. Bonaventurae, vol. 10 (Quaracchi: Typographia Collegii S. Bonaventurae).

Pius XI (1926) *Encyclical Letter of Pope Pius XI to his Venerable Brethren the Patriarchs, Primates, Archbishops, Bishops and other Ordinaries in Peace and Communion with the Apostolic See, on St Francis of Assisi, on the Occasion of the Seventh Centenary of his Death* (London: Burns Oates and Washbourne Ltd, Publishers to the Holy See).

Sabatier, Paul (1894a) *Vie de S. François d'Assise* (Paris: Librairie Fischbacher).

Sabatier, Paul (1894b) *Life of St Francis of Assisi*, trans. Louise Seymour Houghton (London: Hodder and Stoughton).

Sabatier, Paul (ed.) (1928–31) *Le Speculum Perfectionis ou Mémoires de Frère Léon sur la seconde Partie de la Vie de Saint François d'Assise*, 2 vols., vol. 1 (Manchester: Manchester University Press).

Seton, Walter (1926) 'The Rediscovery of St Francis of Assisi', in W. Seton (ed.), *St Francis of Assisi: 1226–1926: Essays in Commemoration. With a Preface by Paul Sabatier* (London: University of London Press), pp. 245–63.

Sorrell, Roger D. (1988) *St Francis of Assisi and Nature. Tradition and Innovation in Western Christian Attitudes toward the Environment* (Oxford: Oxford University Press).

White, Lynn, Jr. (1967) 'The Historical Roots of Our Ecologic Crisis', *Science*; as reprinted in Lynn White, Jr. (1968) *Machina ex Deo: Essays in the Dynamism of Western Culture* (Cambridge, MA: MIT Press), pp. 75–94.

V

Science and Religion in the Thirteenth Century Revisited: the Making of St Francis the Proto-Ecologist Part 2: Nature Not Creature

Accounts of 'medieval science', as currently presented by historians of science, feature men of religion—especially bishops and friars—as playing major roles in developing science in the medieval period. The current consensus amongst historians is that the engagement of professional men of religion in the development of science indicates that there was not any great conflict between religion and science in the medieval period, as some historians in the nineteenth and early twentieth centuries had thought. Science and religion, on this new reading, were complementary, not contradictory, pursuits. Quite why it is men of religion—such as Grosseteste, Roger Bacon, John Pecham, Albertus Magnus—who are found playing such leading roles in the pursuit of science in the medieval period is not something that historians of 'medieval science' have enquired into very much. The assumption on the part of such historians seems to be that all intelligent educated men take up science if there are no restrictions on their doing so. Since in the medieval period the majority of literate men were in the Church, and the evidence of their engagement in science indicates that the Church did not provide any impediment to their practice of science, hence (the implicit argument seems to run) they did so. Hence science and religion flourished together.

However, it is completely inappropriate on our part to interpret in terms of 'science' (in our modern meaning of that term) any accounts that we might find written by medieval people concerning natural phenomena, their functioning and meaning. For the enquiries of medieval people into natural phenomena were not

This article was published in *Studies in History and Philosophy of Science* 32.1 (2001), pp. 69–98. © 2001 by Elsevier Science Ltd.

conducted under this category. Indeed, the modern category of science is one that they did not have at all (Cunningham, 1988). And it is equally inappropriate for us to read a 'science/religion' relationship into the intellectual considerations of medieval people, no matter whether we wish that relationship to have been hostile or friendly. The reason is the same: they did not possess our categories, nor our way of opposing these particular categories to one another. On the other hand, they certainly did have their own categories, under which they made their enquiries into natural phenomena. For the most part historians of 'medieval science' have not paid much attention to the categories with which medieval people made such enquiries into natural phenomena as they did.

In the first of this pair of articles I took St Francis as my example of a medieval man of religion to whom has been ascribed in recent times a concern with Nature—Nature, the subject of the scientist's gaze. I showed that St Francis was not, and could not have been, a proto-ecologist, as he has been credited with having been (Cunningham, 2000). I argued there that St Francis did not have, and could not have had, a feeling for Nature or a sympathy for Nature (as has recently been attributed to him) because he lacked the requisite concepts—that is to say, because he lacked one of our most important modern categories, that of 'Nature'. I also showed that, even if he had had the requisite category, nevertheless he would not have had the motivation to have such interest in or feeling for Nature either, since he was busy being a full-time saint with a major mission for the spiritual renovation of the Church and the society of his day. Francis's concern was with *creatures*, because God had created them, not with *Nature*. Thus Francis's interest in and concern with natural things was not only entirely religiously motivated, but he was also entirely religious in his interpretation of what he saw and found amongst creatures. This was the case with his many followers too.

How could we, as historians of science, have come to accept such bizarre misrepresentations of medieval people, such gross misreadings of their intentions and activities, as to ascribe to them interests in Nature and in science, in our present-day meanings of both these terms? And how did we come to take these misrepresentations seriously and make them the basis of our accounts of what we call 'medieval science', and to count such people as contributors to the development of science? How too did we come to project a modern science/religion relationship onto the medieval period at all?

These are large questions, and obviously the traditions of interpretation from which they stem lie largely outside our profession. These traditions of interpretation originated in the nineteenth and early twentieth centuries, and they originated primarily in the history of the Catholic Church as it was in conflict with its enemies (and indeed also with its friends). Strange to say, this is a largely neglected area of study for historians of science. Even historians eager to investigate the historic relations of science and religion have completely ignored the history of the Catholic

Church and its enemies in recent centuries.[1] Yet the conflict between the Catholic Church and its enemies in the nineteenth and early twentieth centuries has profoundly shaped our views as historians of science, both in leading us to believe that there was indeed an enterprise of 'medieval science', and in encouraging our interpretations of that enterprise in terms which invoke the categories of 'science' and 'religion', whether in conflict or in harmony.

The history of the Catholic Church and its inner and outer conflicts in the nineteenth and early twentieth century has affected all areas of our conceptualisation of the domain we think of as 'medieval science' and of the supposed 'science/religion' relationship in the medieval period. Some consideration of that history will begin to show us where some of our prejudices about the shape of the history of science come from, what work they do, and why we try to defend and buttress them when challenged.

So, if we want to understand in an authentic manner the medieval men and women of religion whom we customarily take as having been engaged in one or another aspect of 'medieval science', or see as epitomising the supposed relationship of science and religion in the medieval period, it is necessary to know something of the revolutions in attitudes which occurred in and around the Roman Catholic Church in the nineteenth and early twentieth century. The task will be quite a long one in order to understand the construction of the nineteenth-century historical reputations of all the medieval men of religion who had a supposed involvement with science. Here I can take just one example, St Francis, and look at the pertinent parts of the history of the Catholic Church which will help us to understand how and why St Francis became the Saint of Nature, and hence became available to be portrayed as a proto-ecologist, and how and why subsequently he has been enlisted as a proto-ecologist by historians, by scientists, by campaigners, and by the Catholic Church. But although my study is focussed on the reputations of just one medieval man, the issues of nineteenth-century Catholicism I deal with also lie at the root of the modern reputations of others, such as Roger Bacon, Grosseteste, John Pecham and Albertus Magnus. For they involve the process by which 'creature' came to be read as 'Nature'.

The image of St Francis the Saint of Nature was indeed a creation of the nineteenth century, and was just one of many new reputations the Saint was credited with in this period. I will deal here with just three lines amongst the many new

---

[1]Although, to be fair to them, the following two recent works are not concerned directly with historical constructionism, even so it is almost unbelievable that there is no treatment *at all* of the Catholic Church in the nineteenth and twentieth centuries in Brooke (1991), and only the slightest of references to it in Brooke and Cantor (1998), and then only in connection with a British Catholic convert who was an opponent of Darwinism! It would be nice to think that such lack of interest in the role of the Catholic Church and its troubles in the nineteenth and twentieth centuries in forming an implicit agenda for the history of science in the twentieth century is limited to Anglophone Protestant historians, but this is not so. An Anglophone but non-Protestant exception is Motzkin (1989).

V

reputations given to St Francis in the nineteenth century, though these will not exhaust the varieties of St Francis made available then. They are:

1. Francis the Romantic: poet and minstrel, a liberal and yet loyal Catholic, helping the poor;
2. Francis the Modern: the model of spirituality, rejecting dogma, reproaching the papacy for its wealth and corruption, and devoted to practising poverty;
3. Francis the true Catholic: social reformer and obedient son of the Holy See.

Francis's supposed universal love of Nature will be seen to be an important element in the first two of these lines. Although these are all wrong reputations to give to the St Francis of the thirteenth century, nevertheless there were good nineteenth-century reasons for seeing these as valid reputations for Francis. The new representations of St Francis which were produced in the nineteenth century—including his supposed love of, and sympathy for, Nature—are closely connected to new representations of Christ produced in the period. These in turn were themselves products of different modernising reform impulses within and without the Catholic Church, and equally of the resistance to modernising exhibited by the central bodies of the Church, especially by the papacy itself.

1. The Pope becomes Infallible

As is well known, the Roman Catholic Church in the nineteenth century suffered a series of great crises (Holmes, 1979; Bury, 1930; Coppa, 1979; Hasler, 1981; Reardon, 1985). On the one hand, the pope's position as a temporal ruler continued to be shown to be very fragile, first by Napoleon, who abducted Pius VII and effectively put him under house arrest, then by the Austrians, who obliged Pius IX to flee Rome in disguise in 1848, and then by the Italian nationalists who in 1870 declared Rome, the papal city, the secular capital of the new unified Italy. Even as head of the hierarchy of the Church, the papacy found that its power to appoint bishops was slipping from it and, in a series of concordats, it was obliged to share this power with lay rulers. On the other hand the pope's position as universal spiritual leader began to look equally fragile. Lay rulers of states increasingly repudiated the papacy's claims to jurisdiction in matters of dogma or discipline. Moreover, there was a great crisis of faith as the Church was repeatedly confronted with the challenges of the modern world. What—if anything—was to be the role of the Catholic Church in the modern world? This question particularly confronted two successive popes, who between them held office for the second half of the century: Pius IX (1846–78) and Leo XIII (1878–1903). We shall see that the most important writers on St Francis in the nineteenth century were themselves directly involved in the struggles for the soul of the Catholic Church. What they have to say about St Francis, how they represent him and his beliefs, directly reflects their own positions in this most difficult era for Catholicism.

Everywhere in western Europe in the nineteenth century there was something of a religious revival, both Catholic and Protestant, though often limited to the middle classes. But it arrived hand in hand with some quite new ideologies, including liberalism, nationalism and secularism, which the Catholic Church found highly problematic. There were many positive acts made by the papacy to promote Catholic devotion and religious practice in the course of the century, including the revival and reform of the monastic orders, and the promotion of new devotions (especially that of Our Lady, and of the Sacred Heart), missionary activity, and the cult of saints. The Social Question troubled the papacy greatly: how to create, or recreate, and then put into practice some properly Catholic political principles which would render nugatory the political ideologies of industrial workers and the middle class, such as socialism and liberalism, which would obviously destroy society if allowed to flourish. By the end of the century the (Catholic) Christian Democracy movement was well established, aiming at achieving the solution of the Social Question through direct political involvement, and in 1890 Leo XIII issued the celebrated bull *Rerum novarum*, which supported workers' rights to form associations, to fight for a just wage, and to go on strike, thus conceding that the solution of the Social Question must involve social justice.

However, except to those Catholics most loyal to the papacy, the ways in which the popes had for so long and without exception opposed the popular and ideological movements of the age—such as democracy (anywhere) and popular sovereignty, liberalism, socialism, and the freedom of the press—made the papacy look like a negative force in the modern world. The case has been made that these wholesale condemnations of modern political aspirations were actually only the responses of the popes to very local conditions, writ large.[2] But the broad new claims that the papacy, especially under Pius IX, made about the papal role, and about new errors, indicate that the papacy still had extensive and very conservative ambitions for the role of the Catholic Church in modern society as a whole. Pius IX issued the *Syllabus of Errors* in 1864. Eighty errors of belief of the modern age were denounced including the opinions:

That God was merely nature, that human reason was the sole arbiter of truth and falsehood, good and evil, that all religious truths were derived from human reason, that Christianity contradicted reason or that revelation hindered the perfection of man, that biblical miracles were poetic fictions and Christ himself was a myth . . . that the Church should be separated from the state and *vice versa*; that it was no longer necessary that the Catholic religion should be held as the only religion of the State to the exclusion of all others. (Holmes, 1979, p. 146)

Most striking of all for our present purposes was error number eighty: it was an error to maintain that 'the pope can and ought to reconcile himself to and agree

---

[2]The popes 'universally condemned movements in the world at large which seemed similar to those causing problems within their own dominions' in Italy, according to Holmes (1979), p. 86.

with progress, liberalism, and modern civilization'.[3] Then, at the first Vatican Council, opened by Pius IX in 1869, the doctrine of papal infallibility was declared: it was not asserted as a novel claim, but it was declared that it had always been the case that the papacy had been infallible. The culture wars between Catholicism and the modern world are usually taken to have been epitomised by the *Kulturkampf*, the battle against the Catholic Church which Bismarck waged in a newly united Germany for a few years from 1872; but our attention here needs to be on the culture wars of German Catholicism much earlier in the century, and on France.

In the German states the Catholic Church was allowed to function again after the Napoleonic interlude. Under the influence of Romanticism, German Catholics returned to the faith and developed it in particular ways. As Madame de Staël wrote in *De l'Allemagne* in 1811: 'The revolution which has occurred in the minds of thinking men in Germany during the last thirty years has brought almost all of them back to feelings of religion' (as quoted in Dru, 1963, p. 23). This Catholic revival was not led from Rome, but was autonomous, with the senior and junior members of the theology faculties of Göttingen and Munich having much influence in establishing new ways of thinking about the nature of faith, about personal religious experience, about the nature and historic role of the inner life of the Church, and about the role of tradition in giving authority to the Church and its teachings. To touch on just one aspect of this romanticized Catholicism, we can find one theologian, Johann Michael Sailer, writing in 1809: 'What makes the inner life of a man outward, or expresses that inner life outwardly, is art in the widest sense of the word, and what makes the inner religious life outward is sacred art in the widest sense of the word' (as quoted in Dru, 1963, p. 46). This link between Romantic aesthetic theory and theology was very close. According to a modern historian of the process:

> In aesthetics, as in other fields, the romantic movement was a conscious return to the primacy of the inner life, regarded not as an end in itself involving withdrawal into an ivory tower (art for art's sake) but as the spring of action, the source of poetic creation, the realm in which experience becomes conscious . . . The romantic aesthete grasped the correspondence between the inner life and experience, calling for an openness to 'inspiration' and by analogy to the transcendent . . . The [Catholic] religious revival was thus implicit in the romantic aesthetic. It was . . . the rediscovery of the primacy of the inner life. (Dru, 1963, pp. 50–2)

Out of German philosophy came one of the most troublesome books of the century for the papacy, *Das Leben Jesu Kritisch Bearbeitet*, the 'Life of Jesus Critically Examined' which appeared in 1835, written by David Friedrich Strauss (1808–74). A 'young Hegelian' and former Protestant of the Tübingen school, deploying some of the advances in scientific historiography (themselves also pro-

---

[3]My translation. The text is: 'Romanus pontifex potest et debet cum progressu, cum liberalismo et cum recenti civilitate sese reconciliare et componere'; as printed with French translation and commentary in Pius IX (1939), pp. 2890–2910.

ducts of the Romantic movement) which were promoted by Hegel and his followers (Reardon, 1985, ch. 3), Strauss for the first time gave the Christ story the close scrutiny of the new German scholarship. It was a 'critically examined' life because Strauss used a number of historical criteria by which to assess what can and what cannot be trusted in the Jesus story as recounted in the Bible. One of these was that an account cannot be historical when it is irreconcilable with the known and universal laws of Nature which govern the course of events. The miraculous, prophecies, apparitions, devils and angels do not belong to the world of credible experience. Such things have mythical or poetic origins and functions. Everything that can be included in a factual history has to be natural, not supernatural. So in Strauss's account Christ had no miraculous birth, no miraculous resurrection, and performed no miracles between these two non-miraculous events. As a Hegelian, Strauss seems to have been able to interpret the Christ story not as the story of a (God-) man, but as that of the human race (Weinel and Widgery, 1914, pp. 77–89).

Of particular concern to the papacy was the Catholic church in France. Though restored to being the state church through the concordat of 1801, as a Gallican church it did not look to Rome for leadership. The most loyal French Catholics, however, looked over the Alps (ultramontanes) and trusted Rome for their dogma and discipline. But the Vatican turned out to be an authority unable to offer Catholic solutions for the modern French world, some ultramontanes finding relations with it very vexing; and in the event the Vatican failed to realise any of their hopes. In the first decades of the nineteenth century the most important Catholic intellectual was the abbé Hugues-Félicité de Lamennais (1782–1854), and the trajectory of his life encapsulates these frustrations and disappointments. Initially a Catholic of ultramontane sympathies, he was highly conservative, seeing the Catholic Church and its morality as the essential basis of social order. 'Re-establish authority!' he cried in 1817 in his *Essai sur l'Indifférence en Matière de Religion*, and he sought that authority in the tradition of the Catholic Church. But by 1830 he had become in effect a liberal, and he began a famous short-lived periodical, *L'Avenir* ('The Future'), whose program was to call for 'God and liberty', as the two necessary ingredients for social stability. It was a liberty which would re-empower Catholics who, like Lamennais, felt oppressed by the collusion of church and state: freedom from state control of education, freedom to form religious associations, freedom of the Catholic Church from state control, freedom of conscience, freedom of the press. When this was opposed by the French clergy, Lamennais appealed to the pope, who (to the surprise of Lamennais, as a loyal Catholic) also opposed it. In time Lamennais took his own principles to their logical conclusion, and embraced the goal of universal suffrage and the cause of the industrial poor, believing 'the spirit of the New Testament to be capable of guiding and sustaining the masses . . . in their aspirations towards freedom and justice' (Reardon, 1985, p. 190). The papacy condemned his passionate book of 1834, *Paroles d'un Croyant* ('Words of a Believer'), in an encyclical, describing it as, 'if small in size, immense in perversity'. By 1840 Lamennais had broken with orthodox religion: he rejected what he now called 'the pope's Christianity' to follow that 'of the human race' (Reardon, 1985, pp. 195, 197). Abandoned by the Catholic Church and his own followers, Lamennais entered journalism, and with the outbreak of the revolution in Paris in 1848 nailed his colors boldly to the republican mast. He had progressed from ardent ultramontane authoritarian Catholic, via liberalism to a sort of Christian socialism, to a sort of religionless Christianity and republicanism; and all these changes were forced on him, he believed, by the failure of the Catholic Church to give proper spiritual and social leadership in modern society.

Another intellectual French Catholic, this one a young man training for the priesthood, also had his world turned upside down by the events of 1848. This young intellectual was Ernest Renan (1823–92), whose role in transforming St Francis would be crucial. He lost his faith in 1848, but devoted his life to studying religion in general and the history of Christian origins in particular from a historical and philological position which did not take the miraculous and supernatural elements of Christianity at face value. Renan was one of the monumental scholars of France in the nineteenth century, Professor of Hebrew at the Collège de France, with a literary style so accessible that he had great popular influence. But it was primarily the subject-matter of his most notorious book which gave him an enormous audience both in France and abroad. For it was another life of Christ—and there is no more direct way of questioning the bases of a religion or a cult than by questioning what it is that one can trust about the life and intentions of its founder. Renan, in his *Vie de Jésus*, published in 1863, presents Jesus as the preacher of a liberal ideal of reform.[4] This book, as will be seen, is of great importance indirectly for the most important and enduring image that we have of St Francis. Renan conceived the idea for his life of Christ on a visit to the Levant, and it was 'with the eyes of a French poet of Nature', according to a modern historian, 'that Renan had himself looked upon the land and people of Palestine' (Weinel and Widgery, 1914, p. 150). And Renan himself writes Romantically about Nature as affecting Jesus's thought pattern:

> A beautiful external nature (*une nature ravissante*) tended to produce a much less austere spirit—a spirit less sharply monotheistic, if I may use the expression, which imprinted a charming and idyllic character on all the dreams of Galilee . . . Galilee was a very green, shady, smiling district, the true home of the Song of Songs, and the songs of the well-beloved. During the two months of March and April, the country forms a carpet of flowers of an incomparable variety of colors. The animals are small, and extremely gentle . . . In no country in the world do the mountains spread themselves out with more harmony, or inspire higher thoughts. Jesus seems to have a peculiar love for them.[5]

[4]This is the title of a chapter in Weinel and Widgery (1914) which deals with Renan.
[5]As translated into English in Renan (n.d.; p. 81), in chapter 4, 'The Order of Thought which Surrounded the Development of Jesus' ('Ordre d'Ideés au Sein duquel se Développa Jésus').

Renan presents Jesus as moving between oppositions which Renan himself had experienced: on the one hand there was 'the young Galilean prophet devoted to everything good, open, and frank in the noblest sense, with a true love of Nature and of women', while on the other hand were 'the gloomy priests, theologians, and lawyers of Jerusalem' (Weinel and Widgery, 1914, p. 151). According to Heinrich Weinel in his study of different literary images of Jesus created in the nineteenth century,

> Renan created that tender type of Jesus, full of mild melancholy, to whom at that time charming women could render homage and love ... Renan looked upon Jesus as a reformer in the realm of religion and morality, a man with definite principles, which it was his hope to establish in the life of the people. [Thus, for Renan] 'A pure worship, a religion without priests and external observances, resting wholly on the feelings of the heart, on the imitation of God, on the close communion of the conscience with the heavenly Father, were the results of those principles' ... 'The abolition of the sacrifices ... the suppression of the impious and haughty priesthood, and in a general sense, the abrogation of the law, seemed to him [i.e. Jesus] absolutely essential'. (Weinel and Widgery, 1914, pp. 157–8)

Renan treated the gospel miracles as tricks, explaining the resurrection of Lazarus, for example, as 'an unpardonable deception played by the sisters upon the Master' (Weinel and Widgery, 1914, p. 157). Again, as with Strauss, Renan's Christ is a man, not a God, a character with natural gifts to a very high degree, but without supernatural ones. This Jesus rightly mistrusts the organized church of his day and its priests and theologians who bring only gloom and oppression. Renan was dismissed from his professorship for publishing this book.

The papacy's rejection of the modern world, as epitomized in the *Syllabus of Errors*, and its continued failure to provide effective Catholic leadership to make Catholicism relevant and responsive to modern industrial society, provoked a distinctive movement of inner reform. Catholic intellectuals, especially Francophone ones, joined together in a loose movement to which the Jesuits and then pope Pius X—all of them its sworn enemies—first gave the name 'Modernism', a name its devotees then adopted. It really began in the 1840s, with the unfortunate experiences of Lamennais, crowned by the revolutions of 1848, though it did not appear as a public movement until the 1890s, and lasted in the public domain until about 1910. Renan's outlook—including the importance of modern scholarship, freedom of thought, liberal politics, freedom from restrictions of dogma—encapsulates the values this movement came to stand for, and which had a crucial role in shaping our dominant view of St Francis, including his supposed love of Nature.

We can characterize Modernism briefly as a movement, both clerical and lay, primarily within the Catholic Church, seeking to relate the Church to the modern world: to modernize it, in other words. These are three of the major points of the Modernist position:

1. *Science*: science in general, and especially the doctrine of evolution, needs

to be accommodated by Catholicism, not denied, and the claims of science as offering an alternative legitimate view of truth need to be recognized.

2. *History*: the study of history has a crucial role in order for us to appreciate that the Church has taken a succession of dogmatic and doctrinal positions over time, in response to particular historical circumstances, and that the current dogmatic and doctrinal positions of the Church are similarly not absolute and fixed for all time.

3. *Today*: the Catholic Church today is still developing, as it has done in the past, and faced with new circumstances it needs to face them in new ways.

These positions, with their relativistic underpinning, were taken (as might be expected) as virtually heretical by the papacy, especially by Pius XI, who described Modernist doctrines as 'pernicious diseases', showing 'the deplorable spirit of *insubordination* and *independence* among young priests' (Pius XI, 1926). Some Modernists were equally outspoken in their criticism of the pope. For instance Paul Sabatier, a pupil of Renan's, and himself a Modernist, criticized the pope in the Jowett Lectures he gave in 1908 on 'Modernism':

> The Church, whose duty was to give to our civilisation a spirit of truth, of scientific exactitude and humility, and of manly sincerity, has fallen into the hands of a [papal] government which does all it can to stifle this spirit . . . She it is who rebels against those signs of the times which she so completely misunderstands. (Sabatier, 1908, pp. 40, 46)

In 1910 the pope demanded an oath against Modernism from all future clerics, and this seems to have stopped the movement from growing within the clergy. The Modernist movement failed, but it produced much scholarly work of a high standard which outlived it.

In the attacks on and defences of the Catholic Church in the nineteenth century, and in the responses of Catholics and other Christians, Nature keeps cropping up all over the place, and is appealed to on both sides of most arguments. It is a concept, or more properly a set of concepts, formed in ways we would recognize today. It is Nature with both scientific and spiritual meanings and resonances, simultaneously.

First, it is Nature as a term which embraces the material universe and the regularities of its functioning, and this concept precludes God from any role at all, whether in creating the world and its constituents, maintaining the laws of Nature, or intervening to suspend them. Nature has laws of her own; and, some claimed, credence should be given to any claim about the present or the past, and even about the divine and sacred, only if such a claim conforms to the established laws of Nature. This concept of Nature had been constructed by this date in such a way as to be the primary criterion of truth. Thus this Nature is a scientific concept, as we would today understand both the terms 'Nature' and 'scientific'.

But secondly, Nature is also simultaneously held to be a spiritual entity as well as a material one. It is not a matter of 'Nature' in this sense being witness to, or evidence of, the divine because it is itself the creation of God. Indeed, insofar as God and the divine are spoken of under this sense of 'Nature', they do not necessarily refer to the Christian God or to the Christian sense of the divine. Rather, Nature is taken to be divine *of itself.* Moreover, Nature is considered to be a self-regulating organism, in a new sense of that old term, 'organic'. Man, who is now seen as 'part of Nature', is considered to be the more spiritually fulfilled the more he is in harmony with Nature and with both her known and her mysterious laws and ways. The more inspired, imaginative and creative a person is, the more he is taken to be in harmony with Nature, and through this he reaches his self-realization. Thus this Nature is a Romantic concept (Cunningham and Jardine, 1990).

These two new concepts of Nature derive from the great secularization of European intellectual life at the end of the eighteenth century on the one hand, which we owe primarily to France, and from the Romantic movement of the same period on the other, which we owe primarily to the German states. Hence Nature is now both scientific *and* Romantic and, while secularists and Romantics fought to control the meaning of the term, many people used both senses of the term simultaneously and without consciousness of their differences, as most of us still do today.[6]

2.   Three New Images of St Francis

2.1.   Francis the Romantic: poet and minstrel, a liberal and yet loyal Catholic, helping the poor.

By shaping new concepts of Nature, the Romantics also helped shape new characterizations of St Francis. We find this Romanticism first in the German states, and then from the 1820s in France and elsewhere, and of course it has persisted through to today. This Romantic sensibility is significant for our present concerns in a number of ways. First, because of the image of Nature that it presented, as an organic being, a self-regulating entity, fecund and dramatic in herself, and with her own life, destiny and mysterious forces. Second, because spirituality has been redefined in Romanticism as meaning, in effect, 'harmony with Nature'. Thus if an individual is in harmony with Nature, or intimate with Nature, he or she is seen as undertaking an exercise in spiritual fulfillment. And while Romantic aesthetic spirituality is not quite the same as Christian religious spirituality, it is counted as an equally desirable condition. The third significance of the Romantic sensibility for our purposes is that, in the Romantic view, from such spirituality follows artistic creativity and sensitivity. The artist, above all the poet, was taken to be the most spiritually aware of people. Poetry, for example, was considered as analogous to prayer. The fourth significance is that Romantics harked back to a supposedly glorious Middle Ages. As Lord Acton put it in a much-quoted sentence: 'The

---

[6]For a very brief history of these modern concepts of Nature, see Williams (1976).

romantic reaction which began with the [French] invasion [of Germany] of 1794 was the revolt of outraged history. The [German] nation fortified itself against the new ideas by calling up the old, and made the ages of faith and of imagination a defence from the age of reason' (Acton, 1907, p. 346). The Middle Ages had a special relevance for German Catholics, for they 'appeared in their totality as an achievement as complete as that of the Greeks, in which the Christian harmony of natural and super-natural had led to the creation of the corresponding forms—not only in poetry and architecture, painting and sculpture, but in the social and political spheres' (Dru, 1963, p. 53). In the early decades of the nineteenth century German Romantics thus promoted a vision of the medieval period as having been particularly beautiful, sensitive and artistic. Finally, the German Romantic movement was one of the sources for the development of the writing of the history of ideas, that is of ideas as disembodied but almost alive entities. Romantic historians of ideas used organic imagery about ideas: for them ideas are born, they develop or unfold, they come to their full development and fructify, they produce seed, they wither and die. This organic imagery for the history of ideas has become very common, and of course is still with us. Romantic images of St Francis created in the nineteenth century will contain most or all of these characteristics.

The first version of a Romantic St Francis seems to have been put forward by a lay theologian, Johann Joseph von Görres (1776–1848), in 1826, in a book called *The Holy Francis of Assisi a Troubadour* (Görres, 1826). It was very much a product of the line of the Catholic revival in Germany in the early years of the century under the influence of the Romantic movement as represented by the Munich school. 'If it were necessary to name the greatest figure in the annals of German Catholicism' of this period, a modern historian has written, 'one could only answer, unfortunately, Görres'.[7] Görres was first a liberal, who established a journal to oppose the Napoleonic occupation: 'He became the spokesman of the nation in the War of Liberation ... Görres spoke for Germany in the name of freedom'. But after 1819 he moved back towards Catholicism, seeing the Catholic church as

the only power which would survive the revolution. In religious matters, he had come to feel, it was wiser to build one's house in the old building which had been constructed long before the oldest monarchies. This was to become the conventionally romantic view [in Germany] of the Church ... The Church was a 'power', a bulwark against unconditional revolution, a safeguard for freedom; and in the name of freedom Görres became more conservative.

In Munich from 1827, at the time ruled by Ludwig I, Görres had a clear platform, where 'the Catholic revival germinated its feudal sentiments and its latent conservatism. The myth of the Middle Ages began to take effect ... The conservative tendencies of the romantic movement appeared as soon as it became involved in politics, and were in the end to master the liberal elements'. Görres was extremely

---

[7]Dru (1963), p. 65. Quotations in this paragraph are from pp. 65–71.

productive: 'There was in fact no aspect of the romantic movement which Görres did not touch upon and none which he did not adorn with his authority ... He was the epitome of his age without being original'.

He wrote his book on St Francis at the height of his conservative Catholic and Romantic phase, in 1826, as he was becoming interested in Christian mysticism. These positions are all evident when Görres gets into full flow on the virtues of St Francis and the world of Nature:

> Thus did the holy man walk about in the world of Nature, and wherever his foot stepped, the ancient curse was lifted from the earth; in the radiance which surrounded him, the dark blemish transfigured itself, like the dark clouds in the dawn; the animals played around him trustingly, the flowers looked up to him with loving eyes; even the elements raised up their heads drunk with sleep from their dark dream world, and blinked, amazed in the unaccustomed brightness which awoke them. Captivated by the higher magical heavenly power, which flowed out of him, all did his bidding willingly, and only once he had walked past, and the last ray [of his power] had ceased to glow, did the curse reassert itself anew; Paradise vanished, life again hid itself behind the bark and the cherub stepped back with its flaming sword, guarding the gateway. With him [Francis] the first stanzas of the Song of the Sun stepped out of this secret intercourse with the powers of Nature. When, as we have already heard, he had returned to consciousness out of this joyous ecstasy, the after-effect of the delight that had filled him in this ecstasy, combined with that naïve intuition of Nature that had seized him (an intuition that we also find in his contemporaries, the Minne poets), made him say—'O Highest One, omnipotent, good Lord, Yours is the praise, the glory, the honour, all blessing ... ' [i.e., the Canticle of the Creatures]. (Görres, 1826, p. 28)

The Romantic image of St Francis owes much more, however, to its development in France by Frederick Ozanam (1813–53). When Romanticism was imported into France it became generally more sentimental. A Frenchman, initially a follower of Lamennais, a passionate admirer of Pius IX in the pope's early, more liberal, days, and devoted to restoring ultramontane Catholicism in France, Ozanam was involved with a group of other young Catholic men in Paris who wished to do something practical to demonstrate their concerned Catholicism. The outcome was the creation of the Society (or Conferences) of St Vincent de Paul in 1833. This was a new form of charity for the sanctification of its members and of society as a whole through good works, involving the visiting of the needy poor of the parish in their own homes and helping them directly. As a proselytizing Catholic Ozanam wanted to write a history of medieval civilization and Catholicism, primarily to fight atheism: 'Never was a history of human religion more imperatively called for by social need', he wrote (quoted in O'Meara, 1876, p. 31). As a Lamennaisian he sought to argue that the historical source of liberty and freedom of thought was Catholicism—and not Protestantism, atheism, materialism or Saint-Simonism, as others might reasonably expect. To make this highly challenging case he became a medievalist, and of a Romantic kind. On a trip to Italy as a twenty-year old young man, he first fell in love with Dante and his poetry. His early book on the debt of Dante to Catholic philosophy of the thirteenth century (Ozanam, 1839) led him back to earlier poets who had helped shape Dante, and this brought him to Jacapone, a Franciscan mystic poet, and thence to Francis himself, the Orpheus of the Middle Ages, as he called him. Ozanam's very popular book *The Franciscan Poets in Italy of the Thirteenth Century*, published in 1852, set the Romantic image of Francis for generations, and was probably the most influential of the publications Ozanam produced in his short but very busy life (Ozanam, 1852).

Ozanam portrayed the thirteenth century as full of chivalry, courtly love, knights and ladies, tournaments, minstrels, fools, poems and songs, and as having a high appreciation of beauty, both man-made and natural. For Ozanam the Romantic, 'poetry at its highest is an intuition of the infinite, the perception of God in creation, the unchangeable destiny of man represented amid the vicissitudes of history',[8] and for Ozanam Francis was above all a poet, and the more spiritual for being so. Ozanam's Romantic leanings come out in his own attitude toward Nature, and in his claims about what St Francis's relation to Nature had been. Ozanam credits Francis with an 'amour de la nature', which (for Ozanam) is one of the things which reveal Francis to have had a great creative mind: 'The love of Nature is the common bond of all poetry', Ozanam writes (Ozanam, 1852, p. 78). In Ozanam's account, Romanticism meets Catholicism:

> Christianity, so often accused of trampling Nature under foot, has alone taught man to respect her, to love her truly, by making apparent the divine plan which upholds her, illumines and sanctifies her. It was with such a clear vision that Francis contemplated the creation; he searched in every corner of it to discover traces (*vestiges*) of his God; he recognised Him Who is superbly beautiful in beautiful creatures ... Where other eyes would perceive only transitory beauties, he recognised, as if by second sight, the lasting bonds which link the physical with the moral order, the mysteries of Nature with those of faith. (Translation as in Ozanam, 1914, pp. 70–1)

Half of this short passage is correct Catholicism and good history, and the other half is Romantic sentimentality. Unfortunately the two are thoroughly mixed up, and Ozanam switches between one and the other from one phrase to the next, the same being true of his book as a whole. It is a curious sight: Ozanam, the devout and theologically informed Catholic, who knows about Francis the *religious* man and the thirteenth-century view of the *relation of creature to creator*, yet who is so swept along by his Romantic commitments that he constantly also expresses this as Francis, a *Romantically* inspired man, in the nineteenth-century view of *the relation of inspired people (poets) to Nature*, with Nature conceived as autonomous, independent of God. Religious spirituality on the one hand and Romantic artistic-Nature spirituality on the other are conflated. The excesses of Görres apart, it is our first sighting of this gross confusion between thirteenth- and nineteenth-

---

[8]'La poésie à sa plus haute puissance est une intuition de l'infini: c'est Dieu aperçu dans la création, l'immuable destination de l'homme présentée au milieu des vicissitudes de l'histoire', Ozanam (1839), p. 70, my translation.

century concepts and attitudes, still so frequently met with in interpretations of St Francis and elsewhere, including in the history of science.

With respect to Francis's commitment to a life of poverty, Ozaman describes such a life as 'hard beyond conception', but claims that it was 'also the freest, and, in consequence, the most poetic . . . Freed from all servitude, from all trivial occupation, Francis lived in the contemplation of eternity, in a devotion which exalts all the faculties, and intimate relationship with the whole Creation which offers the most poignant charms to the simple and the humble' (Ozanam, 1914, p. 69). In other words, the life of poverty allowed Francis to be spiritually fulfilled as a supreme creative artist. This spirituality of Francis, in Ozanam's hands, is simultaneously Romantic and Christian.

Ozanam is probably the writer who in modern times first transformed the 'Canticle of the Creatures' into a song of thanks *to* God for creating a beautiful world, which seems to be how it is generally interpreted today, even amongst the best scholars. 'Loué soit Dieu, mon Seigneur', his translation into French runs, '*à cause de* toutes les créatures, et singulièrement *pour* notre frère le soleil . . . ': 'God my Lord be praised, *on account of* all the creatures, and especially *for* our brother the sun . . . '.[9] This is quite different from the canticle's original meaning, where the creatures were called on to praise God, their Creator, according to their duty as creatures.

This Romantic view of Francis, with its conflation and confusion of thirteenth- and nineteenth-century concepts and attitudes, has proved very popular as an interpretation, having been taken up even by Protestants, such as Karl August von Hase (1800–90), Professor of Theology at Jena, who devoted a book-length section of his 1856 history of Christianity to St Francis (Hase, 1856). The highly popular writer Mrs Oliphant also adopted this view in her life of Francis published in 1868, which seems to be the first modern life of Francis in English. Mrs Oliphant's parents had been members of the Scottish Free church, a Calvinist sect, which she does not seem to have deserted for another church. As someone who lived by her pen and had to support a family of adult spongers, she probably chose the theme of St Francis for its general popular appeal;[10] and for her information on St Francis she depended on Hase, whom she read in German, and Ozanam, whom she read in French. As a result, her St Francis was as confused as Ozanam's. He was

---

[9]Ozanam, A.-F. (1852), p. 88; the English translation of the Canticle in Ozanam (1914) is by Matthew Arnold, and it performs further transformations: 'Be Thou blessed, O Lord, with all things created,/ Especially my Lord and Brother the Sun'. God is here presented as being blessed, rather than praised, and along with all things created (which is not quite the same as all creatures), rather than by them. Further on in this translation of the poem, the Lord is actually praised, but for the creatures, not by them! But compare footnote 11 in the first of the present series of articles.

[10]Goad (1929), pp. 29–30, suggests that Mrs Oliphant was probably inspired by Mrs Jameson's book *Legends of the Monastic Orders*, of 1850. On Mrs Oliphant, see the *Dictionary of National Biography*, vol. XXII, Supplement, 1901, pp. 1102–6.

> a man overflowing with sympathy for man and beast—God's creatures—wherever and howsoever he encountered them. Not only was every man his brother, but every animal . . . And by this divine right of nature, everything trusted in him. The magnetism of the heart, that power which nobody can define, but which it is impossible to ignore, surrounded him like a special atmosphere . . . With Francis the sympathetic power was universal . . . It is difficult to know where to begin in the many stories of this description. We can select at hazard some of those which are the most significant of the sympathetic mind, open to all the influences of nature, with which we have to deal . . . [11]

Sympathy—a very Romantic personal attribute—on the part of Francis, and trust on the part of other creatures, have here replaced power and submission, as the supposed bases for the relation of St Francis and other creatures.

So pervasive has the Romantic view of St Francis become that in the twentieth century even someone so otherwise theologically acute as the Capuchin friar Father Cuthbert can think it appropriate to present St Francis in the extreme sentimentality of the Romantic view, claiming that 'Francis was from the beginning to end an idealist and a poet', and that 'the story of St Francis in its temperament and idealism is peculiarly akin to medieval chivalry' (Cuthbert, 1912, pp. 394, 19; Cuthbert, 1915).

Once established, the Romantic image of Francis, Nature and animals has persisted throughout almost all other approaches to him—for we are all children of Romanticism—including the view of him to which we now turn.

2.2. Francis the Modern: the model of spirituality, rejecting dogma, reproaching the papacy, and devoted to practicing poverty.

The main proponents of this second new view of St Francis were Ernest Renan and his pupil Paul Sabatier. This Francis was a product of the same attitudes which went to make up the Modernist movement within and around the Catholic Church. The life of St Francis was one of the important moments Renan intended to cover in his history of Christianity. Although even his remarkable scholarly energy did not allow him to do this, Renan took the opportunity of writing a review of Hase's book, in 1866, to give his own views on St Francis a quite extensive airing (Renan, 1866). It will be little surprise to hear that Renan's Francis was like his Jesus: pure through his poverty, morally uncorrupt, distant from the priests and obscurantism of the Church, opposed to the corruptions of the Church, rich in spirit, inspirational, who operated only in the world of the natural and who loved Nature and lived in harmony with it. The supernatural has no place in such an account: thus the stigmata (for instance) were, for Renan, a lie started by Brother Elias in order to gain

---

[11]Oliphant (1868), pp. 118, 119, 122. Her confusion between thirteenth- and nineteenth-century attitudes is perhaps best shown by her curious observation that the people of the thirteenth century 'could not realize the possibility of mere animal life without consciousness, without some fashion of intellectual or emotional existence; and were not aware of so great a gap between themselves and the so-called lower creatures as we are'. It would be difficult to be more wrong.

the leadership of the Order after Francis's death. And Brother Elias, of course, was conspiring with the papacy to dilute and distort the teachings of Francis to fit the aims of the papacy.

The dominant characteristic of Renan's Francis is that he is *simple*. His simplicity ensures his purity which, in turn, puts him in touch with Nature:

> Francis does not follow the Bible, and he hardly reads it. He has not studied scholasticism; he is neither priest nor theologian ... He disdains nothing; he is indifferent to nothing; he has joy and a tear for everyone; a flower sends him into raptures; in nature he sees only brothers and sisters; everything has for him a sense and a beauty. The admirable canticle is well known which, they say, he himself called the *Song of the Creatures*, the most beautiful piece of religious poesy since the Gospels, the expression of the most complete modern [*sic*!] religious sentiment ... In the whole of nature he sees nothing hostile or too humble; he collects the worms from the road and places them safe from passers by; he schemes to save a lamb from death or from the bad company of she-goats and billy-goats; he conspires to help the animal caught in the trap, and gives it good advice to avoid being caught. The great sign by which one recognizes spirits preserved from vulgar pedantry—love for and understanding of animals—was present in him more than in any other man. (Renan, 1866, pp. 922, 924, my translation)

This last sentence, claiming that the capacity to relate to, or commune with, animals is a mark of a person with a great spirit, expresses a very Romantic attitude, that those people most spiritually gifted have an instinctive feeling for Nature. According to Renan, in the Franciscan Order itself 'one senses the naivety of men who know nothing but nature and what they have seen and heard of the Church, who then combine all that in the most free way. It is a thousand miles from scholasticism; Francis of Assisi is virtually the only man of the Middle Ages who was completely free from this leprosy, who was in no way tarnished by the false discipline of the spirit which the subtleties of the Schools had introduced. He had no other theological instruction than that of a simple believer' (Renan, 1866, pp. 929–30). Renan's assertion that 'Francis does not follow the Bible, and he hardly reads it', is completely wrong: Francis and the early Franciscans built their speech and writings out of Biblical words and phrases, and allude constantly to the Bible in word and even in deed, reliving the events and moral messages they find there. It may not be too much to say that they knew their Bible off by heart, and were constantly reading it out loud to each other. And 'it is one of Francis' most troubling habits', writes John V. Fleming, a better historian of St Francis than Renan, 'that he reads the Scriptures with utmost seriousness' (Fleming, 1982, p. 82; Fleming, 1977, pp. 22–3 *et seq.*). One can appreciate why Renan the proto-Modernist wanted Francis to exhibit a spontaneous and intuitive spirituality, but one can only hope that he made this erroneous remark about Francis's ignorance of the Bible in order to claim that Francis was free from the learned scholasticism of his time, which is not the same thing at all. But one can see how, via such overstatements, it was possible to represent Francis as a holy innocent, for whom open-hearted relation to animals and plants would have been the most natural thing in the world.

All these elements of Renan's little sketch of Francis in 1866 were expanded by his pupil Paul Sabatier (1858–1928) into the most important scholarly work on St Francis to appear in the nineteenth century, which was published in French in 1893 and immediately translated into many other languages. Sabatier himself recalled, forty years on, how his master, Renan, had handed the task of being Francis's modern biographer to the young pupil. In December 1884, at the end of a lesson in Hebrew at the Collège de France, Renan spoke to a group of students:

'When I began to work, I dreamed of devoting my life to the study of three periods. Blessed be the illusions of youth! Three periods: the origins of Christianity with the history of Israel; the French Revolution; and the marvellous renewal of religion realised by St Francis of Assisi. I have only been able to carry out a third of my programme. But you, M. Leblond,' said he to a young man who then seemed healthy enough but who died shortly after from overwork, 'you must become the author of the religious history of the Revolution, and you,' he said to another [Paul Sabatier], putting his hand on his shoulder, 'You will be the seraphic historian. I envy you: St. Francis has always smiled on his historians. His initial work and his action through the following centuries have never been completely understood. He saved the Church in the thirteenth century and his spirit has remained strangely living ever since. We have need of him. If we know how to wish for him, he will return.'[12]

Evidently Sabatier was being entrusted with writing the biography the way that Renan would have written it, a task he accomplished perfectly. He was also, at least as far as his own recall of the incident goes, being entrusted with summoning up the spirit of St Francis to save the Catholic Church of the late nineteenth century, which is certainly something he tried to do.

Sabatier's biography was extremely controversial in its day, but has become the basis of all subsequent imagery of St Francis. Sabatier's credentials for writing this life are very unexpected. For he was not, and never had been, a Catholic. Rather, he was—of all things—a Calvinist pastor, who later taught at the University of Strasbourg. His time as a pupil of Renan seems to have not only set his course as a historian of St Francis, but also committed him, though a Calvinist, to seeking the Modernist reform of the Catholic Church. In the event it was the fact that Sabatier was a Calvinist that helped make his St Francis acceptable to people of all Christian denominations, for he was concerned to downplay any specifically Catholic elements in Francis's story. And it was Sabatier's commitment to the Modernist reform of the Catholic Church which made that Francis into someone who transcended the dogma of any particular sect, and indeed someone who rejected all kinds of dogma, preaching instead a simple faith of the spirit, in love with humanity, God and Nature—the kind of St Francis who, for a Modernist,

---

[12]Sabatier's inaugural lecture at the University of Strasbourg, 1924, printed in Sabatier (1932), pp. 69–70, as translated in Engelbert (1950), pp. 29–30, with modifications by me. On Sabatier see in general the excellent article by Brooke (1971), pp. 197 *et seq.*, which is very good on his Romanticism; strangely, however, Brooke completely misses Sabatier's dedication to the ideals of Modernism, ideals which are even more prominently reflected in Sabatier's characterisation of Francis than the Romanticism.

could save the Church of today. Sabatier himself wrote presciently (for my present purposes, bless him) in the introduction to his book: 'Objective history is then a utopia. We create God in our own image, and we impress the marks of our personality in places where we least expect to find it again' (Sabatier, 1894b, p. xxxii). We certainly do, and I am sure Sabatier did not intend to make his Francis a Modernist, but that is exactly what he did.

Sabatier's main message about Francis was that his piety was inspired by, and was, the purest *love*: it had nothing in it of the formal or of observance of rituals, and was thus quite distinct from the piety—the false piety—that the Catholic Church had promoted in Francis's day and that it was still promoting, Sabatier believed, in the present day. Francis's piety

> proceeds from the secret union of his soul with the divine by prayer; this intuitive power of seeing the ideal classes him with the mystics . . . Besides this his piety had certain peculiar qualities which it is necessary to point out. And first, liberty with respect of observances: Francis felt all the emptiness and pride of most religious observance. He saw the snare that lies hidden there, for the man who carefully observes all the minutiae of a religious code risks forgetting the supreme law of love. (Sabatier, 1894b, pp. 194–5)

And it is precisely this love, this true piety, according to Sabatier, which makes Francis at one with Nature, and able to respond to Nature:

> We see how Francis's love extended to all creation, how the diffused life shed abroad upon all things inspired and moved him. From the sun to the earthworm which we trample under foot, everything breathed in his ear the ineffable sigh of beings that live and suffer and die, and in their life as in their death have a part in the divine work . . . Francis's sympathy for animals, as we see it shining forth here . . . is only a manifestation of his feeling for nature, a deeply mystical, one might say pantheistic, sentiment, if the word had not a too definitely philosophical sense, quite opposite to the Franciscan thought . . . We should never have done if we were to relate all the incidents of this kind, for the sentiment of nature was innate with him; it was a perpetual communion which made him love the whole creation. He is ravished with the witchery of great forests; he has the terrors of a child when he is alone at prayer in a deserted chapel, but he tastes ineffable joy merely in inhaling the perfume of a flower, or gazing into the limpid water of a brook . . . The thirteenth century was prepared to understand the voice of the Umbrian poet; the sermon to the birds closed the reign of Byzantine art and of the thought of which it was the image. It is the end of dogmatism and authority; it is the coming in of individualism and inspiration; very uncertain, no doubt, and to be followed by obstinate reactions, but none the less marking a date in the history of the human conscience. Many among the companions of Francis were too much the children of their century, too thoroughly imbued with its theological and metaphysical methods, to quite understand a sentiment so simple and profound. But each in his degree felt its charm. (Sabatier, 1894b, pp. 177–82)

Sabatier's Modernist Francis and his Nature-loving Francis are thus two sides of the same coin.[13] Sabatier's interpretation of Francis has dominated all work for a

---

[13]Elsewhere Sabatier even claimed that Francis 'took possession of Nature'. For Sabatier, the *Canticle*, which is one of the final recorded events of Francis's life, marks this 'final stage of Francis's

century, but it is necessary to remember that it is only an interpretation.[14] There is simply no factual basis in the legends for a single one of the statements in the above passage, and the categories in which it is presented are nineteenth-, not thirteenth-, century ones.

St Francis's popular modern reputation as 'the little poor man', the *povorello*, was largely constructed by Renan and Sabatier. Instead of poverty being a virtue since it provides a route to a higher spirituality (as it had been for Francis in the thirteenth century, according to the legend writers), to Renan and Sabatier the virtue of poverty in the case of Francis was that it made him free from the oppressions and abuses of centralised dogmatic and authoritative institutions, especially the papacy. But because he was poor in wealth, this Francis is rich in things of the spirit. And that means that Francis self-evidently *has to be* at one with Nature and his fellow creatures (for that is the Romantic way for the spirit to be rich). 'Poverty not only permitted the Brothers to mingle with the poor and speak to them with authority', writes Sabatier,

> but, removing from them all material anxiety, it left them free to enjoy without hindrance those hidden treasures which nature reserves for pure idealists. The ever-thickening barriers which modern life, with its sickly search for useless comfort, has set up between us and nature did not exist for these men, so full of youth and life, eager for wide spaces and the outer air. This is what gave St Francis and his companions that quick susceptibility to Nature which made them thrill in mysterious harmony with her. Their communion with Nature was so intimate, so ardent, that Umbria, with the harmonious poetry of its skies, the joyful outburst of its spring-time, is still the best document from which to study them . . . (Sabatier, 1894b, pp. 126–7)

thought, the taking possession of Nature': 'This taking possession of Nature (*cette prise de possession de la nature*) was realized, just like the taking possession of the Bible, slowly, profoundly, by the Church and for the Church. For a long time Francis had been impressed by the charm of Nature, seized with pity for animals, now he had found the explanation and the justification for the sentiments which hitherto had been more instinctive . . . He himself felt the life, the hope, which animates all creatures, and in his position, according to his means, he considered himself the collaborator of this universal work . . . Francis tasted with delight all the beauties, all the grandeurs and appearances of Nature. When he contemplated Nature, those who saw him found that he seemed to be in ecstasy not on earth but in heaven, so much did joy rise up from his whole person. Himself drawing continually on this source, he wanted to teach all men, his brothers, to console themselves from it, to renew themselves from it, and to at the same time become aware of the power and beauty of God . . . The Canticle is a hymn to the beauty of the exterior world, at one and the same time an enrichment and a refinement, an exaltation of the senses and of the soul, of the inner and of the outer man. In this way Francis prepared for the Renaissance and gave it a divine goal.' Undated and unpublished notes, printed posthumously in Sabatier (1932), pp. 63–5; my translation.

[14]With respect to the interpretation Sabatier offers, it needs to be noted that he treated as his most trusted authority a work (the *Speculum Perfectionis*) which he believed to date from immediately after the death of Francis, but which is now known to have been composed a century after his death. It is a Spiritualist work, and hence Sabatier was able to treat the Spiritualist message as being the true message of Francis, and to consider the 'official' legend writers (Celano and Bonaventura) as having corrupted this message. Hence Sabatier's Francis is essentially a 'hermit' Francis—which is what Sabatier would have wanted, as a Modernist—but his reading of the historical record is unfortunately back to front. As scholarly opinion during his lifetime inclined more and more to seeing the *Speculum Perfectionis* as a late work, Sabatier continued to try and find arguments to date it as an early one. Given that his whole interpretation of Francis rested on this dating, this is not surprising.

Let me give just one final example to show the general trend of Sabatier's biography in giving roles to Nature in the story of Francis. Sabatier is discussing the *Fioretti* or 'Little Flowers of St Francis', that Spiritual writing of the early fourteenth century, which is the largest source of stories about Francis and his relation to other creatures, especially animals. As a comparatively late document, the *Fioretti* does not have the same status as the legends of Celano and Bonaventura for the deeds and sayings of Francis. But Sabatier, as a Romantic with a Modernist agenda, sentimentalizes this problem: legends—in the modern sense, that is as unconfirmed or super-natural stories that accrue round historic figures—surely must epitomize the truth, he claims.

> We must not exaggerate the legendary side of the *Fioretti*: there are not more than two or three of these stories of which the kernel is not historic and easy to find. The famous episode of the wolf of Gubbio, which is unquestionably the most marvellous of all the series, is only, to speak the engraver's language, the third state of the story of the robbers of Monte Casale mingled with a legend of the Verna ... Scientific history is trying to react, to mark the relative value of facts, to bring forward the important ones, to cast into shade that which is secondary. Is this not a mistake? Is there such a thing as the important and the secondary? How is it going to be marked? The popular imagination is right: what we need to retain of a man is the expression of countenance in which lives his whole being, a heart-cry, a gesture that expresses his personality (*un geste qui a exprimé la personnalité*). Do we not find all of Jesus in the words of the Last Supper? *And all of St Francis in his address to brother wolf and his sermon to the birds?*[15]

Romantic sentimentality wins out over historical understanding: the whole question about Francis's relation to other creatures is begged in favor of finding 'a heart-cry', a *crie de coeur*, or a single sentimental 'gesture that expresses his personality'. Do we not find all of Sabatier in his deployment of the Romantic concept of 'personality' for St Francis, and in his Modernist view of Francis as an inspired and undisciplined Catholic?

Had Sabatier turned Francis into a proto-Protestant, someone whose ideal of Christian life had no need for the institutions of the formal Church, and who could commune with Christ without intermediaries? He was accused of this, and he denied it: indeed, in order to deny it thoroughly he wrote a piece in 1912 on 'The Originality of Francis of Assisi', in which he claimed that Francis's great originality lay, paradoxically, precisely in his Catholicism. Francis, it appears from Sabatier here, took 'le chemin de l'orthodoxie la plus stricte', and he went further down that route than anyone before, thus making his orthodoxy infinitely more rich, more fecund, than other people's (whatever that might mean). He had complete submission (to the Church) and complete liberty: this is what is remarkable about him, 'la soumission dans la liberté, la liberté dans la soumission' (Sabatier, 1912, pp. 3, 15). But despite Sabatier's apparent concession here to the Catholic Church it

---

[15]Emphasis added. This is from the critical study of the sources, which prefaced the French edition; see Sabatier (1894b), p. 418, and Sabatier (1894a), pp. cx–cxi.

is clear, as in his equating orthodoxy in Francis's case with richness and fecundity of religious thinking, that Sabatier's view of a proper Catholic orthodoxy was a Modernist one. His Francis was certainly a Modernist: he did not need to be also a Protestant. It is little surprise that Sabatier's *Life* was immediately put on the Index of Forbidden Books.

Given the 'hermit' or Spiritualist view of St Francis that Sabatier put forward in his *Life*, and which proved so popular, it is no surprise that this *Life* appeared to support the image of St Francis put forward in the fourteenth century Franciscan Spiritualist compilation, the *Fioretti*, 'The Little Flowers of St Francis'. It is this book, perhaps more than anything else, which has spread this Nature-loving view of St Francis so widely in the twentieth century. It has received many editions in many languages, and its many animal stories have been seen as particularly appropriate for children. A scholarly edition of the text was produced by Sabatier himself in 1902.

2.3. Francis the True Catholic: social reformer and obedient son of the Holy See

This third image of St Francis brings us into the beleaguered camp of Vatican Catholicism. We can expect three things, I think, of any image of St Francis that might emerge from here, all of them characteristics conceived in deliberate opposition to Modernism and other ideological enemies of the papacy. This Francis will show total obedience to the pope and the Church (unlike the Modernist version). He will be firm in support of Catholic dogma (again, unlike his Modernist counterpart). And he may offer some means of resolving the Social Question in a Catholic direction (unlike any of the new nineteenth-century ideologies). Any role that this Vatican St Francis has as a 'lover of Nature' will be secondary.

Pope Leo XIII had an independent interest in St Francis and in promoting what he took to be Franciscan ideals in the modern world. Although the future Leo was educated by the Jesuits in the restored Collegio Romano, his mother was a member of the Franciscan Third Order, and was associated with a restored Observant house of the Order. One of his biographers has suggested that the future Leo had wanted to follow the example of the Franciscans and devote himself to a life of self-denial and sacrifice (O'Reilly, 1887, p. 75). In 1875, as a Cardinal, he was appointed by Pius IX as Protector of the Franciscan Third Order, and he took this as an opportunity to publicly defend what he took to have been the ideals of Francis, since 'no man has been held up to so much contempt by French Voltaireanism, by the Revolutionists, Radicals, and Socialists, who so clamorously profess their love for democratic simplicity, equality and liberty'. According to the future Leo, these, of course, were the very things Francis had been trying to promote in his time, but in an authentically Catholic way! The future Leo's dream,

> his aim, the object of his entire life was to bring back the Christendom, the society of the thirteenth century to that democracy, that society of all mankind become children of God and living on earth, according to Christ's doctrine and example, in the

practice of all brotherly virtues ... To revive the spirit and rule of St Francis of
Assisi, to propagate this *Secular Third Order* among all ranks of the Christian people,
would be the providential means to renew the face of the earth. (O'Reilly, 1887, pp.
272, 175)

Thus, a return to St Francis and the Christendom of the thirteenth century, if it
were possible, would indeed solve the Social Question of the nineteenth century.

With respect to Francis's obedience to the Church and his orthodoxy, we can
turn to another Cardinal laying out the position quite clearly. In an Introduction,
written in 1894, to a life of St Francis dedicated to Leo XIII as pope, Cardinal
Vaughan of Westminster stated that the great characteristics of the Franciscan
movement were that 'It was social. It was educational. It was religious, founded
in the charity and humility of Christ. It was Catholic. And, as all things truly
Catholic must be, it was conceived and carried out in the spirit of devoted obedience to the Holy See' (Vaughan in Le Monnier, 1894, pp. viii–x). This characterisation is clearly in opposition to the Modernist interpretation. Vaughan may have
already seen Sabatier's *Life* when he wrote this, since he characterises its position
so perfectly:

> Nothing could be more mistaken than the conjecture of certain writers, who, appar-
> ently anxious to read the Franciscan movement in the light of their own predilections,
> have sought to give the work of the Friars a colour of 'undenominationalism', and
> to represent the drift of their preaching as not only rather moral than dogmatic, but
> as one in which the value of dogma and orthodoxy was discounted to make room
> for a fuller presentment of the precepts of morality. On the contrary, we find that it
> was precisely at the hands of the Friars that dogmatic Theology received its most
> brilliant exposition and its most systematic development. We have only to look at
> the literature of the time to be convinced that the Friar was quite as much a dogmatic
> professor as a popular teacher ... And as to orthodoxy, the position of the Friars was
> luminously conspicuous even in an age when men shrunk from the taint of heresy as
> from a thing of shame. S. Francis, as we might expect, had placed loyal adherence
> to the Catholic faith and obedience to the Pope in the forefront of his observance,
> and had made both to be the very Alpha and Omega of his rule. (Le Monnier, 1894,
> pp. viii–ix)

Whatever one might think of the motives for putting this Vatican image of St
Francis forward, one has to concede that it bears a greater resemblance to the
historic St Francis than does the Modernist image of him.

The success of the Modernist vision of St Francis, and the threat that all Modernist propaganda held toward the papacy, meant that a new Francis had to be constructed, a Francis who was a reformer, but who was not hostile to the papacy.
Catholic scholars produced several new lives of Francis in the next decade. We
can take one of these as our example—that produced in 1907 by a Danish Catholic,
Johannes Jörgensen, who has been given the dubious compliment of being called
'the Catholic Sabatier' (Jörgensen, 1907, pp. 115–7). This characterization of
Francis is written in as accessible and imaginative a style as Sabatier's book, and
in that sense is a rival to it. Jörgensen's work is under the following broad headings:

Francis the Church Builder; Francis the Evangelist; God's Singer (that is, Francis as preacher); Francis the Hermit. These all tend to show Francis as a good Catholic and son of the Church, and his message as one of 'peace as the greatest good for man, peace with God by keeping his commandments, peace with man by righteous conduct, peace with oneself by the testimony of a good conscience' (Jörgensen, 1912, p. 61). As for the image of Francis as nature-lover, a single chapter is devoted to the topic, where Jörgensen says that 'Francis' standpoint as to the conception of nature is entirely and only the first article of faith—he believed in a *Father* who was also a *creator*' (Jörgensen, 1912, p. 309). He has only a little to add in Romantic vein as to Francis's 'love of nature'. So we can see that the image of Francis as a 'lover of Nature' was not of much immediate consequence to the papacy and its supporters. It is not something they particularly cared to embellish or give new meanings to. They had more important things for their St Francis to do. Decades later, by contrast, the Vatican was to have a very positive use for the nature-loving image of St Francis, since it would enable them to put him on their own agenda and link him with ecology and environmentalism.

3. Ecology Before and After the 1960s

It was ecology of which St Francis was to be given the spiritual guardianship by the pope in 1979. Ecology is, in the first place, a science which originated in the second half of the nineteenth century, built on Darwinian evolution (Kingsland, 1985; Acot, 1994; Worster, 1992; Crosby, 1994). The individual who coined the term in the 1860s, and who most set the new science on foot in the last decades of the nineteenth century, was Ernst Haeckel. Sharon Kingsland, historian of this science, writes:

> The science of ecology, as it was understood both by botanists and zoologists at the turn of the century, signified a dynamic, experimental approach to the study of adaptation, community succession, and population interactions. For botanists, in particular, the ecologist was a kind of 'outdoor physiologist', someone who studied adaptation and community evolution in the field using the same rigorous methods that the physiologist employed in the laboratory. (Kingsland, 1991, p. 2)

From the 1930s onwards this new science has concerned itself with the relation of species to their environment, and among its organizing concepts are those of niche, biosphere and ecosystem. Since the same period the science has been highly mathematical, especially in its work on populations and population genetics. One of the concepts that ecologists developed was that of 'environment', since they were very concerned with the relation of species and populations to their particular local environments. However, there is no evidence in the contemporary literature that ecologists were concerned with '*the* environment', in the global sense in which we so readily use that term today. This science of ecology continued to be practised undisturbed into the middle decades of the twentieth century. If one were to characterize the goal of this science of ecology in a phrase, it was about *how Nature manages itself*, with Nature conceived entirely within the scientific paradigm. It was not this ecology to whose protection and heavenly oversight the Vatican would want to assign St Francis.

In the 1960s this science suddenly found its name taken over by campaigners concerned with the plight of the world as a whole and with mankind's failure to manage it wisely. The new ecological movement was part of the popular radical politics of the West in the 1960s, and its activists have continued to volunteer to act as the conscience of the scientists, technologists and politicians, with notable success. The new campaigning ecologists depended, and have continued to depend, on the ecological scientists for their facts, and have made technical concepts of the science of ecology, such as niche, biosphere and ecosystem, into household words. Again, if we want a phrase to characterize this new ecology, we could say that it is about *how man manages (or mismanages) Nature*. Modern campaigning ecology, by contrast with the older science, is at base a neo-Romantic movement, with similar attributions of organism status to the planet and to the universe, the whole being seen as functioning as a (living) unity.

The new campaigners for ecology have given the ancient term 'organic' yet more new meanings, and they have also managed to make the expression 'the exploitation of Nature' into a term of abuse. But above all they have made us all familiar with their new meaning of the term 'environment'. This is so much so that theirs is now the dominant meaning of the term for us, the one we take for granted. And as with all new concepts and terms which we take for granted (as second nature, that is), we forget that we have recently coined the term or invented the concept, and that it was created to meet new, modern, needs—the needs of *our* present-day society. Instead, we spontaneously apply the concept to peoples in other cultures and times, firm in the belief that they, like us, must have had attitudes towards the environment and towards ecology. Our unself-consciousness in this respect enables us to put 'the environment' or 'ecology' into the same sentence as 'St Francis', and to ask what was St Francis's view of the environment, without laughing at how ridiculous our behavior is in doing so.

4. The *Aggiornamento* of the Catholic Church

The decade of the 1960s was also the time when the Catholic Church undertook its greatest series of reassessments of doctrine and discipline. Under pope John XXIII and pope Paul VI, the Church began to receive its *aggiornamento*, its 'bringing-up-to-date'. All that had been resisted so strenuously because it was opposed to the Catholic Church and the papacy when the Modernists called for it at the turn of the century was now, in the 1960s, on the agenda of the popes themselves. The great instrument of the *aggiornamento* was 'Vatican II', the second Vatican Council, announced in 1960 and in session from 1962 to 1965. The achievements of the Council can be described as initiating 'The Gospel of Peace and Justice', according to the title of a book on the Council and its outcomes (Gremillion, 1976).

Environmental concerns were relatively small at the Council, for the creation of the modern environmental conscience was just beginning. But the atmosphere of renewal that came from the Council has enabled environmental concerns to flourish in Vatican circles subsequently; and they were first broached, it seems, in the 1971 papal letter of pope Paul VI, *Octogessima Adveniens*. This letter marked eighty years since Leo XIII's famous encyclical *Rerum Novarum*, and like its predecessor it was concerned with the Social Question, but in a new light. Under the heading 'New Social Problems', paragraph 21 (the last in this section) deals with 'The Environment'. The entire text is as follows:

> While the horizon of man is thus being modified according to the images that are chosen for him [by the media], another transformation is making itself felt, one which is the dramatic and unexpected consequence of human activity. Man is suddenly becoming aware that by an ill-considered exploitation of nature he risks destroying it and becoming in his turn the victim of this degradation. Not only is the material environment becoming a permanent menace—pollution and refuse, new illnesses and absolute destructive capacity—but the human framework is no longer under man's control, thus creating an environment for tomorrow which may well be intolerable. This is a wide-ranging social problem which concerns the whole human family. The Christian must turn to these new perceptions in order to take on responsibility, together with the rest of men, for a destiny which from now on is shared by all. (As printed in Gremillion, 1976, pp. 495–6)

Amongst the organizations the Vatican created to keep itself up to date was one called the 'Planning Environmental and Ecological Institute for Quality of Life', which was chaired by Cardinal Oddi. One of the responsibilities it took on, following Lynn White's call, was to seek a patron saint for these new issues; and hence Cardinal Oddi came to put the case for St Francis. Pope John Paul II duly declared Francis patron saint of ecologists on November 29, 1979. The official *Act* of the event says that

> Among the holy and outstanding men who cared for Nature (*qui rerum naturam coluerunt*) as a miraculous gift given by God to the human race, St Francis of Assisi is justly counted. For he felt exceptionally deeply about all the works of the Creator and, inspired by a certain divine spirit, he composed that most beautiful 'Canticle of the Creatures' through whom (*per quas [creaturas]*), especially through brother sun and sister moon and the stars of heaven, he gave the praise, glory, honor and all blessing due to the most high, omnipotent and good Lord.

The *Act* continues to say that Cardinal Oddi has asked the pope that St Francis be declared the Patron before God of the cultivators of ecology (*Patronus apud Deum oecologiae cultorum ediceretur*),[16] and the pope has agreed. While St Francis was,

---

[16]'Inter sanctos praeclarosque viros qui rerum naturam veluti mirificum donum a Deo humano generi datum coluerunt, Sanctus Franciscus Assisiensis merito recensetur. Namque universa Conditoris opera singulariter ille persensit ac, divino quodam spiritu inflatus, pulcherrimum illud cecinit "Creaturarum Canticum" per quas, fratrem solem potissimum illud ac sororem lunam caeliquem stellas, altissimo, omnipotenti bonoque Domino debitam tribuit laudem, gloriam, honorem et omnem benedictionem. Peropportuno igitur consilio Venerabilis Frater Noster Silvius S.R.E. Cardinalis Oddi, Sacrae Congregationis pro Clericis Praefectus, nomine praesertim sodalium Consociationis Internationalis vulgo "Plan-

strictly speaking, thus made into the patron saint of ecologists and not an ecologist as such,[17] nevertheless, as we have seen, his credentials for being the patron saint were based on nineteenth- and twentieth-century views on his attitudes to Nature and his concern for it, which have become views typical of the modern campaigning ecologist. The environment and ecology have continued to be a concern of the papacy and of Catholics more widely in the years since.[18]

With St Francis now having been given the heavenly patronage of the cultivators of ecology, it is time to conclude. I have shown that while the representation of St Francis as someone with ecological and environmental concerns is pregnant with meaning for our *modern* society and its ills, as a *historical* ascription about Francis himself it is completely without foundation and a nonsense. I have also demonstrated that Francis could not have related to, nor thought about, Nature in any of the ways that other recent scholars have suggested, both because of his lack of the appropriate modern concepts, and also because he was extremely busy doing other things, such as being a Second Christ and the Angel of the Sixth Seal. He did not have, nor could have had, an ecological awareness, let alone an ecological conscience. He did not see other creatures as man's equal, nor care about them because we are all on this fragile planet together.

I said at the beginning of the first of these two articles that while the attribution of roles such as proto-ecologist to a medieval saint may be intended in a merely sentimental and symbolic way by historians of science, nevertheless such attributions perform work for certain interested parties, For instance, the Catholic Church has adopted St Francis as patron saint of ecologists, and is able to support this claim if necessary by referring to works by historians and historians of science which first made the claim that St Francis was a proto-ecologist. With respect to the Catholic Church and its view of the historic relation of science and religion, we can ask: what work does it do, what does it mean, for the Catholic Church to adopt a patron saint of ecologists? What does this act say about the Catholic Church, and what is it meant to say? It seems to me that what it does is claim that the Catholic Church is a socially responsible church and cares for the planet; that the best stewardship of the planet is a Christian and Catholic stewardship. Most important of all, it announces that ecological concern is a long historical

---

ning environmental and ecologycal [sic] Institute for quality of life" nuncupatae, petivit ab hac Apostolica Sede ut Sanctus Franciscus Assisiensis Patronus apud Deum oecologiae cultorum ediceretur. Nos quidem de sententia Sacrae Congregationis pro Sacramentis et Cultu Divino, harum Litterarum vi perpetuumque in modum, Sanctum Franciscum Assisiensem oecologiae cultorum Patronum caelestem renuntiamus, omnibus adiectis honoribus congruisque liturgicis privilegiis . . . .' Latin text as given in *Acta* (1983), pp. 1509–10, my translation. Thomas (1983, p. 23n.) claims that this happened on Easter Day (April 6) 1980, presumably in the pope's speech in front of St Peter's, but I can find no trace of this.

[17]'Oecologiae cultores', in the Latin text in the previous footnote, is probably intended to mean 'practitioners of ecology', i.e., ecologists. A work I have been unable to track down, but which I take to have been issued from a Catholic publishing house in Chicago, goes further in its title in this respect: Hansen (1971).

[18]See for instance the pope's address on the World Day of Peace in 1989. For an account of the increasing importance of the environment in American Catholic circles, see Allitt (1998).

tradition within the Catholic Church, best exemplified by the life and example of St Francis of Assisi. In other words, it makes a statement about the concern for Nature shown by the thirteenth-century Catholic Church and subsequently, and about the seemingly eternal values that the Catholic Church exemplifies and practises. It is thus an act which helps to write—or rather to rewrite—history, and in particular the history of mankind's attitudes to Nature. It links the most popular of Catholic saints with the most popular cause of our time, the defence of the environment, and gives St Francis the guardianship of that cause. It is an act of approval, adding papal blessing and sanction to a movement which had started quite outside the Church. But it is also an act of appropriation, claiming the ideals and idealism of that *modern* movement as exemplifying true *historic* Christian values. And by implication it also claims that the Catholic Church has, and has always had, positive, not negative, attitudes towards science. That is to say, it is a further act in the dispute about science and religion—a *modern* dispute, not a medieval one.

This has been a case-study of how and why one medieval man of religion has been appropriated for nineteenth- and twentieth-century concerns, to do battle for a modern science in the role of predecessor, and for the modern Catholic Church as a saint with scientific leanings and a modern social conscience. Closer inspection of what they were really up to in their own time and in their own categories might, in time, release other medieval men and women of religion from the burden of having also been early men or women of science; but that depends on further research. The success of that research will depend on us historians of science taking care to ensure that we do not import nineteenth- and twentieth-century concepts and concerns into the medieval period—and thus create a science/religion dispute (or harmony or interaction) where there was no such thing. St Francis can only be a proto-ecologist by virtue of our imposing our own values, categories and concepts onto him. It is a curious fate for a man dedicated to getting the creatures to praise their creator because He had made them and they were utterly dependent on Him.

Acknowledgements—I am grateful to Professor Margaret Osler for encouraging me to write this up, and to the audiences in Cambridge, Barcelona and Leeds who have heard it in earlier forms and have offered constructive criticism. I am also grateful to Dr Harmke Kamminga for advice on the history of ecology and discussions on historiography; to Dr Silvia De Renzi for searching out some documents in Italy for me, and with the English rendering of the 'Canticle'; to Dr Martin Kusch for assistance with rendering Görres' impossibly Romantic German; and to Dr Yoko Mitsúi for valuable advice on the structuring and titling of the two papers. The comments of an anonymous referee for *Studies* led me to recast the argument to some extent in the hope of avoiding unnecessary misunderstanding.

References

Acot, P. (1994) *Histoire de l'Écologie*, Que Sais-Je? (Paris: Presses Universitaire de France).
Acta (1983) = *Acta Apostolicae Sedis* (1983), vol. 71, 2 (Rome).
Acton, Lord (1907) 'German Schools of History', in J. N. Figgis and R. V. Laurance (eds), *Historical Studies and Essays by John Emerich Edward Dalberg-Acton, First Baron*

Acton (London: Macmillan and Co.), pp. 344–392; first published in *The English Historical Review* (1886).

Allitt, P. (1998) 'American Catholics and the Environment, 1960–1995', *Catholic Historical Review* **84**, 263–280.

Brooke, J. H. (1991) *Science and Religion: Some Historical Perspectives* (Cambridge: Cambridge University Press).

Brooke, J. H. and Cantor, G. (1998) *Reconstructing Nature. The Engagement of Science and Religion*, The Glasgow Gifford Lectures for 1995–6 (Edinburgh: T & T Clark).

Bury, J. B. (1930) *History of the Papacy in the 19th Century (1864–1878)* (London: Macmillan).

Coppa, F. (1979) *Pope Pius IX. Crusader in a Secular Age* (Boston: Twayne Publishers).

Crosby, A. W. (1994) *Germs, Seeds and Animals. Studies in Ecological History* (New York: M. E. Sharpe).

Cunningham, A. (1988) 'Getting the Game Right: Some Plain Words on the Identity and Invention of Science', *Studies in History and Philosophy of Science* **19**, 365–389.

Cunningham, A. and Jardine, N. (eds) (1990) *Romanticism and the Sciences* (Cambridge: Cambridge University Press).

Cunningham, A. (2000) 'Science and Religion in the Thirteenth Century Revisited: The Making of St Francis the Proto-Ecologist: Part 1: Creature not Nature', *Studies in History and Philosophy of Science* **31**, 613–643.

Cuthbert, Father [L. A. Hess] (1912) *Life of St Francis of Assisi* (London: Longmans, Green and Co.).

Cuthbert, Father [L. A. Hess] (1915) *The Romanticism of St Francis and Other Studies in the Genius of the Franciscans* (London: Longmans, Green and Co.).

Dru, A. (1963) *The Church in the Nineteenth Century: Germany 1800–1918*, Faith and Fact Books: 103 (London: Burns and Oates).

Engelbert, O. (1950) *Saint Francis: A Biography*, trans. E. Hutton. (London: Burns Oates). First published in French in 1947.

Fleming, J. V. (1977) *An Introduction to the Franciscan Literature of the Middle Ages* (Chicago: Franciscan Herald Press).

Fleming, J. V. (1982) *From Bonaventure to Bellini. An Essay in Franciscan Exegesis*, Princeton Essays on the Arts, no. 14 (Princeton, NJ: Princeton University Press).

Goad, H. E. (1929) *The Fame of St Francis of Assisi. The First Walter Seton Memorial Lecture* (London: University of London Press).

Görres, J. J. von (1826) *Der Heilige Franziskus von Assisi, ein Troubadour* (Strasburg: Ludwig Le Roux).

Gremillion, J. (ed.) (1976) *The Gospel of Peace and Justice. Catholic Social Teaching since Pope John* (Maryknoll, NY: Orbis Books). Reprinted 1980.

Hansen, W. (1971) *St Francis of Assisi: Patron of the Environment* (Chicago).

Hase, K. (1856) *Franz von Assisi. Ein Heiligenbild* (Leipzig: Breitkopf und Härtel).

Hasler, A. (1981) *How the Pope Became Infallible. Pius IX and the Politics of Persuasion*, trans. P. Heinegg (New York: Doubleday).

Holmes, J. D. (1979) *The Triumph of the Holy See. A Short History of the Papacy in the Nineteenth Century* (London: Burns and Oates).

Jörgensen, J. (1907) *Den Heilige Frans af Assisi* (Copenhagen).

Jörgensen, J. (1912) *Saint Francis of Assisi. A Biography*, trans. T. O'Connor Sloane, Ph.D. (London: Longmans, Green and Co.).

Kingsland, S. E. (1985) *Modeling Nature. Episodes in the History of Population History* (Chicago: Chicago University Press).

Kingsland, S. E. (1991) 'Defining Ecology as a Science', in L. A. Real and J. H. Brown (eds), *Foundations of Ecology. Classic Papers with Commentaries* (Chicago: University of Chicago Press), pp. 1–13.

Le Monnier, A. L. (1894) *History of S. Francis of Assisi. With Preface by H. E. Cardinal Vaughan, Archbishop of Westminster*, trans. 'A Franciscan Tertiary' (London: Kegan Paul, Trench, Trübner & Co. Ltd).

Motzkin, G. (1989) 'The Catholic Response to Secularization and the Rise of the History of Science as a Discipline', *Science in Context* **3**, 203–226.

O'Meara, K. (1876) *Frederic Ozanam, Professor at the Sorbonne. His Life and Works* (Edinburgh: Edmonston and Douglas).

Oliphant, Mrs (1868) *Francis of Assisi*, The Sunday Library for Household Reading (London: Macmillan and Co).

O'Reilly, B. (1887) *The Life of Leo XIII, from an Authentic Memoir furnished by his Order, written with the Encouragement, Approbation and Blessing of His Holiness the Pope* (London: Sampson Low, Marston, Searle and Rivington).

Ozanam, A.-F. (1839) *Dante et la Philosophie Catholique au Treizième Siècle* (Paris: Debécourt).

Ozanam, A.-F. (1852) *Les Poëtes Franciscains in Italie au Treizième Siècle* (Paris: Jacques Lecoffre et Cie).

Ozanam, A.-F. (1914) *The Franciscan Poets in Italy of the Thirteenth Century*, trans. A. E. Nellen and N. C. Craig (London: David Nutt).

Pius IX (1939) 'Syllabus des Erreurs', in *Dictionnaire de Théologie Catholique*, vol. 4 (Paris: Libraire Letouzey et Ané), pp. 2890–2910.

Pius XI (1926) *Encyclical Letter of Pope Pius XI to his Venerable Brethren the Patriarchs, Primates, Archbishops, Bishops and other Ordinaries in Peace and Communion with the Apostolic See, on St Francis of Assisi, on the Occasion of the Seventh Centenary of his Death* (London: Burns Oates and Washbourne Ltd, Publishers to the Holy See).

Reardon, B. M. G. (1985) *Religion in the Age of Romanticism* (Cambridge: Cambridge University Press).

Renan, E. (1866) 'François d'Assise', in *Oeuvres Complètes de Ernest Renan*, ed. Henriette Psichari, vol. 7 (Paris: Calmann-Lévy, 1955), pp. 919–935. Originally published as 'Saint François d'Assise, Étude Historique d'après le Dr Karl Hase', in the *Journal des Débats*.

Renan, E. (n.d.) *The Life of Jesus* (New York: Dolphin Books). First published in French in 1861 and in English in 1863/4.

Sabatier, P. (1894a) *Vie de S. François d'Assise* (Paris: Librairie Fischbacher).

Sabatier, P. (1894b) *Life of St Francis of Assisi*, trans. Louise Seymour Houghton (London: Hodder and Stoughton).

Sabatier, P. (1912) 'L'Originalité de Saint François d'Assise', in A. G. Little (ed.), *Franciscan Essays (I) by Paul Sabatier and Others*, British Society of Franciscan Studies, Extra Series, vol. 1 (Aberdeen: The University Press), pp. 1–17.

Sabatier, P. (1932) *Etudes Inédites sur S. François d'Assise. Editées par Arnold Goffin* (Paris: Librairie Fischbacher).

Thomas, K. (1983) *Man and the Natural World: Changing Attitudes in England 1500-1800* (London: Allen Lane).

Weinel, H. and Widgery, A. G. (1914) *Jesus in the Nineteenth Century and After* (Edinburgh: T. & T. Clark).

Williams, R. (1976) *Keywords. A Vocabulary of Culture and Society* (London: Flamingo-Fontana).

Worster, D. (1992) *Nature's Economy. A History of Ecological Ideas* reprint (Cambridge: Cambridge University Press).

VI

Aristotle's Animal Books: Ethology, Biology, Anatomy, or Philosophy?

This is a piece about labeling and identity. Aristotle's books on animals are the earliest extensive writings on animals that we have in the western tradition, and they were written by one of the most important philosophers who ever lived. My central question is this: How are we best to label these works, what is their identity? What is best for us? What is best for Aristotle? Are they instances of ethology, of biology, of anatomy, of philosophy, or of what? I will be maintaining that, in trying to understand the past in the realm of the history of ideas, the answers to such questions are critical, and they apply equally to our considerations of all philosophers whose works we might study who lived before the nineteenth century.

In modern science at any given time, one of the sciences of the organic dominates the field. Such an "umbrella-science," as I shall call it, provides the paradigm (in many of Kuhn's senses) for the questions, modes of proceeding, and type of answers pursued in the other sciences it is taken to embrace, include, or subsume. It specifies what is important in the field. As a generalization one could say that, at the end of the nineteenth century, natural history was the umbrella-science of the organic world; then until the 1960s or so it was zoology; from the 1960s it has been biology; perhaps soon it will be ethology. An umbrella-science can be identified in part by looking at the naming, and the renaming, of university science departments.

From *Philosophical Topics* 27, no. 1 (1999): 17–41. Copyright © 1999 by the Board of Trustees of the University of Arkansas. Used with the permission of the University of Arkansas Press, www.uapress.com.

Biology seems to have triumphed in this way due to the success of cell biology and molecular biology. Zoology, the umbrella-science it replaced, survives as a scientific discipline, but it is no longer the dominant discipline in the area, and it would be possible to show that its questions, procedures, answers, and priorities have all been reoriented by the rise of biology. The dominance of ethology (if it happens) will have similar effects on the other life sciences. Umbrella-disciplines dominate not only our view of the present but also our view of the past, as we shall see.

It is now something of a commonplace amongst historians of science and medicine that creators of disciplines of knowledge often also create a history or prehistory for such new disciplines. It is part of the legitimating of a new discipline. In co-opting history to their cause, the creators of any new discipline are faced with three basic options. Either they can show that the new discipline is (1) brand new, and has a short history, and that it was recently created by a particular inspired individual or group of workers. Or they can show (2) that it has a long and honorable history, and is therefore in fact an old subject now coming into fresh glory. Or, finally, (3) that it has a long history, but one that was dark and misguided for all or some of its existence, and it is only now having light thrown upon it, only now being put on the right path at last. Sometimes more than one of these options is chosen at the same time. Which of these options is chosen depends usually on the local circumstances of the practitioner of the new discipline who takes up the role of historian, and the option or options are adopted unselfconsciously. For these discipline histories are, of course, at least as much attempts to define the discipline according to one view, and to guide the development of the discipline in the future, as they are dispassionate attempts to unravel the question of how the practitioners got where they find themselves today. They are often produced in the course of disputes to establish a new discipline in a particular way, or with a particular ideology, and in such contested situations they function as parts of campaigns to seize the authority of the discipline for one particular interest. The positions such histories attribute to past actors are likely to be positions held by living actors.

ARISTOTLE'S ANIMAL BOOKS AS ETHOLOGY?

In a forum on philosophical ethology, it is appropriate to start with ethology. As is well known, the project of ethology originated in the middle decades of the twentieth century, stemming in great part from the work of Konrad Lorenz and Nikolaas Tinbergen in studying the behavior of animals. It is an area in which philosophers have recently begun to be interested, especially (but not exclusively) with respect to cognition, and to the question of whether animals think. What sort of history should it have? A practitioner of ethology, W. H. Thorpe, wrote a history of it in 1979, under the title *The Origins and Rise of Ethology: The Science of the Natural Behaviour of Animals,* which takes the story back to the nineteenth century, but sees the true origin of the subject in the work of Lorenz and Tinbergen.[1] So at present ethology has only a short history, showing that it was created recently by an inspired individual or group of workers. Lorenz has made an attempt to present himself as the great originator of the area of study.[2] Indeed, the only mention that Thorpe makes of Aristotle is to say that "The Greeks, notably Aristotle . . . made desultory notes on the behaviour of animals" (1). However, should circumstances require, a long history of ethology could be written, and Aristotle's studies of the social behavior of bees and ants would make Aristotle an excellent candidate as a putative "father of ethology." When discussing philosophical ethology, the temptation is always there to enlist Aristotle as an early ethologist. It is a temptation which I believe needs to be resisted, and my attack on the historical construct "Aristotle the biologist" in a moment should be read as standing also for a pre-emptive attack on "Aristotle the ethologist," a reputation he has not (to my knowledge) yet been given. I hope I am in time to save him from it.

ARISTOTLE'S ANIMAL BOOKS AS BIOLOGY?

The study of Aristotle's thought is today divided between different scholarly disciplines. It is as if all of Aristotle's surviving books had been sent to a university library, and the library staff had decided, on the basis of the titles of the individual works, to send them out to the departments which seemed most appropriate for their study. Thus his *Politics* went to the department of political science; the *Nicomachean Ethics,* the *Prior* and the *Posterior Analytics* together with the *Metaphysics* to the philosophy department; the *Rhetorics* and the *Poetics* to the literature department; *On the Soul* either to the theology department or to the psychology department; the *Physics* and *On the Heavens* to the history of science department; and the *History of Animals,* the *Parts of Animals,* and the *Generation of Animals* (the "animal books," as I shall be calling them) to the historians of biology in the same department. While some modern scholars of Aristotle certainly do try to look at Aristotle and his philosophy in the round, even they bow to the experts with respect to the interpretation of some of the books. Of none of Aristotle's books is this more true than of the animal books: it is taken for granted by everyone that the appropriate expert to read, interpret, and translate these is the historian of biology. So much is this the case that the animal books and their contents are today customarily referred to as "the biological works" or, as a group, as being or containing what is called "Aristotle's biology."

In 1913 D'Arcy Thompson (1860–1948), later the author of *On Growth and Form* (1917), presented a talk at Oxford called "On Aristotle as a Biologist."[3] Fresh from translating Aristotle's *History of Animals* into English (1910), Thompson declared, "Aristotle seems to me to have been first and foremost a biologist, by inclination and by training" (11). Thompson argued that this biology was not carried out in Aristotle's old age, but when he was a young man, living in or near Mitylene on the Asiatic coast, before he taught Alexander and before he taught in the Lyceum. Thus, Thompson wrote, "[I]t follows for certain, if all this be true, that Aristotle's biological studies preceded his more strictly philosophical work; and it is of no small importance that we should be (as far as possible) assured of this, when we speculate upon the influence of his biology on his philosophy" (13–14). Thompson waxes lyrical on the similarity between Aristotle's views and those of today:

> When he [Aristotle] treats of Natural History, his language is our language, and his methods and his problems are well-nigh identical with our own. He had familiar knowledge of a thousand varied forms of life, of bird and beast, and plant and creeping thing. He was careful to note their least details of outward structure, and curious to probe by dissection into their parts within. He studied the metamorphoses of gnat and butterfly, and opened the bird's egg to find the mystery of incipient life in the embryo chick. He recognized great problems of biology that are still ours today, problems of heredity, of sex, of nutrition and growth, of adaptation, of the struggle for existence, of the orderly sequence of Nature's plan. . . . It would take more than all the time I have, to deal with any one of Aristotle's theories— of generation, for instance, or of respiration and vital heat, or those still weightier themes of variation and heredity, the central problems of biology, or again the teleological questions of adaptation and design. (14–15, 17)

D'Arcy Thompson then lists some of the more striking observations about animal structure, life, and behavior which Aristotle first noted, drawing special attention to observations which had been remade by modern observers only in very recent times. He concludes that Aristotle's biology did have an influence on his philosophy. For instance,

> in his exhaustive accumulation and treatment of political facts, his method is that of the observer, of the scientific student, and is in the main inductive. Just as, in order to understand fishes, he gathered all kinds together, recording their forms, their structure and their habits, so he did with the Constitutions of cities

and states [creating] a Natural History of Constitutions and Governments. (25)

Thompson's characterization of Aristotle as a *biologist* is unusual for this date, and has to do with the fact that this speech was the Herbert Spencer Lecture, and Thompson was celebrating Spencer as much as Aristotle. Spencer had wanted biology to be one of the departments of his great *Synthetic Philosophy,* and the two-volume work which filled this role, *The Principles of Biology,* had appeared in 1864 and 1867. Moreover, Thompson was a pupil at Cambridge of Michael Foster, who was particularly keen to claim that the umbrella-science of the organic should be biology. It would otherwise have been more expected at this date for someone to promote the Aristotle of the animal books as having been a natural historian or a zoologist. However, Thompson's choice of "biologist" to describe Aristotle was almost prescient, and thus has not gone out of date; rather, it has come into date. Spencer himself was not concerned at all with history in his *Principles of Biology.* Thompson is one of the first people volunteering to write a history of biology which traced it back to Aristotle and forward to Spencer. Thus Thompson's view of biology is of a subject which has a long if intermittent history, with Aristotle as its founder. And, one might ask rhetorically, who better to recognize Aristotle as the earliest biologist than Thompson, who was himself a professor of biology (1884–1917, a title later changed to natural history, at the University of Dundee), and then a professor of natural history (1917–48, at the University of St. Andrews), and who had learnt his Greek at the knee of his father, who had been a professor of Greek? But since this is an instance of disciplinary history by a practitioner of that discipline, we might ask also: Was Thompson attributing to a past actor, Aristotle, a position and practice held and pursued by a living actor— himself?

Thompson's characterization of Aristotle as a biologist is gloriously explicit. This Aristotle saw "problems in biology well-nigh identical to our own"; "he recognised great problems of biology that are still ours today, problems of heredity, of sex, of nutrition and growth, of adaptation, of the struggle for existence, of the orderly sequence of Nature's plan. . . ." Aristotle today still gets represented as a biologist or protobiologist in popular works, in a manner following the kind of analysis offered by D'Arcy Thompson. In the scholarly community, calling Aristotle a biologist has become virtually second nature for scholars studying the animal books. Yet we do not employ a precise characterization of biology today in our Aristotle studies which could serve as a point of comparison or assessment with respect to what Aristotle was doing. We would shrink from saying anything as explicit as D'Arcy Thompson for fear of being guilty of whiggism. Moreover, scholars interested in the animal books are usually classical scholars before they are biologists, and are not particularly concerned with the celebration of biology itself, as Thompson was. But in using the term "biology" so freely for Aristotle's project in the animal books, we do not even ask a question such as "In what ways did Aristotle's enterprise differ from our practice of biology, and in what ways might it be said to resemble biology?" which might yield some interesting information. We do not even open this question when we call Aristotle a biologist; in fact, we open no questions at all. Rather, we use the term in a knee-jerk sort of way: If these books are about animals, then they are obviously instances of biology, and Aristotle was therefore a biologist when engaged in the work reported in them.[4] Our tacit working definition of biology is, in fact, so loose as to be of no practical use: it gives us no key to learning more about these books as instances of biology in practice.

Why do we do this? Why do we call Aristotle a biologist, without reflection? The answer seems to lie simply with the dominance of biology today as an umbrella-science. We are brought up to think it the dominant (and best) way of investigating the organic world, and this message is reinforced every day by the press and other media. Given the mind-set that this dominance creates in us and for us, it seems *self-evident* to us that Aristotle's animal books are instances of biology or its subsciences. There simply does not appear to us to be a question there to be raised!

And yet a little reflection will remind us that biology, as a term, as a discipline, as a domain of knowledge, is of very recent construction. Had Aristotle needed to characterize as "biology" the enterprise he was engaged in that is recorded in the animal books, then it was open for him to coin the term, especially since Greek was his native language and the term "biology" has respectable Greek roots. But he did not. Indeed, the term was not coined by a Greek either ancient or modern. Rather, it was coined by a German (Karl Friedrich Burdach, as *Biologie*) in 1800, and then again, virtually simultaneously, by a Frenchman (Jean-Baptiste Lamarck, as *biologie*) and by another German (Gottfried Treviranus, as *Biologie*), both of whom first used the term in print in 1802. "Biology" was a term coined when it was first needed, as soon as it was needed, and not a moment before—and certainly not two millennia before. As we have already seen, a few investigators in the nineteenth century (such as Huxley, Spencer, Foster, and Thompson in Britain, and others on the Continent) tried to promote this into the umbrella-science of the organic sciences, though with only limited success at the time; the modern discipline of biology is of twentieth-century origin.

It is prima facie unlikely that a relatively recently coined modern term in a modern language is going to be the one we need to characterize an ancient and lost enterprise which had been conducted in a dead language.

This very recent coining of the term indicates that we must be very cautious in using it to describe whatever it was that Aristotle was up to that is recorded in his animal books. For we run the danger of anachronism, of ascribing to Aristotle an enterprise—and thereby of ascribing to him intentions—that he could not possibly have had in his time. We run the danger of ascribing to Aristotle the practice of biology not only *avant la lettre,* but before the concept or the practice was available to be practiced. This is the first and greatest problem arising from the handing over of Aristotle's animal books to the historians of biology for interpretation, however appropriate such a handing-over might seem at first sight, since they looked as if they were the scholars with the requisite expertise. The problem is that the historians of biology will turn Aristotle's practice as evidenced in the animal books into biology (or something closely resembling it), whatever it actually had been.[5] Deploying a modern term ("biology") for Aristotle's practice as evidenced in the animal books, simply obscures from our own sight what that (ancient) practice might have been.

But before going farther down this road we need to ask: Are there any *advantages* to using the terms "biology" and "biological" with respect to Aristotle's work as represented in the animal books? How does such usage help us? How does it help Aristotle? The only advantages I can see are (i) we are using a familiar word, whose basic meaning we all roughly understand, so it does not need to be explained, and (ii) since that term is the name of a modern scientific discipline, Aristotle thereby gets to sound like a modern, someone like us.

The *disadvantages* seem to me to greatly outweigh these apparent advantages, and to stem directly from them. They are the disadvantages of misrepresentation and of anachronism, and while both might help us by making Aristotle sound like ourselves, neither of them helps Aristotle sound like himself. As I have already claimed, to label these aspects of Aristotle's work as biology not only begs the question of the nature of his enterprise in the animal books, but it also (whether this is intended or not) draws Aristotle's work into the twentieth century and allies it to the modern practice of biology, a practice and a discipline not available to him, and in these ways thus *misrepresents* what he was up to. I know that some scholars believe that we can still call Aristotle's enterprise in the animal books "biology," and evade this problem of misrepresentation, for (they claim) we are doing no more than saying that these books deal with "life" (in Greek, *bios*) and its study (in Greek, *logos*). But such a claim that our personal use of the term "biology" with respect to Aristotle's animal books is quite distinct from the use of the term for a modern scientific discipline, is mere naive solipsism. Words such as "biology" and "biological" cannot just be given any private meaning we, as individuals, choose. However much we may think we are using the term "biology" in an innocent way with respect to Aristotle, we cannot avoid being heard to be using it with all the weight of the full-blown scientific discipline that this term was coined to characterize. But the issue goes deeper than this, for I think the real problem here is that historians of the animal books *want* Aristotle to have been doing biology or some early version of it: not only does this make him a "father" of the modern discipline of biology, but also the ascription to Aristotle of the practice of the extremely important modern discipline of biology, in its turn, is taken to make *Aristotle* more important historically.

With respect to the anachronism inherent in using the term "biology" for the enterprise evidenced by Aristotle's animal books, it might seem that we can avoid this by claiming that we are not referring to the enterprise, but only to the individual pieces of information that we might find in those books. Surely we can claim that a simple statement or piece of information is "biological," without implying or being taken to imply that Aristotle acquired it through the practice of the modern discipline of biology? Surely we are just saying it is a statement about a phenomenon of life (*bios-* and *logos,* again)? Surely all we are offering is simply a description of the status or nature of the information in the modern world? My answer to this would be twofold. First, that we have already committed the anachronism by offering a description of the status or nature of the information *in the modern world*. Second, I would say that it is certainly the case that in common speech we frequently take terms which have precise and strict meanings, such as "biology," and use them with loose meanings. In this loose sense, yes, discrete items of information about life phenomena from Aristotle's writings could today well be called "biological." But if it were to be contested whether or not such a discrete piece of information about life phenomena *counts as* biological, we would instantly abandon the loose meanings typical of common speech and instead turn to the expert to adjudicate. The expert we would turn to would be a modern biologist; for it is he or she who has the ultimate say in what "counts as" biological or not. And that means that the adjudication would be being made according to the criteria of the modern discipline of biology. In these circumstances, if something does not "count as" a biological phenomenon, or does not "count as" correct by the criteria of modern biology, then it simply is not biology or biological. So underneath *all* our uses of such terms as biology and biological, lies the modern discipline of biology, and the authority of the modern practitioner of biology.

The other side of this issue is that if we go into our studies of Aristotle's animal books without inspecting the disciplinary, activity, and intentional terms we are ascribing to him, then we will believe that everything we find confirms our assumption. It would be nice to think that extended immersion in the historical documents would somehow spontaneously lead us to appreciate and recognize the nature of the activities (especially the intellectual activities) that past people were engaged in, but experience does not show many examples of this happening. If we go into our studies thinking Aristotle was practicing biology, we're likely to come out thinking the same, however confused or inadequate a biologist we might then think him to have been.

The case is the same with those other apparently innocuous terms, comparative anatomy, embryology, (experimental) physiology, zoology. As disciplinary labels, they too are all creations of the late eighteenth and early nineteenth centuries, and were coined in order to characterize new domains of study which were originated in that period. The use of any of them with respect to Aristotle turns him, whether we like it or not, into a modern, indeed into a modern scientist. But he was not a modern but an ancient, not a scientist but a philosopher. He could not practice modern disciplines, only ancient ones.

But there is an even bigger problem which arises from our spontaneous knee-jerk classifying of the animal books as works on biology. For once we have separated these books from Aristotle's other books, we become obsessed with questions about the *relationship* of his "biology" to his "philosophy." As we saw, it was a question which struck D'Arcy Thompson after he had separated these books from the others. We agonize over questions such as "Did Aristotle's biology influence his philosophy, and if so, how?" or its reverse. And when we have finished *contrasting* or *opposing* the supposed biology to the philosophy (and vice versa), we sometimes move on to trying to *reconcile* the two, usually coming to the unremarkable conclusion that they are related. Questions like these have dominated the scholarly literature on Aristotle's animal books in recent years.[6] But all these questions arise from *our* action in separating the animal books from the rest. Yet, these questions assume that it was *Aristotle* who made this separation. We seek to reconcile in Aristotle's mind and practice things which may never have been separated there. (The same issues would arise with respect to ethology, if that was how we chose to characterize Aristotle's animal books.)[7]

What we need to do first is to ask an open, not a closed, question about what Aristotle was doing, in his own terms. To *start* our investigations by calling this activity "biology," etc., is to close the question at the very moment it needs to be opened.

I would like to propose a moratorium on the b- word, biology, with respect to Aristotle's animal books. But it might seem as though before doing so we need to call in an expert. For what I have not done here yet is actually compare the *contents* of the animal books to the concerns and questions of modern biology. Perhaps they do cover the same area? Perhaps they really are historically related? Perhaps, strict terminology apart, Aristotle was a biologist after all and these are books of (very early) modern biology? As I asked above, what better expert could do this than a biologist with excellent ancient Greek? But that has already been done, by D'Arcy Thompson and others. The experts have been called in, and they have produced an anachronism. What we need to do is not look at the contents of the animal books, item by item, sentence by sentence, proposition by proposition. What we need to do is look at the *argument* presented in them, and at the *enterprise* that that argument serves or served. This has not been done, and biologists are probably not the right people to do it, however good their ancient Greek.

ARISTOTLE'S ANIMAL BOOKS AS ANATOMY?

Perhaps the correct category for Aristotle's animal books is as instances of anatomy or anatomizing? Anatomy is certainly a modern scientific discipline, a subdiscipline of biology. Perhaps it was an integral, independent field of investigation in Aristotle's day? The term "anatomy" is ancient Greek in origin (rather than modern-style Greek, like "biology") and is the title of one of Aristotle's own books (unfortunately now lost), which is usually rendered in English as *Dissections*.

Anatomy is a *practical and empirical* pursuit, aimed at acquiring *theoretical* knowledge. Without his knife in his hand, the anatomist is helpless: he cannot be an anatomist unless he cuts up and explores, as at least one part of his investigative procedure. But equally, without thinking and reflection, the anatomist gains no more knowledge through all his cutting-up than the butcher has. This theoretical, contemplative, dimension of anatomizing is what distinguishes the anatomist from the slaughterer, butcher, cook, or huntsman, who seek knowledge of the insides of animals for practical purposes, such as preparing meat for table. While anatomical knowledge can certainly be put to practical use in other contexts, such as medicine and surgery (two of the disciplines in respect of which we usually think about anatomizing), and this ultimate practical use may be the reason why any given individual engages in anatomizing, yet in itself anatomical knowledge is not practical but wholly theoretical. This characterization of anatomy applies equally to all the different anatomical projects that there have been over the centuries.

It is clear that Aristotle did perform anatomy, asking questions about the nature (and natures) of animals and their organs, questions which could only be answered by employing the manual practice of dissection and vivisection.

There is extensive evidence in the animal books that Aristotle had personal experience of the dissection of a wide range of animals. Aristotle repeatedly refers his readers to his (now lost) book, *Dissections* (or *Anatomy*). In the *Parts of Animals,* for instance, when speaking of the parts which pass the food, Aristotle writes: "The mouth, then, having done its duty by the food, passes it on to the stomach, and there must of necessity be another part to receive it in its turn from the stomach. This duty is undertaken by the blood-vessels, which begin at the bottom of the mesentery, and extend throughout the length of it right up to the stomach. These matters should be studied in the *Dissections* and my treatise on *Natural History* [i.e., the *History of Animals*]" (650a28–32). When discussing the blood-vessels, he writes: "For an exact description of the relative disposition of the blood-vessels, the treatises on *Anatomy* and the *Researches upon Animals* should be consulted" (668b28–31). Similarly, when discussing the lobster, Aristotle writes, "For an account of every one of the parts, of their position, and of the differences between them, including the differences between the male and the female, consult the Anatomical treatises and the *Inquiries upon Animals*" (684b2–6). And everywhere in the *Parts of Animals* there is material presented which could only have been established by dissection; for instance: "In all cases that we have examined the heart is boneless, except in horses and a certain kind of ox" (666b18–20). In other treatises linked to the animal books there are also references to dissection: in *On Respiration,* for example, and *On Sleep.* It is also clear that Aristotle had prepared diagrams for his *Dissections.* Speaking of the cuttlefish in *History of Animals,* for instance, Aristotle writes: "The cuttlefish has two sacs and numerous eggs in them, like white hailstones. For details of the arrangement of these parts, the diagram in the *Dissections* should be consulted" (525a6–8).

But did Aristotle have an anatomical project? It is certainly the case that at least two anatomists of a much later period believed that he had had such a project, and they set out, with great success, to model their projects on his. This recreation of Aristotle's anatomical project was first attempted in sixteenth-century Italy, by Hieronymus Fabricius ab Aquapendente (Girolamo Fabrici or Fabrizie), professor of anatomy at Padua university from 1565 to 1613,[8] and it was the model of anatomizing adopted by the most famous pupil of Fabricius and the most famous anatomist of the early modern centuries, William Harvey (1578–1657). It is possible to see that Fabricius not only followed Aristotle in his choice of anatomical topics, as Harvey was to do after him (e.g., the generation of animals, respiration, the local motion of animals), but that he also followed him right down to the detail—to the use of the categories as the basis for building a *historia* of the parts of the body of "the animal," and Harvey was to follow him in this too. I have made the case elsewhere that Harvey discovered the circulation of the blood—that

most important of early modern anatomical discoveries—as a direct consequence of studying the operations of the heart as the center of the
Aristotelian "vegetative soul" in the animal (not just in the human), as a
committed follower of Aristotle.[9]

However, the fact that two much later anatomists, one of them a professor of the subject for forty-eight years, the other who practiced it as a
passionate hobby for over fifty years, could model their practice on what
they believed Aristotle's anatomical practice had been, does not in itself
mean that Aristotle himself had actually had such a practice. And even if
they were right, it does not necessarily mean that their revived Aristotelian
practice is the historical forerunner of our modern discipline of anatomy.
Even if we were right in calling Aristotle an anatomist in his animal books,
we might have to surround the term with many caveats lest we confuse ourselves by conjuring up images of the modern scientific discipline of
anatomy, which lacks all reference to the "vegetative soul" that Fabricius,
Harvey, and perhaps Aristotle before them were studying.

Anatomy has two interdependent facets: manual (dissection) and mental (rational and contemplative). We have found that Aristotle engaged in the
manual practice of dissecting animals. But why? What were the rational and
contemplative grounds for his doing so? Usually this question would not
arise in our scholarly work on him. And if the question were raised, it would
seem like a question whose answer is self-evident: it would seem that obviously Aristotle anatomized animals because in conducting the work reported
in the animal books he was being a biologist (and/or a zoologist, or a comparative anatomist, or an experimental physiologist, or an embryologist). In
such a role or roles *of course* he would have anatomized animals. How else
could one sensibly be a biologist (or zoologist/comparative anatomist/exper-
imental physiologist/embryologist)?

However, we have disposed of this line of argument already. So the
question can indeed be asked: What reason did Aristotle have for engaging
in such systematic dissection of animals (for it is not the sort of thing one
engages in for no purpose)? What was Aristotle up to when he turned to animals in the first place as an object of enquiry? For he knew of no predecessors in this enquiry, and we know of no investigations into animals in the
Greek tradition before Aristotle either. What was he looking for? What questions was he asking? What kind of answers was he satisfied with? How did
he demarcate his area of enquiry: quite what did he take as his material for
investigation? What was his enquiry about? And precisely how did he go
about it? The obvious place to start answering these questions is Aristotle's
writings themselves.

The starting point of Aristotle's anatomizing is his book "On the soul,"
known to scholars under its Latin title, *De Anima*. So in order to follow

Aristotle here, we will first have to recall the *De Anima* from the theology or psychology department, or wherever it has ended up in our modern university system, and read it as the necessary preliminary to the animal books. Indeed, it is more than a preliminary: it is the book whose theme and thesis called into existence the animal books themselves.[10]

A work on the soul ought to seem to us to be a mighty odd place for anyone to begin the exposition of a project leading one to engage in anatomy. In the modern way of going about things, there is no relation between the "soul" and the enterprise of seeking anatomical knowledge. Yet, Plato, Aristotle's own teacher, had talked in the *Timaeus* about the human body as divided into certain regions which correspond to the nature and needs of the soul, and talked of the body as "the vehicle of the soul." Indeed, it can be claimed that the philosophical projects of all three of that great triad of philosophers, Socrates, Plato, and Aristotle were *soul-centered,* that their respective philosophical systems are in this sense "about" the soul. Historians have of course been right to point out the great differences between the outlooks of Socrates, on the one hand, and his pupil Plato, on the other; and between Plato, on the one hand, and his pupil Aristotle, on the other: there are indeed differences, and they are highly significant. Yet the things that Socrates, Plato, and Aristotle disagree on, in successive generations, are the same things. For instance, they agree on what the enterprise of philosophy is about, and one of the things it is about is the soul gaining wisdom. They disagree on precisely what kind of thing the soul is, and hence on what kind of wisdom it can properly acquire and how it acquires it.

Thus when Aristotle, first as a pupil of Plato in the Academy, and then as an independent teacher in the Lyceum, set out to engage in the enterprise of philosophizing, he had to start (logically, if not chronologically) with the issue of what kind of thing the soul is, how it acquires knowledge, and to what extent that knowledge is to be trusted. The *De Anima* is where Aristotle starts his own discussion of this topic. Or rather, it is where Aristotle starts his *argument* about the nature of the soul. For Aristotle of course knew the answer, in general outline if not in all its ramifications and details, before he started the teaching or writing which is recorded in this book. What Plato taught (and hence what Plato taught to Aristotle) was a view of the soul as a thing in three parts: an immortal soul located in the head; and a mortal soul, in two parts, located in the thorax and abdomen respectively. This soul applied only to man, and Plato was simply mapping the characteristics of this tripartite soul onto what he knew of the body of man. He had not been doing anatomy. Now Aristotle, for reasons we do not fully know, disagreed with Plato about the nature of the soul: he disagreed with Plato's view of what kind of thing the soul is, what kind of thing it can "know," how it comes to knowledge about things, the status of that knowledge, and the practical consequences for man of the wisdom that the practice of philosophizing produces. Whatever the ground of Aristotle's disagreement with Plato, we can assume that, like philosophers' differences in general, it lay outside the practice of philosophizing itself. At all events, until Aristotle had specified his view of what the soul is and how it works, his attempt to reorient the philosophical enterprise could not get started. For what prompted Aristotle to teach and write was, as much as anything, his desire to correct his teacher: if, as Plato, you have the wrong concept of the soul, you will philosophize incorrectly.

Aristotle's particular view of the soul, of what it is and how it functions, was to lead him to undertake the study of anatomy. Plato's view of the soul had not led him in this direction; nor had anyone else's led them thither either. And the precise "thing" in nature which Aristotle anatomizes is given its identity by the starting point of Aristotle's enquiry: his view of the soul and of what the soul can know. For, in Aristotle's view—and here he is at one with Plato—the soul can only know, can only "be at one with" *universals*. Hence when he turns to anatomizing, what Aristotle anatomizes is not man; nor is it animals; nor is it different kinds of animal. It is, instead, a universal: it is "The Animal." It is clear that we will only understand Aristotle's anatomizing if we see it as he did: as an integral part of his philosophizing.

ARISTOTLE'S ANIMAL BOOKS AS PHILOSOPHY?

One of the things that Aristotle disagreed with Plato about was whether *this* world was of interest and consequence to the true philosopher. Aristotle agreed with Plato that the perfect, the eternal, and the stable were the true objects of the philosopher's quest for knowledge. But Plato had claimed that such things were not perceivable in this imperfect and transitory world, only in the world of Forms; for Plato, "the world of perfect Forms contains all that is truly real."[11] Aristotle, by contrast, claimed that the perfect, eternal, and stable could indeed be seen within the imperfect and transitory things of this world.[12] As has been often remarked, this difference between them is nicely captured in the famous *School of Athens* fresco in the Vatican painted by Raphael in 1509–12, which portrays Plato pointing upwards to the perfect, unchanging heavens and the permanent "Forms," and Aristotle pointing downwards to the things of the changing sublunar world, each thereby indicating that *this* is what the true philosopher should be concerning himself with.

Aristotle talks about his desire to turn philosophizing away from obsessive concern with the eternal divine bodies in the heavens, and to direct it toward things in the constantly changing sublunar region, especially animals, in a celebrated passage in *Parts of Animals,* book 1 chapter 5. Here, he claims, we can find constancy in the midst of change. This passage is sometimes quoted as being Aristotle's rationale for practicing biology.[13] We need to listen to it in the context of his philosophy as a whole.

> Of the beings such as are composed by Nature—the ones we call ungenerated and imperishable for all eternity, and the others we say partake of coming-to-be and passing-away—it so happens that while the first ones [i.e. the heavenly bodies] are worthy and divine, there are very few views [*theoria*] of them available for us because, if we try to investigate these beings— even those which we long to know about—that which is clear to perception is very scanty. It is easier for us to acquire knowledge about the perishable plants and animals because we live among them, since one can acquire much knowledge about each kind that exists if one cares to take enough trouble. And each of them has its charm. For even if we grasp only a little bit about the heavenly bodies, yet by the excellence of the knowledge it is more pleasurable than all the things around us (just as seeing a fleeting glimpse of our loved ones is more pleasurable than seeing many other things large and clearly), nevertheless with respect to the others [i.e., the earthly beings] since we know more and better about them they have superiority in knowledge [over what we know of heavenly things]. And furthermore, by being nearer to us and more similar in nature, they make some compensation for the philosophy about heavenly things.
>
> As we have already treated the heavenly bodies, giving our opinion about them, it remains to speak about animal nature leaving out nothing as far as possible, neither nobler nor less noble. For even with animals which are not pleasant to look at, nevertheless the Nature at work in them holds extraordinary pleasures for those who are capable of recognising the causes [*tas aitias*] and are philosophers by nature. After all, it would be unreasonable and absurd if looking at pictures of them we rejoice that we see [both the representation of them and] at the same time an art-at-work (such as the art of painting or the art of sculpture), and yet we did not love more the contemplation [*theoria*] of the things themselves composed by Nature since we are capable of discerning the causes.
>
> Therefore one must not be childishly disgusted by the study of less worthy animals, because in all the things in Nature there is something marvelous. And just as it is said that Heraclitus said to some visitors who wanted to meet him and stopped when they entered and saw him warming himself by the stove—he ordered them to come in and not be afraid, "because even here there are gods," so one must go about the investigation of every animal without expressing distaste, since in all of them there is something of Nature and of The Beautiful. For the non-random, the for-something's-sake is in the works of Nature most of all; and the thing because-of-which it is composed, or

has come-into-being—its purposes [*telos*]—are part of The Beautiful. And if anyone thinks that the investigation [*theoria*] of the other animals is unworthy, then he must think the same way about himself too, for it is not possible without much disgust to see the things of which the human kind is composed, such as blood, flesh, bones, veins, and other such parts. Equally, one must recognise that he who discusses any one of those parts or equipment is not speaking about the material [*hyle*] itself and not for its own sake, but for the sake of the entire form [*morphe*]—just as one discusses a house but not the bricks, mortar, timber. In the same way one must recognise that discourse [*logos*] about Nature is about the composite-thing and the entire being as a whole, but not about those things which never occur separately from those beings. (644b22–645a37)

First of all, our business must be to describe the attributes found in each group; I mean those "essential" attributes which belong to all the animals, and after that to endeavour to describe the causes of them. (645b1–4)

Now, as each of the parts of the body, like every other instrument, is for the sake of some purpose, viz., some action [*praxis*], it is evident that the body as a whole must also exist for the sake of some complex action. Just as the saw is there for sawing and not sawing for the sake of the saw, because sawing is the using of the instrument, so in the same way the body exists in some way for the sake of the soul, and the parts for the sake of those functions for which each of them has been formed. (645 b 15–20)[14]

With this claim that "the body exists in some way for the sake of the soul," we can turn back to Aristotle's argument in the book on the soul itself, the *De Anima*, and in order to see how this book is the key to understanding Aristotle's program of anatomizing and its relation to his philosophy, I will paraphrase some of his argument early in that book.

The *De Anima* opens with Aristotle's disingenuous claim that in undertaking philosophy, an investigation into the nature and properties of the soul is of the greatest importance. It will be of especial use, he says, to the understanding of Nature since "the soul is in a sense the *principle of living things* [*arche ton zoon*]." Aristotle's aim here, he says, is to discover (1) the nature and essence of the soul, and (2) its attributes/properties. The first problem will be to distinguish between (a) such properties as are characteristic of the soul by itself, and (b) such properties as cannot be separated from their presence in living creatures. The thrust of the whole book is, of course, to refute the position of Plato and others that the soul is something which can be, or which can be considered to be, detached from the living body. By contrast Aristotle is arguing that it can be known only through its manifestations, that is, through the actions of its instruments or organs: viz., through the body. In short, Aristotle's position was that the body is the soul-in-action. It was in pursuit of making this case that Aristotle was led to the study of anatomy.

Aristotle claims that ideally one knows something when one can offer a definition [*logos*] of it. When it comes to things in the world which really exist—things unlike mathematics—then one cannot hope to start with a definition (as one can in mathematics) and then show how the properties or attributes follow from that definition, because one simply does not yet know what the attributes or properties of those things are. So, in inquiries which concern things which actually exist in the world, such as the soul, one has to start by exploring the attributes they have. Until one has done this, one stands no hope at all of arriving at an authentic definition. And it will be necessary to look at every possible case, because we intend our definition ultimately to deal with, to be true of, *every case* (410b20): we want our definition to be such that we can demonstrate that *all* the important attributes that actually occur do indeed follow logically (necessarily) from it. Aristotle is saying that only when one has looked as thoroughly as possible at all the instances one can find, will one have a sufficient clue or hint such that one will be able to create an adequate definition. But going from the attributes of the particular instances to the definition will still involve a mental leap. So, long before one can show the causes of a given thing demonstratively, it is necessary to become acquainted as thoroughly as possible with the properties or attributes of the things themselves.

Aristotle wants to reach a proper definition of the soul, which will be a demonstration of its essence (of *what it is*); he will then be able to *demonstrate* what attributes the soul has, in all its particular instances, from that definition. This demonstration will itself be an account of the *causes,* the logical causes, why the soul has those attributes *(why it is as it is).* But he is going to start his investigation not at the definition, but at the *attributes* that the soul actually has in its manifestations. And that brings him back to his original question of establishing which attributes of the soul are characteristic of the soul itself, and which ones depend on, and cannot be separated from, its presence in living creatures.

Aristotle's answer to this question is built into his formulation of it: if any function or affection of the soul is peculiar to it, it can be separated from the body. But if there is nothing peculiar to the soul, it cannot be separated. Yet, Aristotle claims, in most cases it seems that none of the affections, whether active or passive, can exist apart from the body. This applies to anger, courage, desire, and sensation generally, though possibly thinking is an exception. Thus (Aristotle concludes) the soul does not have any attributes of its own: there are no attributes of the soul *except when the soul is present in a living body.* He has already shown that in the investigation of the soul, it is necessary to work *from* its properties or attributes, *towards* its ultimate definition. It is now clear that the soul has attributes only when it is, quite literally, "embodied" in living creatures. Hence any definition one might give of any of the attributes of the soul must take this fact into account—and any attribute of the soul must be defined as a movement of a part of the body, or of a part or faculty of a body, in a particular state, aroused by a particular cause, with a particular end in view.

No one, Aristotle is claiming, can investigate the soul properly unless he or she starts the enquiry from its actual manifestations in living creatures. The nature of the soul is the object of Aristotle's enquiry here: and it is this enquiry which leads him to look at living creatures. Aristotle looks at *animals* because he is interested in the *soul*. He looks at them solely with respect to what light they shed on the essence and attributes of the soul. He looks at animals because they are the soul-in-action.

Does the soul have parts or not? And is there more than one kind of soul? The distinctive feature of the things which have soul is, of course, that they are living; and living involves (Aristotle says) the presence of mind, sensation, movement (or rest) in space, and the movement involved in nutrition, decay, and growth. In that they have in themselves a capacity to feed, to decay, and to grow, *plants* are alive. These capacities Aristotle describes as the "nutritive": "we call 'nutritive' that part of the soul of which even plants partake." This "nutritive" aspect of the soul, what we might call its "nutrivity," was later often called the "vegetative" soul, or vegetivity. All animals have this set of capacities *plus* sensation (the sense of touch); some animals have both of these *plus* movement; and at least one animal (man) has all of these *plus* thought. The soul is the origin of—what is responsible for, the source of—all these faculties; nutrition, sensation, appetite, movement, and thought are all *aspects* of the soul. Of these, thought alone may perhaps be separable. But the other faculties cannot be thought of as "parts" of a single soul (or as separate souls), for the evidence of sense is against it. For one can successfully take cuttings from a plant; and one can cut certain insects into two, and both parts will continue to live. The evidence appears to show that somehow the soul is a "unity" which can be potentially divided into many little "unities"; it is not a series of "parts" joined and held together. The soul does not have parts, therefore.

This way of approaching the question of whether the soul has parts or not has brought Aristotle to the position where he claims that life consists of various "faculties" [*dynameis*], and the possession of even one of these indicates the presence and workings of soul; he has also claimed that they are simply *aspects* of the *one* soul. But what he has established in addition is that there is a hierarchy of these faculties: "of the faculties [*dynameis*] of the soul which we have mentioned, some living things, as we have said, have all, others only some, and others again only one" (414a29–30). Given this variety in the extent to which the soul is present in different living things, it is clear that one definition of soul which covers them all is going to be too general: it would fit them all, but would not be the figure [*schema*] of any particular one of them (414b23).

It is necessary therefore, he suggests, to reformulate the question. Given that the soul has several aspects, or "faculties"; given that living things share to a greater or lesser extent in these faculties; given, too, that a definition of soul which covers all of these would be too general to be worth seeking; then, what is needed instead is to seek definitions of *each of these faculties*. Then these definitions, taken together, will be the full definition of the soul: they will cover all aspects of the soul as it actually has incidence in living things. We have noticed (Aristotle claims) that these faculties are in a hierarchy, with plants at the lowest end, possessing just the vegetative faculty, and man at the highest end, possessing all of the faculties. This raises a further question which must now be asked: Why are living things (the different instances of the soul-in-action) thus arranged in such a series (415a2)?

Once it has been recognized that the soul-in-action is a number of faculties arranged in a hierarchy, the tasks ahead of one in pursuit still of a full definition of the soul, says Aristotle, are these:

1. To explore each of these *faculties* (nutrition, sensation, appetite, motion, thought, and any others).
   To do this
   (a) in general; and
   (b) in particular, by exploring the soul (the particular set of faculties) of individual types of living creatures, "for instance of the plant, the man, and the beast."
2. to investigate why these faculties are arranged in a series.
   As Aristotle writes: "It is clear that the account of each of these faculties is the most relevant account that can be given of the soul" (415a13–14). In all stages of this exploration one will be starting from the actions of the soul: from the faculties of the soul as actually exhibited in (particular) creatures.

Aristotle has now established, at least to his own satisfaction, that there are five major faculties of the soul: (i) the nutritive/vegetative; (ii) the sensitive; (iii) the appetitive; (iv) the motive; and (v) the thinking. They are in a hierarchy, rising from the nutritive to the thinking. In the rest of *De Anima*, he deals with these faculties in a general way and, following his own considerations on the appropriate method of procedure, he starts in each case with things such as food, color, and sound.

The first faculty that he deals with is the lowest, the one which all living things share, the *nutritive/vegetative*. It is through the presence in them of this aspect of the soul that all living things have life: "the nutritive soul is present also in the others and it is the first and most universal power of the soul on account of which life is present in all" (415a24). This aspect of the soul has two major functions: to assimilate food and to generate other living things of the same kind. After all, as Aristotle says, it is because of this that all living things do what they do according to nature.

In mentioning "for the sake of," Aristotle is now talking about cause; and thus he diverts for a moment to discuss all the senses in which this term can be taken. Hitherto Aristotle has been talking about living creatures, because his quest is for the soul, and they are (for him) the soul-in-action. Now he turns the issue around and asserts that the soul is the *cause* [*aitia*] of living creatures. It is the cause in three major senses of the term:

(i) It is the cause in the sense of *what-makes-them-what-they-are*, that is, *living*. The presence of the soul is the cause why any potentially living creature is an actually living creature. This sense of cause is often translated into English as the "essence" or "essential nature" of a thing. In the case of living creatures, their "essence" or "essential nature" is their possession of soul.

(ii) The soul is also the cause in the sense of being the *first principle* of living creatures: it is the (internal) *source* or *origin* of motion in the living body. Motion is characteristic of living creatures for Aristotle, and is of three kinds: (a) change of *quantity* (growth and decay); (b) change of *quality* (as in sensation); (c) change of *place* (locomotion).

(iii) The most important sense in which the soul is the cause of living creatures is cause in the sense of *that-for-the-sake-of-which*. All living creatures exist for-the-sake-of being instruments of the soul. This is, of course, a very important concept for Aristotle: the "goal" [*telos*], the "end," the "for-the-sake-of-which" something exists. It is evident for Aristotle that the soul is *prior to* the bodies of living creatures, which are simply its instruments. He constantly draws a parallel with mind: humans act with some purpose in view, and Nature is just the same. Recognizing this is the clue to understanding why Nature is as it is. A living creature does not exist "for its own sake": it exists "for the sake of being" the instrument of the soul. The "goal" of serving the soul thus controls and determines how any living creature *must be,* what parts it *must have*—subject only to the particular requirements of its life, its particular activities, its habits, and its other parts.[15]

It is thus the requirements of the soul which specify and control what forms or forms living creatures can have. When he is looking at living creatures, Aristotle is assured that their characteristics are determined by the needs of the soul. And it is because the soul is "that for the sake of which" living creatures exist that Aristotle's project of finding out about the soul by looking at animals is both possible and logically sound. For animals are not just a convenient means of looking at the soul-in-action: animals exist *simply to be* instruments of the soul. They have, for Aristotle, no interest except as being instruments of the soul. Philosophers should look at animals because they are the instruments of the soul, and everything the philosopher finds out in this enquiry will tell him more about the soul and its operations—because the soul is here the object of the enquiry. This is beginning to sound very much like Plato's claim that the (human) body is simply the "vehicle of the soul," and his insistence likewise that the body exists "for the sake of" the soul. Aristotle is indeed dealing with this issue because he has inherited this way of thinking from Plato, his teacher. But Aristotle is turning the issue upside down. He is saying that instead of *assuming* what kind of thing the soul is, what its requirements are, and how the body must therefore be structured and function to fulfill these requirements, what the philosopher needs to do is to investigate and hence *discover* the nature of the soul from the soul-in-action. This central concept, the "for the sake of which" will be the grand clue to enable the investigator to understand why animals are the way they are, why they have the parts they have, and why those parts function the way they do. Investigating animals is the necessary means to the end of understanding "The Animal"—which is "The Soul in action."

To accomplish this plan, and to gain an understanding of "the soul in action," it was essential for Aristotle to systematically cut up animals—that is, to anatomize them. The use of the knife was central to his investigations, because one cannot reach the internal organs without cutting animals open, either alive or dead, and the organs are the instruments of the soul. Hence Aristotle turned to the work which is recorded in the animal books.[16]

Aristotle is informative in a number of places in the animal books about what his program and his procedure are. A full characterization of Aristotle's project in the animal books would take up at least another article. For the sake of brevity, we can here gather together some more of the programmatic statements (that is, in addition to those passages I have already quoted) that he makes in *Parts of Animals,* a book whose title ought properly to be rendered *Of the Causes of the Parts of Animals.* Aristotle asks: "[S]hould the student of Nature follow the same sort of procedure as the mathematician follows in his astronomical expositions—that is to say, should he consider first of all the phenomena which occur in animals, and the parts of each of them, and having done that go on to state the reasons and the causes . . . ?" (639b6).To which the answer is yes, "we ought first to take the phenomena that are observed in each group, and then go on to state their causes" (640a15). "The best way of putting the matter would be to say that *because* the essence of man is what it is, *therefore* a man has such and such parts, since there cannot be a man without them" (640b1). "We have to state how the animal is characterized, i.e., what is the essence and character of the animal itself, as well as describing each of its parts; just as with the bed we have to state its Form. Now it may be that the Form of any living creature is Soul, or some part of Soul, or something that involves Soul" (641a17). "[The student of Nature should] inform himself concerning Soul, and treat of it in his exposition; not, perhaps, in its entirety, but of that special part of it which causes the living creature to be such as it is. He must say what Soul, or that special part of Soul is; and when he has said what its essence is, he must treat of the attributes which are attached to an essence of that character" (641a25). And finally, for the present, "Now the body, like a hatchet, is an instrument; as well the whole body as each of its parts has a purpose, for the sake of which it is; and the body must therefore, of necessity, be such and such, and made of such and such materials, if that purpose is to be realized" (642a11).

We have now seen that Aristotle was an anatomist: someone who systematically dissected or vivisected animals to find out certain kinds of things about their inner structure and functioning for particular theoretical and contemplative purposes. But we have also seen that he had a very precisely delineated program of dissection, leading him to look at particular animals and particular parts of animals, and to wield the knife in particular ways. So he anatomized *as a philosopher.* Anatomy was an integral and necessary part of his philosophical program.

Today we think of Aristotle as a philosopher who could wear many hats, and we do not remark that the extension of his concept of philosophy and of the role of the philosopher is matched by no modern category of knowledge and by no modern scholar—and, indeed, is not even captured by all the faculties of a modern university and all its academics, taken together. As with some of the scientific disciplines we have been discussing, the modern discipline of philosophy is itself also a creation of the eighteenth and nineteenth centuries.[17] The point, concerns, topics, and identity of old philosophy were all transformed and reshaped then, and were greatly restricted compared with their former range and extent. The philosopher was now obliged to yield his expertise with respect to Nature to that new person, the scientist. The new philosopher now had to pursue his philosophy not in the world but in the ivory tower of academia: philosophy was no longer to be the pursuit which enabled one to live the life of the good man, but was restricted to a set of discrete topics for discussion and argument.

In order to appreciate Aristotle's view of the role of the philosopher and of the goals of philosophizing, and of his view of the role of the study of animals, the first thing we have to do is to abandon the modern discipline boundaries and definitions. For it is only our commitment to these which leads us to characterize his animal books currently as "biology" and therefore to contrast, oppose, or even seek to "reconcile" them with his supposedly distinct and supposedly contrasting "philosophical" works. *Our* discipline of philosophy has no role for anatomizing. But it is clear that *Aristotle's* understanding of philosophizing meant that anatomizing, undertaken in order to understand the nature and functioning of the soul, *had to be* an integral and crucial part of philosophy. Similarly, for us to talk of "philosophical etholog" in the case of Aristotle would also seem to demand this enforced separation of Aristotle's supposed "biology" from his supposedly separate "philosophy." But only when we can see the animal books and Aristotle's anatomizing as an *integral* part of his particular view of what philosophy is and how it should be lived and practiced will we have properly understood them.

It may be that scholars of Aristotle's animal books will believe that my position differs from their own only in form of words, and will maintain that they, too, of course view the animal books as instances of anatomizing and see the anatomizing in turn as part of Aristotle's larger philosophical project, but that for convenience they choose to refer to this part of that larger project as "Aristotle's biology." Are we only arguing over words? I certainly think that some of these modern scholars have done excellent work in recognizing and laying out the details of Aristotle's goals and procedures in the animal books, even though they have done so while insisting on regarding Aristotle as engaged in biology.[18] But I do think the words matter, especially ones which describe intentional activities and modern disciplines, and that if we choose the wrong words, we not only make life quite unnecessarily difficult for ourselves, but we end up turning Ancients into Moderns, which is bizarre.

NOTES

I thank the editors, Prof. Barbara Massey and Prof. Gerald Massey, for their invitation to contribute to this forum. I am most grateful for encouragement and advice to Dr. Harmke Kamminga, Dr. Yoko Mitsui, and Prof. Heinrich von Staden.

1. See William Homan Thorpe, *The Origins and Rise of Ethology: The Science of the Natural Behaviour of Animals* (London: Heinemann Educational, 1979). See also Donald A. Dewsbury, "Rhetorical Strategies in the Presentation of Ethology and Comparative Psychology in Magazines after World War II," *Science in Context 10* (1997): 367-86, and Robert A. Hinde, *Ethology: Its Nature and Relations with Other Sciences* (Oxford: Oxford University Press, 1982).

2. Karl Z. Lorenz, *The Foundations of Ethology,* trans. Konrad Z. Lorenz and Robert Warren Kickert (Vienna and New York: Springer-Verlag, 1981).

3. D'Arcy Wentworth Thompson, *On Aristotle as a Biologist. With a Prooemion on Herbert Spencer. Being the Herbert Spencer Lecture Delivered before the University of Oxford, on February 14, 1913* (Oxford: Clarendon Press, 1913).

4. We make a similar knee-jerk identification with respect to the sciences of the organic which come under biology, when we find Aristotle saying things which to us seem to be obviously like those modern-day sciences: physiology, embryology, comparative anatomy, and the like.

5. The same will be the case if we choose to label Aristotle's animal work as examples of one of the subsciences of the umbrella-science of biology, such as zoology, comparative anatomy, experimental physiology, or embryology, and hand it over to the historians of those subjects.

6. See, for instance, Allan Gotthelf and James G. Lennox, eds., *Philosophical Issues in Aristotle's Biology* (Cambridge: Cambridge University Press, 1987); Wolfgang Kullman, *Die Teleologie in der aristotelischen Biologie: Aristotles als Zoologe, Embryologe und Genetiker,* Sitzungsberichte der Heidelberger Akademie der Wissenschaften. Philosophisch-Historische Klasse (Heidelberg: Winter, 1979), and Wolfgang Kullman and Follinger, eds., *Aristoteliosche Biologie: Intentionen, Methoden, Ergebnisse, Akten des Symposions über Aristoteles' Biologie vom 24.-28. Juli 1995 in der Werner-Reimers-Stiftung in Bad Homburg* (Stuttgart: F. Steiner, 1997).

7. The concept of "philosophical ethology," as I understand it, as applied to past philosophers, is also built on this separation of the "biological" from the "philosophical," and their relations. Applied to Aristotle this would involve this same separating-off by us of his supposedly "biological" books from his supposedly distinct "philosophical" ones.

8. See Andrew Cunningham, *The Anatomical Renaissance: The Resurrection of the Anatomical Projects of the Ancients* (Aldershot: Scholar Press, 1997), chap. 6.

9. Andrew Cunningham, "William Harvey: The Discovery of the Circulation of the Blood," *Man Masters Nature: 25 Centuries of Science,* ed. Roy Porter (London: BBC Books, 1987).

10. One or two other scholars have made points similar but not identical to this. See, for instance, G. E. R. Lloyd, "Aspects of the Relationship Between Aristotle's Psychology and his Zoology," in *Essays on Aristotle's De Anima,* ed. Martha Nussbaum and Amelie Oksenberg Rorty (Cambridge: Cambridge University Press, 1992), 147–67. Unlike me, Lloyd is not here concerned with anatomy.

11. Francis Cornford, *Before and After Socrates* (1932; reprint, Cambridge: Cambridge University Press, 1979), 64.

12. Marjorie Grene explains their relative positions thus: "the core of Plato's doctrine, the Forms, recollection, the dualism of soul and body, Aristotle was at pains to refute. . . . [Aristotle agreed with Plato that] it is forms that the knowing mind properly and rightly knows. Further, form is causal, it is the reason why things are what they are and the reason why our minds can know them. Further still, form is the source of unity: it is one as against the multiplicity of the informed, or of the unformed. Moreover, the mind in knowing form is somehow like it, at one with it. All this is common ground. But Aristotle found form, intelligibility, definiteness, where Plato had never found it: in the limited, recurrent but orderly processes of nature itself" (Marjorie Grene, *A Portrait of Aristotle* [London: Faber, 1963], 65).

13. As, for instance, by Allan Gotthelf: "With these words Aristotle introduced his students to the study of biology"; see vii. Similarly, Geoffrey Lloyd writes that this chapter "provides a fascinating insight into the resistance that Aristotle had to overcome among some of his contemporaries in order to get biology accepted as a worthy subject of the philosopher's investigations" (Geoffrey Lloyd, *Aristotle: The Growth and Structure of his Thought* [1968; Cambridge: Cambridge University Press, 1980], 71).

14. Here and elsewhere, I follow the Loeb translation by W. S. Hett, modified according to the advice of Christine Salazar, whose assistance I am most grateful for.

15. On this see *History of Animals* 487a10.

16. This claim does not have any consequences for establishing the chronological order in which Aristotle went about his work; it is a claim about the logic of his argument only.

17. For a preliminary discussion of this transformation, see Gilbert Ryle, ed., *The Revolution in Philosophy* (London: Macmillan, 1956). This transformation still awaits proper historical study.

18. For instance, Allan Gotthelf and James Lennox might reasonably think that they have

said all this themselves, as in the introduction to the section "Biology and Philosophy: An Overview" of *Philosophical Issues in Aristotle's Biology,* where they write: "But 'biological' is *our* label: Aristotle has no such term, and speaks rather of the general study of nature [*phusike*], and within that, of the study of plants, or of animals, or of the capacities of soul. Nor must we assume that we can straightforwardly map these studies as Aristotle conceived them onto *portions* of our own general biology, or botany, or zoology (or embryology, or comparative anatomy, etc.)—or even assume that they are *science* rather than philosophy (or philosophy *rather than* science). We need instead to approach them fresh, on their own terms to come to see what their aims are, and their methods, and their contents, and what their relation is to (and how they might be of use in the understanding of) Aristotelian philosophy itself" (5–6). However, the last part of the last sentence reveals that they are still contrasting the animal books to "Aristotelian philosophy itself," and this caveat paragraph appears in a volume which they have chosen to call *Philosophical Issues in Aristotle's Biology.* Moreover, I can find no discussion of soul in the main Introduction to the book, and the first essay in the volume, by David Balme, is called "The Place of Biology in Aristotle's Philosophy" and contains no mention of soul at all.

VII

Paracelsus Fat and Thin:
Thoughts on Reputations and Realities*

I Paracelsus contradictory

[53] I am proposing the historiographical category of the fat and the thin. I hope it will be recognised that while my account of this category begins in Paracelsus, the concept of the fat and the thin is a general historiographical one which could perhaps be put to useful work to help understand the relation between reputations and realities in the case of other historical figures. The fat and the thin is a concept which closely concerns the transformations of Paracelsus and of Paracelsianism.

I was relieved to discover that the elusiveness of Paracelsus is not something that has only been experienced by me. Forty years ago, that most eminent historian of medicine, Owsei Temkin, published an article 'On the elusiveness of Paracelsus', and in it he pointed out that 'two of the greatest historical figures of the medical past are also amongst its most elusive ones'.[1] These two are Hippocrates and Paracelsus. This elusiveness is indeed quite remarkable, especially since I believe we can say that Hippocrates and Paracelsus are two of the only three characters from the history of medicine whose names are known universally, to the man in the street (the other, I hazard, would be Pasteur).

Was there one man called Paracelsus? I asked myself. Or were there many men called Paracelsus? For I found that Paracelsus was variously claimed as:

A SUCCESSFUL PHYSICIAN	A FAILED PHYSICIAN
A PRACTICAL PHYSICIAN	A SPIRITUAL PHYSICIAN
A CHEMIST	AN ALCHEMIST

---

[1]   Owsei Temkin, 'The elusiveness of Paracelsus', *Bulletin of the History of Medicine* 1952, 26, 201–17, see 201.

* Originally published in *Paracelsus: The Man and his Reputation, his Ideas and their Transformations*, ed. O.P. Grell (*Studies in the History of Christian Thought* 85). Leiden: E.J. Brill, 1998. This article has been reset with a new pagination. The original page numbers are given in square brackets within the text.

[54] A MAGICIAN OR MAGUS	A SCIENTIST (*or* someone turning magic into science)
A PROPHET	A MADMAN
AN OPPONENT OF THE ANCIENTS	A HUMANIST (that is, a Renaissance follower of the Ancients)
THE HIGH POINT OF MEDIEVAL NATURAL PHILOSOPHY	THE BEGINNING OF MODERN SCIENCE (*or* the bridge between the two)
A FULLY POLITICISED SOCIAL-REVOLUTIONARY	AN OTHER-WORLDLY MYSTIC
A MAN OF HIS TIME	A MAN FOR ALL TIMES

And in addition to these roles, he also had reputations among historians as an *astrologer*, as an early *biologist*, *mineralogist*, *anthropologist*, even *psychiatrist*.

What is particularly striking about these roles is not just that they are multiple, but that so many of them are mutually contradictory. We get to the ridiculous position where, in the classification still used by the history of science journal *Isis*, Paracelsus supposedly contributed at the same time, *and with the same writings*, to Science and Pseudo-science: he was simultaneously both Scientist and Pseudo-Scientist!

Paracelsus's motto is said to have been: 'Let him not be another's, who can be his own' (Alterius non sit qui suus esse potest). This sounded to me like a statement about preserving one's personal identity, and this question of Paracelsus 'being himself' led me (eventually) to ask: Who was this man? What did he really think and teach? Why, given how incomprehensible I found his writings and the writings of some historians about him,[2] did he have such a persistent reputation? And what was it 'really' a reputation for? The 'real' Paracelsus and his supposed achievements all eluded me. I should stress that this was not a problem about the notorious difficulty or obscurity of Paracelsus's *language*, but about getting a rounded view of Paracelsus from

---

[2] I should mention that the fundamental work on Paracelsus with which I had to wrestle at the time was that of Walter Pagel, *Paracelsus: An Introduction to Philosophical Medicine in the Era of the Renaissance*, Basel 1958.

the writings of historians, even if one also had recourse to the writings of Paracelsus himself.

[55] I did at one time think that the problem was one of my own limited vision, rather than something to do with either Paracelsus, or with other historians. Perhaps his multiple historical reputations simply reflected the multi-faceted nature of Paracelsus's personality, the universality of his message, the profundity of his doctrines, the diversity of his achievements? Perhaps this is the point of an enigmatic historical character such as Paracelsus, this is what makes him so 'great', that he is available to and seemingly 'speaks to' later generations? Perhaps it requires the work of many historians for us to be able to see the diffcrent facets of his reality? Hence the historian interested in the political history of the sixteenth century can discover the political Paracelsus who is not evident to the historian of chemistry who, by contrast, finds Paracelsus the proto-chemist, someone not evident to the political historian? Perhaps only if all such accounts are taken together can we perceive the true Paracelsus in all his complexity?

But I realized (again, eventually) that in fact this is not the case: issues of anachronism render some of our current representations of Paracelsus completely invalid. For in the centuries subsequent to Paracelsus we have separated-out, or marked off, disciplines of knowledge in a way alien to Paracelsus's time. Hence he has become a supposed 'contributor' to a range of fields which, by our modern categories of knowledge, are contradictory, and which did not exist as categories of knowledge in his own time. And he has been given separate reputations in each of these histories. That is, it is us, with *our* subject-divisions, who have created a multi-faceted and contradictory picture of Paracelsus, and one which in many respects simply cannot be authentic. Amongst the many important new distinctions and contrasts in the domain of knowledge that have been made since Paracelsus's time, there are those between religion and science, as incompatible and mutually exclusive areas; between science and magic, similarly seen as incompatible; between the supposed 'rational man of the Renaissance', and the mystic of the sixteenth century; the fundamental distinction between 'medieval' and 'modern', with all the judgemental force involved in making such a contrast; the distinction between science and (popular) politics, and our customary belief that these are properly distinct and unrelated domains; the distinction between chemistry and alchemy, and the view that the history of chemistry must show a new rational science here replacing an old mystical art – that sense must oust nonsense. Moreover, a host of new sciences have been created since Paracelsus's time, [56] including a number to which Paracelsus has been described as a contributor: such as mineralogy, biology, anthropology, all created in the early nineteenth century; psychiatry, created at the end of the nineteenth century; and biochemistry, created in the twentieth century. In particular, we have completely lost *natural philosophy* as a coherent and meaningful category of knowledge and enquiry, which is one within which Paracelsus was working, at least some of the time.

In addition, the multiple reputations of Paracelsus are in part due to our perfectly natural desire, especially when creating a *new* discipline, to have founding fathers. The desire of historians of chemistry in the 19th century, for example, to have a suitable transitional figure to bridge between 'alchemy' and chemistry, between medieval and modern, drew their attention to Paracelsus, and created for him the role of father of modern chemistry.[3]

So the world of knowledge, the map of the disciplines, has completely changed since Paracelsus's day. Yet historians and others have been eager to give Paracelsus some roles as a proto-practitioner of modern disciplines, disciplines quite unknown in Paracelsus's own time, logically impossible though such roles are. And many writers on Paracelsus have generally ignored the categories of knowledge which existed in Paracelsus's day, and not sought to characterise him as a man of his time.

The problem about this attribution of 19th and 20th century activities to Paracelsus, is not just that we thereby have only a partial view of the historical rounded figure, but that we produce for ourselves a view of Paracelsus that logically could not have been the case. We distort the history if we try and make it fit inappropriate categories – not matter how we bolster our work with direct citation of Paracelsus's own works and words. If, for instance, we seek the chemist in Paracelsus, we find in him a 19th or 20th century man, [57] something which, no matter how talented he was, he could not possibly have been, because he was a sixteenth century man.

---

[3]   This view of the historic role of Paracelsus in the history of chemistry continues today. J.R. Partington in his multi-volume work, *A History of Chemistry*, London, 1961, says 'From our point of view, he [Paracelsus] represents a step forward from alchemy. He administered a rude shock to the conventional alchemists, and by his blustering profusion of abusive rhetoric he pushed aside their unintelligible jargon by one even less comprehensible but more modern and in nearer relation to reality' (vol. 2, 123). Similarly, in his recent survey of chemistry (*The Fontana History of Chemistry*, London, 1992), W.H. Brock sees the mid-sixteenth century as the time during which 'alchemy had been transmuted into chemistry' (28), and for his story Paracelsus still plays the role of intermediary between (medieval) alchemy and (modern) chemistry, by employing 'an empiricism that was controlled by Christian and Neoplatonic insights' (45).

So the real issue here is that of recognising which questions about Paracelsus are legitimate and historically valid, and which ones are not. All of which leads me to the fat and the thin.

II The fat and the thin

There are two parallel iconographic traditions for Paracelsus, which originate from his lifetime. One shows him thin, hatless, bald, simply dressed, and with an expression which is neither smile nor scowl, but looks like the face at rest. This is the portrait by Augustin Hirschvogel of 1538, and claims to show Paracelsus at age 45. The other shows him fat, hatted, with hair showing beneath the hat (and presumably continuing under the hat), richly apparelled, and with an expression which might be a smile. This derives from a painting at the Louvre, now attributed to the school of Q. Metsys, produced c. 1528–30. Later representations of Paracelsus come primarily from these traditions, taken as models.[4] Strikingly, over the centuries, in subsequent representations, the thin Paracelsus seems to get thinner, while the fat Paracelsus gets fatter.

---

[4] My information here is built on the twenty-one engravings and other portraits held in the Wellcome Institute, London, as described in Renate Burgess, *Portraits of Doctors and Scientists in the Wellcome Institute of the History of Medicine*, London 1973, 271–2. The portraits listed there, deriving from originals from the period of Paracelsus' lifetime, appear to divide into four traditions, which might be built on just two representations:

> 1. *thin* A. (Burgess numbers 10–11), bald, from original in the State Library, Vienna, 1538, by Hirschvogel; this is my 'thin' Paracelsus. B. (Burgess numbers 12–16, and 20) bald with sword, earliest listed appearance 1587; this looks to me like a development of number 1A, so I count it also as in the tradition of the 'thin' Paracelsus.
> 2. *fat* A. (Burgess number 19) with a hat and furred coat, derived from a drawing, sometimes known as 'a young man with slouch hat' by the younger Holbein, 1526; this might be the origin of the tradition of the 'fat' Paracelsus (see Pagel, *Paracelsus*, 7 n. 13). B. (Burgess numbers 1–9) with hat, from a Louvre painting, c. 1528–30, attributed to school of Q. Metsys; also sometimes known as 'the Hollar version': my basic 'fat' Paracelsus. This has also been attributed to Jan Van Scorel (1495–1562); see Pagel, *Paracelsus*, 90.

Burgess numbers 17–18 are a quite different representation of Paracelsus, with a clipped beard, and which dates from the 17th century. Number 21 is a twentieth-century image of Paracelsus. I have heard that there is an article by Walter Pagel on this general theme, but I have not been able to locate it.

Paracelsus. Etching by Augustin Hirschvogel (1538). Courtesy of Wellcome
Institute Library, London

Phyſick Proffeſsor at Basil.
Philip Theophraſtus PARACELSUS *He died at*
Saltzburge Añ. Dom: 1540. aged
47 yeares.
W. Marſhall ſculpſyt.

Paracelsus. Line engraving by W. Marshall after J. Payne. Derived from School of Q. Metsys (c. 1528–30). Courtesy of Wellcome Institute Library, London.

This is interesting in itself and for what it says about historical [**58** and **59** illus.] [**60**] perceptions of Paracelsus. But here I want to use the existence of these parallel iconographical traditions as a motif for a treatment of Paracelsus's historical reputations, and the relation between his reputations and his realities.[5] That is, we can take the existence of an iconographical tradition of a 'thin' Paracelsus as representing Paracelsus stripped of all his accreted reputations – the 'real' Paracelsus, as it were, the one who lived. Given the nature of the picture itself one might risk a joke, and regard this as 'the bald truth'.[6] This thin Paracelsus must, of course, be constructed basically from the original (Latin and German) printed writings of Paracelsus himself, in all their abundance, and from related contemporary materials, and it is an image still under construction by historians. There *was* a historical Paracelsus who lived, and this Paracelsus is being reached by proper historical exegesis and is turning out to be a very complex figure, though not a self-contradictory one.

The parallel existence of an iconographical tradition of a 'fat' Paracelsus, can be taken by us to represent the Paracelsus-es of all the different historical reputations. Again, to risk another joke, given the picture of the fat Paracelsus, we could regard these fat Paracelsus-es as presenting 'the embellished truth', as long as we bear in mind that while clothes embellish and present a certain image of a person, they at the same time conceal the body beneath. And hats are ideal for concealing the bald truth.

We all of us come to the study of Paracelsus initially with some image or other about him in our minds: that he was, for instance, an alchemist, a magician, a scientist, a popular political figure, a man of the middle ages, and so on. These are the fat Paracelsus-es. Fat Paracelsus-es are all later creations. These images we bring to our work are ones which have been created, unconsciously and [**61**] unselfconsciously, by people later than Paracelsus. Most of the historical literature of over three hundred years on Paracelsus is doing such work: making fat Paracelsus-es of one kind or another. Even the most scrupulous historians can bring to their work on Paracelsus (or any

---

[5]   For this vocabulary of 'fat' and 'thin' I am initially indebted to Johanna Geyer-Kordesch's comments on my original paper at the Glasgow conference, and to the subsequent table-talk of Allen Debus. The present considerations have been developed since then.

[6]   Jokes in academic papers are very risky so, to avoid misunderstanding, I must stress that I am *not* saying that the illustration of the thin Paracelsus is a true likeness, and the fat one is not; I am not at all concerned with the issue of the relative authenticity (if any) of these images to the historical Paracelsus. Perhaps neither of them correspond to his true likeness. For present purposes it doesn't matter. What matters is what images they were *meant* to convey, and what images they *did* convey, to the viewer (these not necessarily being the same), especially the later viewer who could not have seen Paracelsus in person.

other historical figure) an agenda which leads them to find *in* Paracelsus (or whoever) what they brought *to* him.

Every time people write about Paracelsus, the Paracelsus they think they are describing is the thin Paracelsus, the Paracelsus who really lived. But in fact this is not necessarily the case. For finding the thin Paracelsus is very difficult indeed, for the simple reason that fat Paracelsus-es keep getting in the way and making it not only difficult but very often impossible to see the thin one. And new images of Paracelsus are unwittingly being constantly created by ourselves as historians, in the very act of seeking to find something in Paracelsus and his work which is relevant to the agenda of historical research we ourselves are engaged on.

Of course it does not feel as though we are creating new fat Paracelsus-es as we labour to read and understand the writings of Paracelsus and his contemporaries. Indeed, the act of reading Paracelsus's own writings and deploying them in our writings about him, make us feel that the opposite is true: that we are reading and thus listening to the true Paracelsus, and in our writings we are revealing the true, thin, Paracelsus to our audiences. Yet always there is the problem of the selectivity we unwittingly practise: our eagerness to see in what Paracelsus wrote what we want to find, plus our willingness to believe that Paracelsus must have meant what we want him to have meant.

These historical reputations, the multiple reputations Paracelsus has had in all the periods since his day, including today, are just as valid reputations as the thin Paracelsus, and are also historical Paracelsus-es. For these varied Paracelsus-es had and have important roles within the collective imaginations of those societies in which they are created, roles which are usually far more important to those societies than the figure of the 'true', thin, Paracelsus.

Each fat Paracelsus does important ideological and psychological work in the society or group which promotes and projects him, and in this sense any fat Paracelsus is more important generally than the thin one. These fat Paracelsus-es are not false Paracelsus-es: they are just different from the 'thin' Paracelsus, Paracelsus the historical figure [62] who actually lived. Moreover, of course, the reputations are *also* themselves historical realities.

The existence of fat Paracelsus-es raises questions of *why* Paracelsus was chosen to be 'fattened up' at a particular time and by particular people? And these questions are, of course, not in any way questions about the thin Paracelsus and his time, but about the time when, and the society in which, each particular fat Paracelsus was created.

The basic historiographic points at issue (and they are very basic) are two. First, unless we are very careful and alert to what we are doing in our historical studies, we confuse one or other fat Paracelsus for the thin one. That is, we attribute characteristics, goals, achievements, beliefs, and so on, to the thin Paracelsus which are in fact from our present, and thus construct fat Paracelsus-es which serve our later purposes – but which we mistakenly take to be true about the historic ('thin') Paracelsus. This can lead to our history-writing being simply self-serving and indeed simply circular, as we reinforce the image of Paracelsus that we bring with us by bolstering it with layers of authentic historical information and quotation from the writings of Paracelsus; such work may simply further obscure from view the thin Paracelsus rather than making him more visible. I should stress that the encounter with the historical texts of Paracelsus will not, in itself, ensure that we are dealing with the thin, rather than our preferred fat, Paracelsus. Quite the reverse: prejudice or preconception can withstand any number of encounters with the evidence and remain unscathed, as my examples will show. One could, I suppose, make the joke that inside every fat Paracelsus there is a thin Paracelsus trying to get out. But in fact we will only find the thin Paracelsus if we go deliberately looking for him, and without preconceptions about his proper shape.

The other basic historiographic point comes from my own experience in trying to understand the historical reputations of Paracelsus. It is this: unless and until we can distinguish between the fat and the thin, then we simply go round in circles asking absurd questions which try and cope with the apparent contradictions in Paracelsus – such as 'how much was Paracelsus a medieval mystic, how much a modern scientist?' – questions which derive simply from historical reputations Paracelsus been given, rather than from Paracelsus the historical figure in his own time.

It is inevitable that around the thin Paracelsus fat Paracelsus-es **[63]** would gather. And this is the case with all historical figures later thought to be important: fattening-up is a recurring, natural, historical process which happens to celebrated figures of the past when they are needed to achieve latter-day purposes, and in the course of which they are given new pasts, new historical reputations.

It may be that the thin Paracelsus, whenever we might succeed in reconstructing him in his completeness, will be ultimately of interest only to historians. But that should be an incentive to us. The picture of the thin Paracelsus is, of course, what the best practice of historians working on him is bringing to light.

III The occult Paracelsus

So: there are many fat Paracelsus-es. Paracelsus today has many reputations *simultaneously*, each of which has its own *historical* origin and its *modern* reason for surviving and being promoted/celebrated by historians and others. I cannot, of course, hope to look at the source and function of all of these, especially since some of them are still in the course of being made, such as Paracelsus the theologian, or Paracelsus the political activist. But I will look at the making of one particular reputation, the most dominant one still current, and at the version of it which has been with us and developing for a century or more now: Paracelsus as occultist/mystic.

My questions here will be: why does Paracelsus have such a large reputation today as an occultist, and especially why has his 'occult' medicine been more studied than his practical medicine? and, why do we have texts (translations) which promote this view? These are questions about fat Paracelsus-es, how and why Paracelsus was historically *given* certain reputations, not questions abut what kind of occultist the thin Paracelsus 'really' was. That would be an enquiry of a different kind. I present two little case-studies.

To illustrate the source and force of Paracelsus's recent reputation as an occultist, I take as my first example late 19th century Britain. As far as Paracelsus studies are concerned, this is when the English-speaking world received its first extensive set of translations of the works of Paracelsus, made from the Latin edition of Geneva of 1658.[7] Two volumes, each of 400 pages of translation, **[64]** were produced, and a hundred years later these two volumes are still the largest source of Paracelsian works available in English, having been reprinted in the 1960s. This project was a product of the occult revival of 19th century England.[8] The Paracelsus that this translation/edition gives us is therefore, as one might expect, an occult Paracelsus.

The co-ordinator of this substantial translation exercise was Arthur Edward Waite (1857–1940). The period c. 1820 to c. 1910 in Europe has been called by James Webb, 'The flight from reason'.[9] In splendidly robust fashion

---

[7]   Something of this story has been told by Charles Webster, 'The nineteenth-century afterlife of Paracelsus', in Roger Cooter, ed., *Studies in the History of Alternative Medicine*, London 1988, 79–88, but the historiographic points he makes are different.

[8]   Despite their date of publication, the appearance of these volumes had nothing at all to do with 1893 being the 400th anniversary of Paracelsus' birth.

[9]   James Webb, *The Flight from Reason* (vol. 1 of *The Age of the Irrational*), London 1971. On these matters see also, for instance, Bruce F. Campbell, *Ancient Wisdom Revived: A History of the Theosophical Movement*, University of California Press 1980. I am grateful to my former colleague Perry Williams for guidance here.

Webb argues that the new views of the world portrayed by the 'new science' of the 19th century elicited, as its counterpart, a 'crisis of consciousness' [sic], a crisis in which resort to the occult, in a myriad of ways, was central:[10] the 'new science' (together with the 'new politics' of bourgeois nationalism and liberalism) had as their counterpart what one might call a 'new occultism'. This occultism included Spiritualism, 'hypnotism, magic, astrology, water-divining, "secret" societies', and others.[11] As far as Britain was concerned such occult concerns began around 1848 and peaked in the 1880s.

All these occult obsessions were lived out, one after the other, by Arthur Edward Waite. In 1880 Waite was in his early twenties. He had left Catholicism as a young man, through the rather unusual route of reading Robert Chambers' *Vestiges of Creation*, but did not abandon the search for the mystical 'One'.[12] Waite was a life-long mystic, looking for Fairy land or 'Faërie' – an early work of his was *The Heart's Tragedy in Fairyland* – and he had 'a bee-filled bonnet'.[13] He always hoped that research into the occult and occult practices [65] and their histories would help in his 'lifelong Quest' (as he called it) for mystical union. Waite was able to make a living out of his concern with the history of the occult and the mystical, for he luckily fell in with two publishers keen to put out occult works (though a lot of their stock went unsold); and when they all failed him, Waite became London manager of Horlick's Malted Milk (itself a drink with proverbial restorative powers!), and wrote their advertisements and pamphlets.

Waite's life-long Quest was intense, but sceptical: he was always hoping that 'another explosive might be placed [by himself] astutely in the camp of fond believers'.[14] In particular what Waite believed was that scholarly work on the *history* of each of these traditions would serve to sift the authentic truths from bogus later accretions in each of these traditions, and this historical work is what he devoted his life to: history will take us to the mystical truth. It should be noted that while Waite is here dealing in terms of the 19th century sense of 'history', in which some development or progress is to be perceived, he is actually expecting *degeneration* over time from some pristine original to have happened.

---

[10] *The Flight from Reason*, xiii.

[11] *The Flight from Reason*, xiv.

[12] Indeed, according to Waite, it was in reading Chambers' book that 'the whole cosmos began to move about me, spelling out living oracles, of which Church, and Rome especially, had never conceived', *Shadows of Life and Thought*, London 1938, 48–9.

[13] *Shadows of Life and Thought*, 128.

[14] *Shadows of Life and Thought*, 100.

His autobiographical work, *Shadows of Life and Thought* (1938), reveals his personal life-route through the mystical and occult: he went via Spiritism (or Spiritualism), which did not satisfy him and which disillusioned him by its rampant trickery; he tried Theosophy, as did many others, for this was the time of Madam Blavatsky and Anna Besant; Occult Science attracted him in turn, including Alchemy; so did Magic, especially the works and approach of the Frenchman Eliphas Lévi (Alphonse Louis Constant, 1810–78); Rosicrucianism lured him in its turn, and here as always he is sympathetic to the 'real' Rosy Cross and its mysteries, but full of doubt about the drivel put forward by its modern day advocates. The Order of the Golden Dawn called him next, and then Esoteric Freemasonry; he found in most cases that after the necessary historical research he could, unlike most members, reach the true mysteries, and receive 'that life in the heart, that secret of light in the mind, which no one tells to another, because it cannot be put into words'.[15] Waite talked to the trees (or rather, he listened while they talked to him),[16] he was **[66]** fascinated by Animal Magnetism and Hypnotism, he earnestly sought out the history of the Holy Grail and toyed with Tarot cards, always believing that one or other of these would be the true key to the mystic route to 'pure being which is the Life of the Soul in God'.[17] By the 1890s (that is, after his work on Paracelsus), he claimed that he was weary of all the false prophets and the widespread ignorance of the occult shown by his contemporary practitioners of the occult. Yet still he cherished

> the vague feeling ... that something was lying *perdu* beneath the cesspools and ashpits of so-called occult science. It was the notion presumably of a Higher magia, with Jacob Böhme and a dozen lesser but reputed names as apologists, if not as champions.[18]

It was in this commitment to occult seeking that Waite enthusiastically directed the translation and publishing of the works of Paracelsus. For Paracelsus was, for him, a potential source of the 'Higher magia'.

The man who personally financed the translations and publication of the Paracelsus edition was Fitzherbert Edward Stafford-Jerningham (1833–

---

15   *Shadows of Life and Thought*, 162.
16   *Shadows of Life and Thought*, 68.
17   *Shadows of Life and Thought*, 195.
18   *Shadows of Life and Thought*, 145.

1913).[19] His was an old Catholic aristocratic family. The Honourable Stafford-Jerningham succeeded his brother as Lord Stafford in 1892; this brother, though officially declared lunatic in 1862,[20] had held the title since 1884. Lord Stafford was another convert to the occult world, and was using up the family fortunes by trying to make gold by alchemical transformation, and in this was obviously an undeclared lunatic. Stafford needed certain texts translated to help his transmutations, initially the *Lexicon Alchemiae* of Martinus Rulandus, and this is why he employed Waite. When Waite had had the translation made, Stafford wanted it printed – in an edition of one, just for his own use. Waite's view of the general project was that

> what was wanted was a working canon of distinction between those
> alchemists who were dealing with a supposed physical operation and
> those who used the symbolical language of Alchemy in a spiritual or
> transcendental sense.[21]

[67] For Lord Stafford he produced English versions of a number of alchemical works, including *The Triumphal Chariot of Antimony* by Basil Valentine. And then, in 1894, appeared 'the *magnum opus* of the whole incredible adventure, being my edition of *The Hermetic and Alchemical Writings of Paracelsus*', and for this Waite had to work against time 'so that the Earl might not wait unduly'. And although, according to Waite, at the time 'seekers were few indeed' for the work of Bombast von Hohenheim, this edition has nevertheless been of great influence in subsequent years.

The first volume is devoted to works of Paracelsus which could fit under the title of 'Hermetic Chemistry', such as *The Aurora of the Philosophers* and *Concerning the Nature of Things*. The themes of the second volume are 'Hermetic Medicine' and 'Hermetic Philosophy', with 'Hermetic Medicine' including texts such as *The Archidoxies* and *Alchemy the Third Column of Medicine*, but also including several texts on how to prepare alchemical medicines. The emphasis on hermeticism and alchemy will be evident, as will the omission of all the practical medical treatises of Paracelsus, such as the surgical works.

What Paracelsus appears from this set of volumes? Obviously this Paracelsus is an alchemist. Thus Lord Stafford the practical alchemist

---

19   His grandfather had succeeded in having the old barony restored in 1829, the previous
baron, Lord William Howard, having been executed for treason in 1680 as one of the 'five
popish lords' involved in the Titus Oates affair.

20   See *Burke's Peerage*.

21   *Shadows of Life and Thought*, 131.

wanting the transmutation of metals had his Paracelsus here, with recipes he could follow. But Paracelsus is also presented as an alchemist concerned (as Waite put it in his Preface) with 'the development of hidden possibilities or virtues in any substance, whether by God, or man, or Nature'.[22] And so Waite also had his Paracelsus here, a Paracelsus who is an alchemist at the spiritual and transcendental level rather than at the material level, indeed who is a spiritual alchemist, using 'the symbolical language of Alchemy in a spiritual or transcendental sense'. And, on the medical side, this edition presents Paracelsus the *spiritual* physician, with his Great Elixir and his Universal Medicine: but deliberately excludes Paracelsus the *practical* physician with his 'many formidable treatises on surgical science, and on the causes and cure of diseases'.[23] For, to Waite, even Paracelsus' prescriptions 'were not to be literally understood, even when they were apparently the ordinary formulae and concerned with the known [68] *materia* of medicine'.[24] *All* was spiritual and to be understood spiritually. To put it bluntly, Waite's Paracelsus was a Paracelsus who had been engaged on a spiritual journey, and not a sixteenth century one but a *19th* century spiritual journey, taking him through Spiritualism, Theosophy, Occult Science, Alchemy, the Magic of Eliphas Lévi, Rosicrucianism, and the Order of the Golden Dawn, and who was about to move on to Esoteric Freemasonry and the lore of the Tarot cards. Late-nineteenth century occultism is built into the selection of the texts, into the interpretation of them, into the very translation itself, and into the characterisation of Paracelsus the historical man. In particular, Waite imported from Eliphas Lévi the view that Paracelsus was restoring an ancient occult medicine.[25]

One can see, thus, how *this* occult Paracelsus was the construction of modern, 19th century British, occultists. A similar state of affairs seems to have obtained in the early years of this century in both France and Germany,

---

[22]   *The Hermetic and Alchemical Writings of Aureolus Philippus Theophrastus Bombast, of Hohenheim, called Paracelsus the Great, Now for the First Time Faithfully Translated into English*, by Arthur Edward Waite, 2 vols., London, 1894, repr. New York 1967, vol. 1. Preface, xxii–xxiii.

[23]   *The Hermetic and Alchemical Writings*, vol. 1, Preface, xxii.

[24]   *The Hermetic and Alchemical Writings*, vol. 1, Preface, xxii.

[25]   Lévi says, 'This truly universal medicine is based upon a spacious theory of light, called by adepts fluid or potable gold ...', *The History of Magic, Including a Clear and Precise Exposition of its Procedure, its Rites and its Mysteries*, translated by Arthur Edward Waite, London, 1913, 340.

where editions of the translated works of Paracelsus were produced under the auspices of mystic and occult publishers.[26]

There is another occult Paracelsus, a different and more famous one, created shortly after, and which is still flourishing into the present day, and this tradition, too, produced a set of translated texts of Paracelsus, seeming to offer unobstructed – if assisted – access to Paracelsus the man speaking. This is the Paracelsus which emerges from the work of Carl Gustav Jung and his circle, and Jung's 'analytical psychology'.

Jung's encounter and fascination with alchemy is very well known, and is integral with his working-out and presentation of his 'analytical psychology'. Jung was the son of a Lutheran pastor. Always religious in outlook from his childhood, and working on psychic (hypnotic) phenomena for his thesis, from around 1910 Jung became completely **[69]** fascinated by the mythical, spiritual, and mystical traditions, and this interest was a significant factor in his break with Freud in 1913.[27] In his autobiography Jung records some telling moments in this respect. In 1910 Freud 'adopted' Jung as his intellectual heir, and urged Jung to defend the sexual theory. Jung said, 'Against what?', and Freud said 'Against the black tide of mud of occultism'.[28] Jung went out and explored the occult. So the period of Jung being made the heir is the period of Jung beginning to split from Freud – and it was precisely over the rational/occult distinction! One can see, therefore, how very important the occult dimension was to Jung and to his analytical theories. Jung, of course, interpreted this event as Freud denying his mystical (feminine) side; Jung says that Freud had dreams that Jung wanted to kill him (Freud), the father figure. So each interpreted the event according to their pre-existent theories.

---

[26]   Hans Kayser, ed., *Schriften Theophrasts von Hohenheim gennant Paracelsus*, Leipzig, 'Bücher der deutscher Mystik', 1921, in which brief passages from Paracelsus are assembled under various headings; Grillot de Givry, *Oeuvres complètes de Paracelse, Traduites pour la Première Fois du Latin et Collationnées sur les Editions Allemandes*, 2 vols., Paris, in the 'Classiques de l'Occulte' series, 1913–4.

[27]   See George B. Hogenson, *Jung's Struggle with Freud*, University of Notre Dame Press 1983, esp. Part 2, 'From Freud to Mythology', 15–40.

[28]   C.G. Jung, *Memories, Dreams, Reflections*, orig. German edition, 1962; tr. by Richard and Clara Winston, 1967; London, reprint, 1983, 173. The claim has been made that Freud's psychoanalysis was deeply influenced by mysticism, but mysticism from the Jewish tradition, not the alchemical tradition as in the case of Jung; see David Bakan, *Sigmund Freud and the Jewish Mystical Tradition*, Princeton 1958.

Astrology Jung took up at a time when it was becoming of great interest in the German speaking world,[29] but his interest in alchemy seems more unusual. Jung's own deep interest in Paracelsus seems to have begun in 1928,[30] and was thereafter very committed, and not only did he seize the opportunity of twice giving major public lectures on Paracelsus (in 1929 and 1941), one of them on 'Paracelsus the physician', but he wrote another important essay on him too, 'Paracelsus as a spiritual phenomenon' (published in 1942). Of this essay Jung wrote in his autobiography:

> Through Paracelsus I was finally led to discuss the nature of alchemy in relation to religion and psychology – or, to put it another way, of alchemy as a form of religious philosophy. This I did in *Psychology and Alchemy* (1944). Thus I had at last reached the ground which underlay my own experiences of the years 1913 to 1917; for the process [**70**] through which I had passed at that time corresponded to the process of alchemical transformation discussed in that book.[31]

So Jung equates his own personal psychic journey and spiritual trans-mutation with the alchemical transmutations of Paracelsus: that is to say, in Jung's view, Paracelsus and his work had, at a higher or more universal level, prefigured Jung himself and *his* work. And in writing so extensively on alchemy in later years, once he had had the initial insight or vision, Jung's general aim was

> to demonstrate that the world of alchemical symbols definitely does not belong to the rubbish heap of the past, but stands in a very real and living relationship to our most recent discoveries concerning the psychology of the unconscious. Not only does this modern psychological discipline give us the key to the secrets of alchemy, but, conversely, alchemy provides the psychology of the unconscious with a meaningful historical basis.[32]

---

[29] Ellic Howe, *Urania's Children: The Strange World of the Astrologers*, London 1967, esp. chapter 5, 'The German revival'. For the British story, see now also Patrick Curry, *A Confusion of Prophets: Victorian and Edwardian Astrology*, London 1992.

[30] 'Light on the nature of alchemy began to come to me only after I had read the text of the *Golden Flower*, that specimen of Chinese alchemy which Richard Wilhelm sent me in 1928. I was stirred by the desire to become more closely acquainted with alchemical texts'; Jung in *Memories, Dreams, Reflections*, ch. 7, 230.

[31] Jung, *Memories, Dreams, Reflections*, 236.

[32] C.G. Jung, *Mysterium Coniunctionis: An Inquiry into the Separation and Synthesis of Psychic Opposites in Alchemy*; first German edition 1955, 1956; English translation by R.F.C. Hull,

With his essentially religious and spiritual attitude to psychological phenomena, and having found in Gnostic writings the prototype of the 'archetype',[33] in Paracelsus and alchemy Jung then found the prototypes of neurosis and transference, much of dream symbolism, and the role of the therapist who 'confronts the opposites with one another and aims at uniting them permanently'.[34] The Jungian therapist is an alchemist. Conversely of course for Jung, Paracelsus the alchemist must have been an early Jungian therapist.

And Paracelsian alchemy is for Jung more than just a metaphor or model for psychic transformations:

> As the alchemists, with but few exceptions, did not know that they were bringing psychic structures to light but thought that they were explaining the transformations of matter, there were no psychological considerations to prevent them, for reasons of sensitiveness, from laying bare the background of the psyche, which a more conscious person would be nervous of doing.[35]

[71] So alchemy for Paracelsus – at least in Jung's view – 'was not simply a chemical procedure as we understand it, but far more a philosophical procedure, a special kind of yoga, in so far as yoga also seeks to bring about a psychic transformation'.[36]

All this is delightfully and necessarily circular in argument. Jung thinks he finds *in* Paracelsus and alchemy certain truths about psychoanalysis, 'truths' he is in fact projecting *onto* Paracelsus and alchemy. And then he uses these 'truths', deriving ultimately from his psychoanalytic thinking, to interpret Paracelsus the historical figure as an alchemist-who-is-really-a-psychoanalyst. This is, of course, the only way one can find a twentieth century psychoanalyst in the sixteenth century: by putting him there, and claiming he was not aware of what he was 'really' doing. 'None of the alchemists ever had any clear idea of

---

Princeton 1963, repr. 1977, xiii.

[33]   The whole concept of the 'archetype', so central to his developed theories, Jung took in 1919 from the *Corpus Hermeticum* (God is 'the archetypal light'). Jung, 'Archetypes of the collective unconscious', *Collected Works*, vol. 9, 4; Jolande Jacobi, *Complex/Archetype/Symbol in the Psychology of C.G. Jung;* original German edition, Zurich 1957, English translation by Ralph Manheim, New York 1959; see 34.

[34]   *Mysterium Coniunctionis*, xv.

[35]   *Mysterium Coniunctionis*, xvii.

[36]   'Paracelsus the Physician', paragraph 28; see the English translation by R.F.C. Hull (1966) reprinted in *The Spirit in Man, Art and Literature*, London, 1966.

what [their] philosophy was really about', writes Jung,[37] who from his superior position knows the truth: they were really psychoanalysts!

Why Jung chose Paracelsus as his model, apart from the fact that both were Swiss and Paracelsus was the most famous of German-speaking physicians in history, I have not been able to determine. But it was, I suppose, important for Jung that Paracelsus was someone from well before the 'atomistic, mechanistic views of the late nineteenth century' which Jung disliked and which he associated with Freud.[38]

The actual set of Paracelsus texts in German which projected this view of Paracelsus, using Paracelsus' own words – and which were then translated from German into English – was produced by Jolande Jacobi. Jacobi was one of the group of loyal female disciples that Jung attracted round himself, and who have been recently well described as his 'Valkyries'.[39] She came from a Hungarian Jewish family, and left them behind her when she went to work with Jung in 1938, at the age of nearly 49. She had religious and mystical inclinations, eventually converting to Catholicism. She became an analysand of Jung, and eventually became a Jungian analyst herself. [72] She has been described as 'the disturber of the peace, the one with the vision of Jung's greatness who wished to present him to the world in a big way; an extrovert who grated on the nerves of the very introverted circle of people around Jung'.[40] Her sharp tongue put her at odds with others of Jung's circle, and her fiery temper led to disputes with Jung himself. The end of one such dispute has passed into the lore of Jungian analysts: to his great surprise, another analyst saw Jacobi being thrown out of Jung's office after she had had a row with Jung, and she went down the stairs on her bottom, bump, bump, bump.[41] But, whatever this apparent rebellion against Jung might mean, when she applied herself to the task of expounding and popularising the works of Jung, Jacobi is found to 'stick closely to Jung's words and ideas, simply clarifying him;[42] she became the great populariser of the Master's works, and her writings on, from and about him are still in print.

---

[37] 'Paracelsus as a Spiritual Phenomenon', 231 of the original German edition; see *Collected Works*, vol. 13, 186, translated by Hull.

[38] The quotation is from Jolande Jacobi, *The Way of Individuation*, tr. R.F.C. Hull, London, 1967, 3.

[39] See the chapter 'Jolande Jacobi: Impressionist', in Maggy Anthony, *The Valkyries: The Women around Jung*, Longmead, Shaftesbury, Devon, 1990, 55–63.

[40] *The Valkyries*, 55.

[41] *The Valkyries*, 59.

[42] *The Valkyries*, 60.

So in this context it is quite significant that in introducing her earliest essay at popularising Jung, *The Psychology of C.G. Jung*, first published in 1940, Jacobi was to quote Paracelsus in her Introduction. Following the 'inner path', Jacobi claims, equips the individual to turn to the 'outer path' to 'fulfil the claims of life in the collectivity':

> Within and without he will be able to affirm his personality. 'The world was created imperfect' says Paracelsus, 'and God put man into it in order to perfect it'.[43]

And her second work promoting the Jungian vision was in fact the book of Paracelsus texts: *Paracelsus: Selected Writings*, which first appeared in German in 1942 with the more revealing title of 'Living Heritage'.[44] This volume is put together from paragraphs and short passages from Paracelsus' writings, illustrating various themes, such as 'Credo', 'Man and the Created World', 'Man and his Body', 'Man and Works', 'Man and Ethics', 'Man and Spirit', 'Man and Fate', and 'God the Eternal Light'. By this technique we get, for instance, [73] a smoothly flowing 'Credo' from Paracelsus, outlining succinctly his beliefs, and using only his own words. But it is a 'Credo' which Paracelsus did not write as a 'Credo': in other words, the 'Credo' is a Jacobi construct. It is also, of course, a Jung construct. How far it represents any 'Credo' that Paracelsus himself might have written, must remain an open question.

And this is where it gets really interesting, because only a couple of years later, in 1945, Jacobi compiled a similar work from the writings of Jung himself, in part as a 70th birthday gift. This is *Psychological Reflections*.[45] Among the headings Jacobi uses here are 'Recognition of the soul', 'Man in his relation to others', 'Between good and evil', 'The life of the spirit', 'The way to God'. The two works are, in some ways, interchangeable (at least in the eyes of a non-believer in either alchemy or Jungian analysis, like myself), and many of the texts within them could be quoted as characterising either Paracelsus or Jung, so close has the identification of the two become in the eyes of Jacobi the Jungian student and, I suggest, in the eyes of Jung himself.

---

[43]   This passage is introduced into the 4th English edition of 1944, and is the only appearance of Paracelsus in this work; this passage was obviously introduced by Jacobi after she had completed her next book, the one on Paracelsus.

[44]   Originally published as *Theophrastus Paracelsus: Lebendiges Erbe*, Zurich 1942; English translation by Norbert Guterman, Princeton 1951, second edition 1958.

[45]   Originally published as *Psychologische Betrachtungen: Eine Auslese aus den Schriften von C.G. Jung*, Zurich, 1945; second edition 1949; English translation, mostly by Hull, London, 1953.

So here are two 'fat' occult Paracelsus-es: one a late 19th British one, one a mid-20th century Swiss one, with different occult visions, serving different occult interests, yet each built authentically, it would appear, on genuine Paracelsus texts, and each producing – making accessible – Paracelsus texts which reinforce that vision and place it at the door of the historical Paracelsus. There is nothing like a translation or edition of texts to have this kind of effect! There are probably other occult Paracelsus-es too: to each occultist his occult Paracelsus.

None of this is to deny that there *is* indeed of course an enormous and enormously significant occult dimension to Paracelsus – to the thin Paracelsus, that is. It is just to point out that the *construction* of the later *reputation* (or, more properly, reputations) of Paracelsus as an occultist was made by people whose first interest was in their own spiritual view of the occult or in their own view of psychical life, not in retrieving the true historical 'thin' Paracelsus. The *only* Paracelsus they were each interested in was *their* 'occult' one. We have seen two occult Paracelsus-es here: someone whose alchemy [74] was a mystical spiritual adventure and whose 'true' medicine was also wholly spiritual, the other someone whose alchemy was truly a process of psychological and psychoanalytical self-liberation. It is perhaps not surprising, therefore, that these occult Paracelsus-es do not mesh particularly well either with each other or with other Paracelsus-es, those constructed in pursuit of different interests.

Incidentally, tracing the origins of these occult Paracelsus-es did help me understand why I had originally found Paracelsus so antipathetic: the answer lay in the fact that I cannot stand modern occultism, so I simply detested the historical Paracelsus presented through the distorting lenses of modern occultisms. It was, of course, before I learnt to distinguish between the fat and the thin Paracelsus-es. What was particularly liberating for me was to discover that even a deeply scrupulous scholar such as Walter Pagel, whose writings on Paracelsus had seemed so authoritative but which I had found so deeply irritating and intellectually impenetrable, had an agenda of his own when he drew our attention to Paracelsus as being a most significant figure in the history of medicine; but once one recognised the agenda, even Pagel's Paracelsus was another fat figure.[46]

---

[46] The nature of Pagel's Paracelsus has been explored by T.M. Stokes, 'Walter Pagel as an historian of medicine', unpublished M. Phil. Essay, Cambridge Wellcome Unit.

IV Thin Paracelsus

In writing about the 'elusiveness' of Paracelsus, Owsei Temkin wrote that 'The literature on Hippocrates and Paracelsus is abundant just because it is provoked by so many unanswered questions', questions, he implied, about the historical Paracelsus. I beg to differ. We have seen a little of how malleable the reputation of Paracelsus has been over the centuries, how open to interpretation, how available the thin Paracelsus has been for fattening. Paracelsus has multiple reputations because of the multiple later interests of others, who have had an interest in co-opting him to their side.

Is there anything about the historical, the thin, Paracelsus which makes, or made, him particularly attractive/available for these many imposed roles, these multiple after-lives, these 'subsequent performances' (as Jonathan Miller might term them)? That is, why have [75] later people so often chosen Paracelsus, rather than anyone else, on whom to foist such varied roles? A possible answer might be that Paracelsus had many roles for his contemporaries. And here, in closing, I want to raise the question of the meaning of certain distinctive features of Paracelsus' reputation in his own day and shortly after – beyond his involvement with magic and his possible pacts with the Devil – and which have needed explaining or explaining away by later admirers intent on characterising Paracelsus in their own images.

I call to attention six aspects of his contemporary reputation:[47]

1. his excessive drinking
2. his excessive travelling
3. his constant raging against authority
4. his dirtiness
5. his madness
6. his lack of sexual reputation – indeed the rumour that he was a eunuch or had been castrated.

I am not concerned here with the validity of these reputations, but with their existence and their association with the Paracelsus who lived: I am calling on the principle that gossip like this must, at some level, correspond to something about the real person, or have been seen as a possible thing which might be true of this person, or it couldn't 'stick', as gossip must do if it is to be believed.

---

[47] Most of these derive from the account of Paracelsus' assistant, Johannes Oporinus, in a letter dating from 1555; the letter is translated in Henry M. Pachter, *Paracelsus: Magic into Science*, New York 1951, 154–6.

The point I am centrally interested in here is this, which I have not been able to find dealt with in the literature: what was the *sixteenth* century meaning of such reputations? Although the tradition of work on Rabelais begun by Mikhail Bakhtin has opened up such questions, and for the very period of Paracelsus,[48] I have not found the meanings of these particular reputations dealt with. We know that today reputations for drunkenness, dirtiness, rootlessness, raging, raving, and not having sex, are not desirable, and they would tend to undermine the validity of what one preached. One or two historians have claimed that these are simply slanders put about by Paracelsus' enemies, and these characterisations would certainly work [**76**] like that today. In a Romantic interpretation – a kind of interpretation which was of course created long after Paracelsus' own day – by contrast, these characteristics serve, in the eyes of admirers of Paracelsus, to make him into a romantic genius. If we take just the drinking, then Eliphas Lévi, for instance, the fully romantic French magician of the 19th century, to whom Paracelsus once appeared in a dream, interpreted Paracelsus' drunkenness in just this way – as revealing the true spiritual and magical gift of Paracelsus. 'The marvellous Paracelsus', he calls him, 'always drunk and always lucid, like the heroes of Rabelais':[49]

> To strive continually against Nature in order to [achieve] her rule and conquest is to risk reason and life. Paracelsus dared to do so, but even in the struggle itself he employed equilibrated forces and opposed the intoxication of wine to that of intelligence. So was Paracelsus a man of inspiration and miracles.[50]

In other words, like a true magical adept, Paracelsus kept himself balanced between the opposing forces by keeping himself drunk.[51]

Paracelsus' reputation as someone immune to sexual temptation, and possibly as castrated, may possibly be linked to either the Christian ascetic tradition or to his magical interests. The Greek Church Father, Origen,

---

[48]   Mikhail Bakhtin, *Rabelais and his World;* original Russian version, 1965; English translation by Helene Iswolsky, Cambridge, Mass. 1968; Carol Clark, *The Vulgar Rabelais*, Glasgow 1983.

[49]   *Transcendental Magic*, 11.

[50]   *Transcendental Magic*, 266.

[51]   Another post-Romantic interpretation puts Parcelsus' drunkenness down to simple excess, typical of a genius: Giordano Bruno supposedly said (according to Henry Pachter):

> Who after Hippocrates was similar to Paracelsus as a wonder-working doctor? And, seeing how much this inebriate knew, what should I think he might have discovered had he been sober?

As quoted by Pachter, *Parcelsus: Magic into Science,* 297, without reference.

castrated himself to achieve greater commitment to the kingdom of heaven,[52] the medieval scholastic and priest, Peter Abelard, also castrated himself. Similarly there is a tradition of a mythical figure, embodied in the character of Klingsor, whose (self-?) castration gave him superior magical powers.[53] Paracelsus and those gossiping about him may not have known about any of these. But was [77] there possibly a *sixteenth* century image of castration as leading to greater purity, holiness or power with respect to nature?

Of course, if we knew the 'real' Paracelsus, he would be firmly placed in his time, historicised, and limited to his time: and hence he would be less famous, less available to be all the things we want him to be. I come back to Paracelsus' supposed motto: 'Let him not be another's, who can be his own'. He has been everyone else's: is it time for him to be his own man?

---

[52]   Joseph Wilson Trigg, *Origen: The Bible and Philsophy in the Third Century*, London 1985, 53–4.

[53]   In Wolfram von Eschenbach's *Parsifal* Klingsor is unmanned by a jealous husband, in Wagner's *Parsifal* he is understood to be self-castrated. I am grateful to Martin Preston for information on these points.

VIII

Thomas Sydenham: epidemics, experiment and the 'Good Old Cause'

QUESTIONS

My central question here will be: *why, with respect to medicine, did Thomas Sydenham see what he saw and do what he did?* Answering this will of course entail answering the question *what it was* that Sydenham did with respect to medicine in the London of the three and a half decades 1656 to 1689.

Thomas Sydenham has posthumously acquired one of the greatest of all names in the history of western medicine. He is celebrated amongst historians of medicine as the inaugurator or reviver of *clinical* or (to express it in English rather than Greek) *bedside* medicine – which we treat as a highly positive achievement – and has been awarded the title of the 'English Hippocrates'. Such was his fame in early nineteenth-century Britain that two successive societies dedicated to the publishing of modern and ancient 'classic texts' in medicine were named after him. The works of Sydenham, in Latin and then in English, were amongst the earliest works issued, and I like to think they were planned to be the first.[1] Naturally enough, therefore, Sydenham has attracted his share of medical historians to celebrate his work and achievement. I trust that nevertheless it does not appear too arrogant of me to claim that all this effort has not yet given us a satisfactory account of either the activity or the achievement of Thomas Sydenham.[2] At all events, it is to this most fundamental of issues that I address myself here.

---

[1] See G.G. Meynell, *The Two Sydenham Societies* (Acrise, Kent, 1985). He numbers the Latin edition as the third of the works published by the first Society.

[2] The main biographers are J.F. Payne, *Thomas Sydenham*, (London, 1900), still very valuable; Payne wrote the article in the *Dictionary of National Biography* too; R.G. Latham, who presented a 'life' in volume 1 of his *The Works of Thomas Sydenham, M.D., translated from the Latin Edition of Dr Greenhill* 2 vols. (London, Sydenham Society, 1848–50); Kenneth Dewhurst, *Dr Thomas Sydenham (1624–1686): his Life and Original Writings* (Berkeley,

To answer my question, what we need to discover and establish is an account which is *about Thomas Sydenham and his world*. An account which locates Sydenham's experience of, and his thinking and decisions about, medicine and its practice in the same place where all his other experiences of the world were registered, and where all his other thinking and decisions took place. In other words, in his mind: one mind, undivided. The mind of a particular Englishman alive, active in and affected by, the specific events of his own time. I am going to look at Sydenham's activity to reform medicine (and that means every last thing about that medicine) as a product of the life he led in the world in which he led it. Approaching the matter this way will necessitate us putting aside from our minds the reputation that we (I speak as a historian of medicine) normally ascribe to Sydenham: what Sydenham intended and achieved in his own world and his own terms may well have been quite different from what we, centuries later and with different interests, usually credit him with having achieved (and hence persuade ourselves that he must have intended!). How and why Sydenham was given his later reputation is a different question, and one to which we shall return. So I want here to argue that Sydenham's medicine was the produce of a person highly *politicized*, and in a particular way; that his attempt to reform medicine was a continuation of politics by other means. And I want further to suggest that this was evident to his contemporaries: the *adoption* of Sydenhamian medicine by other people was thus also a political act. If we do not understand the politics we will not understand the medicine.

First, however, a note on texts. A number of English manuscripts in the handwriting of Sydenham or Locke (and sometimes even of both of them on the same page) have been brought to light. But Sydenham's works were all published in Latin: whether or not (as rumour had it) they were translated into Latin by other hands,[3] these Latin editions must for the present be treated as the 'original' texts. Sydenham's major work (and the focus of our present concern) appeared in two forms. First in 1666 as *Methodus Curandi Febres, Propriis Observationibus Superstructa* (Method of curing fevers, built on my own observations). This was given a second, slightly modified edition

---

1966); Donald Bates, 'Thomas Sydenham: the Development of his Thought, 1666–1676' (Johns Hopkins University Ph.D. thesis, 1975); Bates also wrote the article in the *Dictionary of Scientific Biography*. But all of the writers on Sydenham seem (to my mind) to leave out the medicine: or rather, they deal with the life of Sydenham and, quite separately, with the medicine – as if the medicine owes little or nothing to the life.

3 On the rumours about other people translating Sydenham's works, see Payne *Thomas Sydenham*, ch. 13: 'In what language did he write?'

in 1668; neither of these were ever published in English. Then in 1676 Sydenham published the *Observationes Medicae circa Morborum Acutorum Historiam et Curationem* (Medical observations on the history and cure of acute diseases), announcing it as the 'third edition' of the *Methodus*. However, it is significantly different in structure, size, presentation, title and content from the *Methodus*. It is this 'third' edition which has been the basis of all English translations.[4]

MEDICAL PRACTICE: WHY?

Let us start right in the middle of the story, at the moment when Thomas Sydenham is taking up the practice of medicine. The date is about 1656 or a little later; the place is London. Sydenham is in his early thirties. Let us ask: why does he take up medical practice at all? This is a most important question since, on his own account, the only good medicine is practical medicine; and what he advocated was a particular form of practical medicine. Let us begin our exploration by inspecting the accounts of this event that Sydenham himself published. First, let us take the account of 1676 (the dedication to Mapletoft):

It is now the thirtieth year since, being on the way to London in order to go from thence a second time to Oxford (from which the calamity of the first war had kept me away for some years), I had the good fortune to fall in with the most learned and highly honourable man, Dr Thomas Coxe (who even throughout those years, and right up to today, practised Medicine with great renown), then attending my ill Brother. He, with his well-known kindness and courtesy, enquired of me to what art I was preparing to devote myself, now that I was resuming my interrupted studies, and had reached man's estate [i.e. age 21]. Hitherto undecided, and not even dreaming of the Medical Art, I [now] readied myself seriously for it – greatly stimulated by the urging and counsel of so great a man, and in some way I suppose by my own fate. And certainly, if these my efforts [i.e. the book] turn out to be in the smallest way for the common good (*in publica commoda*), the credit must be thankfully referred to him, with whom as promoter and inspirer I first began those studies. After a few years spent in the academic arena, having returned to London I came to Medical Practice . . .[5]

Thus does Sydenham account for the years and events: they are largely ignored. There is nothing here about the fact that Sydenham comes from a gentry family (the fifth son of nine); nothing about the fact that, like all the other adult male members of his family, he fought in the parliamentary army (1643–5 in Dorset in 'the first war', six months of so of 1651 in the midlands and north in the 'second' war). In

---

[4] It was first translated by John Pechey in 1696, and again by John Swan in 1742. Greenhill's nineteenth-century Latin edition, and Latham's English translation from it, are of the 'fourth edition' of 1685, itself a revised version of the third.
[5] This is my translation from the Latin as printed by William A. Greenhill, *Thomae Sydenham, M.D., Opera Omnia* (London, Sydenham Society, 1844), pp. 3–6.

the above account the civil war simply plays the role of a mysterious calamity which kept him away from Oxford for some years. There is no mention either of the fact that Thomas Coxe, at the time that Sydenham met him, was a physician to the parliamentary army. And then, the period at Oxford, some ten years (*c.* 1647–*c.* 56), is simply time 'spent in the academic arena'. There is nothing here about the medical degree (the M.B.) that he was awarded in April 1648 – a mere year or so after arriving – nor of the medical fellowship at All Souls that he was thrust into (October 1648) by the Visitors set up in the university by the victorious parliamentary authorities. Most important of all, there is no mention of the Commonwealth and Protectorate, nor of the political career that Sydenham had been promoting for himself: he twice stood for parliament, for one of the Weymouth seats (in 1658 the election was abandoned on Oliver Cromwell's death, and in 1659 Sydenham was defeated); he also held the Exchequer post of 'Comptroller of the Pipe' from 1659.

So what, one might say: why should he have mentioned such things? He was, after all, writing to his contemporaries and not for the benefit of the historians, and writing about how he took up medical practice, not how he was detained from doing so. To which the answer must be this: that Sydenham is here offering a fragment of autobiography to explain to his contemporaries the source and nature of the book which follows. Hence what he mentions and what he omits in this account are the result of active choice on his part. But for him to omit mention of the things I listed above, and especially of his political commitment (as exemplified by his service in the parliamentary army), was for him to omit his *raison d'être* for these years. It is Hamlet without the Prince. And it is to leave unexplained why ten years passed between him being 'greatly stimulated by the urging and counsel' of Thomas Coxe to take up medical practice, and him actually doing so – time which was not being spent on a conventional eleven-year university training in medicine. In this context we should note that in this account of how he came to take up medical practice there is actually no direct mention of the circumstances which led him to do so! From this account we are none the wiser: and, more important, nor would his contemporaries have been. Could it perhaps be that Sydenham had good reasons for not mentioning what he did not mention?

We return to our question: why did he take up medical practice at this period? It would seem to have been connected with him giving up some other things. For Sydenham took up medical practice when he (i) left his fellowship at Oxford, and when he (ii) abandoned active politics. His leaving Oxford in 1655 was the price he paid for getting married (to a Dorset girl): resignation from the celibacy of his Oxford fellowship was obligatory. He was now short of an income.[6] Even though he was now the eldest surviving Sydenham brother, there was no chance that Thomas would inherit the family estate; when his father eventually died the estate went to Thomas's eldest nephew. The necessity of seeking stable employment was pressing. He had been holding a 'physic place' at All Souls for some eight years. Medicine it had to be. But, although he now in 1656 did engage in medical practice in London, he did not yet abandon his ambitions in the political arena, for he was to stand (unsuccessfully) for election as an M.P. in both 1658 and 1659. He did not abandon such ambitions until 1660 – until the restoration of Charles II. Could this be what drove Thomas Sydenham to become a full-time medical practioner: the failure of the Commonwealth?

Already we have had a taste of the awful silence in Sydenham's published writings about politics or life outside medicine: Sydenham would, on this evidence, seem to live in a world where the political was as seemingly distinct from the medical as it is in our world today. But there are reasons why our medicine and science are performed and presented as if they are free from social or political content. Sydenham, in his turn, may have had reasons to present his medicine in a similar light. If so, his political concerns can nevertheless be recovered by us from the extant thing that would contain them: the medicine that Sydenham was now to practise, and on whose remarkable properties he was to report in the two versions of his book in 1666 and 1676.

Sydenham's political position and commitment is partly visible in the preface he wrote to the first edition of the book (in 1665). This is how it begins:

Whoever takes up medicine ought to weigh these matters seriously. First, that he will one day have to give to the supreme Judge an account of the life, committed to his care, of his patients. Second, that whatever skill or knowledge he has acquired by Divine favour, is to be devoted above all to the glory of the Highest Numen and to the health of the human race, for it would be unworthy that those heavenly gifts should be in the service of either avarice or ambition. Third, that he has taken on the care not of some ignoble or contemptible creature; for, so that we should acknowledge the value of the human race, the Only-Begotten Son of God was made man, and thereby ennobled by his own dignity the nature he had assumed. Finally he should remember that he is not exempt from the common fate, but that he is subject and liable to those same laws of

---

[6] It may be his shortage of money which led to him making a petition (printed in Payne, *Thomas Sydenham*, pp. 72–4, and elsewhere) in 1653/4 for the arrears of salary due to his dead brothers; it is certainly curious that Sydenham claimed that these arrears were due to him as a *brother*, rather than to his father, who was still living.

mortality, the same accidents and tribulations as everyone else; how much more diligently should he try to bring help to the suffering, and with a more refined compassion (being himself in the same situation).

. . . it behoves all Physicians who want to be and to be considered honorable and prudent, to behave thus; so that, first acknowledging the Divine goodness and then also calling upon it, they may await from it wisdom and a favourable outcome of their cases; and also so that with every effort and application, they should thereto so apply themselves that they procure not only health for the sick but also some greater certitude for the Medical Art which they profess: steering their *experimenta* in this direction so that the procedure [*ratio*] of healing may daily become fuller and more reliable, and ultimately so that the human race might more securely and universally enjoy the benefit of improved medical practice even after they themselves have passed on.

Therefore, aware of my duty [*officii mei conscius*] I offer to the world this *Method of Curing Fevers, Built on my own Observations . . .*[7]

It is clear that Sydenham felt that as a practitioner of medicine he had a moral obligation to take active steps to improve medicine. He refers thus this obligation to a religious (perhaps a Christian) motive. It is also clear, from the fact that he mentions it at all (let alone at such length) that he did not think most of his fellow physicians shared this attitude. It was an unusual attitude, and Sydenham is here preaching it: the sacred obligation on physicians to do 'research' (as I shall call it) to improve medicine.

To see how this expresses a political position we need to consider the interests and concerns of the group of which Sydenham was a member and product. Sydenham and folk like him had by this date been to war with and executed their king, and then set up a form of government alternative to monarchy. For what cause had the Sydenhams and their like been fighting? We have an account of that cause from someone very close to Thomas Sydenham, his eldest brother William, who rose to be a colonel. In a letter of 1645 to a Royalist commander who was threatening an eternal vendetta on him for killing innocent people, William wrote that he had looked upon his own heart

and find written there, in the fairest characters, a true desire of advancing God's honour, maintaining the King's just power, and contending for the privileges of the Parliament at Westminster, and the liberty of the subject; which when I find you so maliciously opposing and despitefully styling treason and rebellion, I am induced to think this age hath produced unparalleled monsters.[8]

God's honour, a monarchy with limited powers, the privileges of parliament, the liberty of the subject: it would be difficult to give a

---

[7] My translation from the Latin as printed by Greenhill, *Sydenham . . . Opera Omnia*, pp. 22–4; this passage is also translated in part by Payne, *Thomas Sydenham*, p. 118, and Latham, *The Works of Thomas Sydenham*, 1, pp. 25–7.

[8] Printed in A.R. Bayley, *The Great Civil War in Dorset, 1642–1660* (Taunton, 1910), p. 245, and elsewhere.

better short account of the cause as seen from the parliamentary side. The 'subjects' whose liberty was to be protected were of course primarily the people whose representatives sat in the house of commons: the English gentlemen (gentry, lesser nobility, burgesses, the 'middling sort'), predominantly rural, exclusively male.

The Sydenhams and their like had, thus, been fighting primarily for the rights of the *free-holder*, the independent land owner who could 'live off his own'; they were not democrats. Seeing themselves as natural leaders of their society, they had been fighting for a greater share of political power for people like themselves. And this is what we should expect. But the predominant political formulation they used – the way that they conceived and experienced it – was that they were fighting to *recover* their historic rights, which had been usurped by 'tyranny', usurped that is by an unlimited, an unjust, monarchy. They came to believe that there had been an 'Ancient Constitution' before the Norman conquest, when all was peace and harmony, royal power was limited and the English gentleman's house was his castle.[9]

But most of them conceived and experienced their current plight also in *religious* terms. In the decades previous to the civil war more and more such people had come to find the attraction of extreme protestantism irresistible. Whether it was the stress on individual responsibility inherent in such protestantism; whether the strict moral and spiritual practices typical of it gave them a means of regulating the details of their own and other people's lives in their own material interest, or whether it was some other feature of it which served this function for them, must for now remain a matter of conjecture. At all events they had taken to it in their thousands. Their enemies called them 'the godly people' and they have come to be especially associated with the description 'puritan'. And because (in the words of no less than their chief enemy, the 'tyrant' Charles I) 'the people are governed by the pulpit more than the sword in time of peace', so 'no concessions will content them [i.e. the rebels, the puritans] without the change of church government, by which that necessary and ancient relation which the church hath had to the crown is taken away'.[10] Hence much of the struggle took place over religious issues and in religious terms. But every religious position was also a political one, with implications, direct and immediate, for the kingdom of this world. Thus we can see

[9] See for instance Zera S. Fink, *The Classical Republicans* (Evanston, 1945); J.G.A. Pocock, *The Ancient Constitution and the Feudal Law* (Cambridge, 1957).
[10] Charles I, in 1646; as quoted in Christopher Hill and Edmund Dell (eds.), *The Good Old Cause: The English Revolution of 1640–60, its Causes, Course and Consequences* (London, 1949), p. 173.

why the principled core of the parliamentary army was constituted by those who were at the same time both 'godly pretious men' and also freeholders and freeholders' sons who 'upon matter of conscience engaged in this quarrel'; such men were the ones that Cromwell chose to have in his own regiment.[11] They were fighting to recover a holy commonwealth; they were 'good commonwealth men'. Such were the Sydenham brothers. And their personal attitudes closely resembled those of Cromwell himself: like him they were 'Independents' in both politics and religion.[12]

This locates Sydenham with fair precision on a clear and familiar political–religious spectrum. But we are still some way from understanding why Sydenham as a member of this group who by chance was obliged to be a practising physician in the 1660s, should have thought his urgent duty lay in improving medical practice by new (and radical) investigations. Fortunately the work of Charles Webster can help us here. Webster, in an astonishing piece of reconstruction,[13] has uncovered the existence, the membership and the activities of an extensive network of such puritans in mid-seventeenth-century England. Around Samuel Hartlib, who offered himself as a general co-ordinator, there appears to have been, in the period *c.* 1626–*c.* 60, a movement to bring about the 'Great Instauration' which Francis Bacon had called for. It was carried out, according to Webster, under the prompting, and to achieve the goals of, millenarian puritanism. Such puritans felt it was their duty – urgently – to follow a text in the Book of Daniel and thus to 'run to and fro, and knowledge shall be increased'. That running, as Webster has shown, was aimed at changing the material (and spiritual) circumstances of the whole of life, and in a highly directed manner. The improvement of education, trade, husbandry, surveying and medicine were among their concerns, and Webster has documented their projects and plans for these areas fully. Webster judges this movement to have been successful in parts, and robbed of final success only by the collapse of the Commonwealth. He also claims that this pre-restoration movement was the source of many of the enterprises that we have customarily put to the credit of the restoration (in particular, the origins of the Royal Society).

The millenarianism embraced by these puritans was of quite recent vintage. To be a millenarian at all one has to believe in and welcome the

[11] Bulstrode Whitlock, as quoted by Brian Manning (ed.), *Politics, Religion and the English Civil War* (London, 1973), p. 99.
[12] On whom see George Yule, *The Independents in the English Civil War* (Cambridge, 1958).
[13] Charles Webster, *The Great Instauration: Science, Medicine and Reform, 1626–1660* (London, 1975).

inevitability of the thousand-year reign of Christ with his Saints. But whether one believes this lies in the distant future, or in the immediate future, or is already taking place, depends on circumstances. The puritans whose activities Webster studies believed that the millenium was imminent and – most important – that they had to work urgently to prepare the earth to be a new Garden of Eden. Why should they have felt this need to work – to run to and fro – why with this urgency, and why promoting these particular things? We can, I think, expect that the form of any future earthly paradise would coincide nicely with the interests (in every meaning of the word) of those proclaiming it: it would reflect their present interests, and its implementation would represent a furthering of their interests too. No one promotes a millenium which has no place for people like themselves, nor one which would be an expression of all the values they despise. Thus the social position, power and aspirations of any group of millenarians should be detectable from their characterization of the millennium.

The millennial vision of the puritans that Charles Webster studies seemed to call on them to be active and urgent in improving the material conditions of the world. In other words (and these are now my words) they created and held up to themselves this particular characterization of the millennium in order to encourage themselves, and everyone else, to work in the interests – the material and power interests – of people just like themselves. The rewards would not have to wait for the hereafter; indeed, if everyone buckled down to it, the rewards were just around the corner. Hence (I suggest) the urgency, the call to be making improvements in whatever one's sphere of activity might be. This is a millennium fit for those who in *this* world are nearly in power. We need (they claimed) to recreate the Garden of Eden, where Nature obviously was for man's benefit. It should take little imagination on our part to guess who were to be the saints ruling at Christ's side when the millennium came: the saints would be the gentry come into their own. *That* is how the millennial puritan movement charted by Webster is a political position: its realization would have changed the world in the interests of those advocating it, changed it into a fulfilment and expression of those interests. Thus it can be little surprise that the people who were millenarian puritans in their religion at this time were so often 'commonwealth men' in their politics. This is why Sydenham's authentically religious call (in the preface of 1665) to other physicians to treat their calling as a sacred duty, a duty to seek out new ways of improving medicine, was also a political call.

Such people, the advocates and entrepreneurs of the millennium, are the very people Thomas Sydenham had been in contact with: it was one of them who had introduced him to and inspired him to take up medicine – Thomas Coxe. Coxe was a friend of Samuel Hartlib. The class that these people represented and came from was the same class that Sydenham himself came from.

But all that happened in the 1640s and 1650s. In 1660 that world suddenly came to an end, as the old order and its symbol (monarchy) were restored.[14] But Sydenham never gave up his political ideals. Instead, once resigned to the full-time practice of medicine, he set out to exemplify the principles of the now-defunct Commonwealth in his medical practice. His medicine was now to be the focus and expression of his politics. And hence the fact of the restoration accounts for Sydenham's silence about his engagement in the war on the side of parliament and about his other political activities since. It was political circumstances (not personal choice nor, as we might assume, some self-evident separateness between the enterprises of medicine and politics) which led Sydenham to write about his medicine as though it had no political content.

In the period of Sydenham's practice of medicine, (c.1656–88), his political position (and hence his medicine) was to be that of the 'Good Old Cause'. This term expresses (and was coined to express) all the things that people like Sydenham had been fighting for – but in defeat. The term came into use in the late 1650s as Cromwell, with his growing ambitions to set up a family dynasty, seemed to be abandoning the last principles for which the pious gentry of the parliamentary army had fought. The cause was 'good' because it was the defence of freedom against tyranny, it was 'old' both because it was (supposedly) pre-Norman and because it had inspired the parliamentarians in the civil war, it was a 'cause' because it still needed to be fought for. The 'Good Old Cause' was the cause of those who had not compromised and who refused to compromise with 'tyranny'. From the 1690s it was to be the rallying cry of those who wanted to keep the 'whig' interest and party true to its roots. To speak of the Good Old Cause was to try and keep alive a particular political position and ambition. Other terms were sometimes used of adherents of this position, such as 'commonwealth men' and (later) 'radical whigs'. The position was indeed radical; but it was also backward looking, and adherents of it could look like living fossils who were unaware of the realities of the new age of restored

---

[14] For a good account of the changes that the restoration brought about, and their effect on the world of learning, see Steven Shapin and Simon Schaffer, *Leviathan and the Air-Pump: Hobbes, Boyle and the Experimental Life* (Princeton, 1985), ch. 7.

monarchy. Sydenham was an adherent of this position. If (as I claim) he now turned his energies from overt public politics to a politically-informed kind of medicine, then that medicine too would represent a view which was both radical and out of date: it would be the medicine, in the 1660s, 70s and 80s, of someone who still lived mentally in and with the aspirations of the 1640s and early 1650s.

MEDICAL PRACTICE: WHAT?

Let us now turn to that medicine and try and trace the way in which Thomas Sydenham conceived and built his new approach to curing diseases. Our question here is: *what* precisely was he studying, and why?

First let us inspect Sydenham's own accounts of these important events. Here he is writing in 1676:

After a few years spent in the academic arena [Oxford], having returned to London I came to Medical Practice; which when I inspected it with a very intent eye and carefully using all diligence, I soon came to that opinion (which has grown with me right up to this very day) that this Art is to be in no way more properly learnt than by the exercise and use of the Art itself; and that it is definitely the case that he who has turned his eyes and mind most accurately and diligently to the natural phenomena of diseases, will be best at eliciting true and genuine curative indications. Therefore I gave myself over wholly to this method, confident enough that if I followed Nature as my leader, even

Wandering through desert places of the earth, trod before
By no man's foot

I would never depart a finger's breadth from the correct path. Guiding myself by this thread, I first applied my mind to a closer observation of fevers, and after much weariness and the most troublesome agitations of mind by which I let myself for some years be fatigued, I at last fell on the method by which fevers could be cured; and this, under pressure from my friends, I long ago [i.e. in 1666] allowed to be published.

He had described that process in print in 1666 slightly differently. He had claimed then (as we will recall) that the physician had a duty to get greater certainty for the medical art,

steering their *experimenta* in this direction so that the procedure [*ratio*] of healing may daily become fuller and more reliable . . . Therefore, aware of my duty, I offer to the world this *Method of Curing Fevers, Built on my own Observations* . . .

Though this earlier account is much thinner (and more pious in presentation), it has additional information for our story; if we put the two together we can produce the following account of the sequence of events:

[1676:] 1. He comes to London and sets up in medical practice (in about 1656).

[1666:] 1a. He believes that the physician has a (religious) duty to do 're-search' to improve the art.

[1677:] 2. From the experience of practice itself (but not before) he now realises that only (i) accurate *observation of the natural phenomena of diseases*, giving the (ii) *true and genuine curative indications*, will enable one to do this.

[1666:] 2a. The route is by performing *experimenta*.

[1666:] 2b. The work should be built on *one's own* observations.

[1676:] 3. He turns to the study of *fevers*, and after a number of years of very hard work he strikes on the *method* by which fevers can be cured.

We have already dealt with 1 and 1a, and located them in his life and experiences as a young man. But why did Sydenham believe that accurate observation and consideration of the true and genuine curative indications was the necessary route to improving the practice of medicine? What does he mean by *experimenta*? Why should one's work be based on *one's own* observations? And, first of all, why in particular does he turn to the study of *fevers*?

He studied *epidemic fevers* because of the nature of the early medical practice open to him. In the years between about 1656 and 1665 when he wrote the first version of the book about the new method of cure, Sydenham lived in Westminster and then in Pall Mall in London. He was one of the first residents of Pall Mall; his neighbours were to include members of the aristocracy such as Katherine Lady Ranelagh, and fashionable people like Mary Beale the painter and Nell Gwyn the actress. Sydenham was in full-time medical practice now, and some of his practice was amongst the reasonably wealthy. He was for instance sometimes called to offer a second medical opinion for certain aristocratic families. But there were really only two such families in his life, and both of them were (at least sometimes) of similar political leanings as himself. One was that of Anthony Ashley Cooper (from 1672 the Earl of Shaftesbury) who had been a republican and was still if necessary prepared to oppose the royal prerogative; and the other was that of the Percys who supported the 'Good Old Cause' and whose famous scion was Algernon Sidney (who had also been a friend and colleague of Sydenham's elder brother on the Cromwellian Council of State). And even to these Sydenham had access to only through his own friends: Cooper's physician from 1668 was Sydenham's close friend John Locke; the physician of the Percy family was John Mapletoft, another close friend.

Such work thus does not seem to have made up even a significant fraction of Sydenham's medical practice.[15] Even as late as 1687 he was aware that the – presumably political – 'scandal of my person' was still

---

[15] Dewhurst, *Dr Thomas Sydenham*, pp. 49–50, has a different opinion of the extent of Sydenham's practice amongst the well-to-do.

a problem for many people. The consequence of this seems to have been that Sydenham had to find most of his medical practice amongst the *poor*, those the conventional physician did not usually treat (except as an occasional favour). In a letter in 1677, for instance, advising his fellow physician Mapletoft about the treatment of the Countess of Northumberland (Algernon Sidney's mother), he writes: 'were she one of *those poor people whom my lot engages me to attend* (for I cure not the rich till my being in the grave makes me an Authority) I would take the following course . . .'[16]

Thus Sydenham based his practice amongst the poor, and it may well be that it was force of political circumstance which obliged him to do so. It takes the fees of many poor people to match the fees obtainable from the rich. Thus in order to gain a reasonable living, Sydenham would have seen probably far more patients than most of his fellow physicians. He already believed that it was the duty of physicians to work to improve medicine, in whatever sphere of practice they found themselves. The poor *en masse* was the particular realm of practice in which Sydenham perforce chose his first topic of 'research'. This topic was highly unusual, to say the least, and probably unique to Sydenham. For his first and primary object of investigation – the 'what' of our question – was *epidemic fevers*: 'epidemic' means '(a disease falling) on the people'. Without an extensive practice amongst the poor such a topic would have been impossible and indeed unthinkable. His experience of practice (see item no. 2 on p. 175 above) was thus the experience of seeing *many* patients and, at times of epidemics, of seeing many patients suffering from the *same* disease.

It is important that we appreciate the radical novelty of this position. It is not just that Sydenham was studying the diseases of greatest virulence, the diseases with the highest mortality, the diseases which are apparently the least curable; it is also that he was studying them as affecting the people that the physician did not conventionally treat – people in the mass. Now, although many conventional physicians were of course concerned with fevers and their cure, yet their approach to fever was locked in the one-patient-one-doctor pattern. They certainly treated a whole range of fevers if their (rich) patients showed the symptoms of them: but always they treated them *as the fevers of individual patients*. It was not the nature of the *fever* which was their starting-point, but the predisposition of their *patient* to suffer from it, and the particular capacity of their *patient* to withstand and overcome

---

[16] My emphasis. Cited from Dewhurst, *Dr Thomas Sydenham*, p. 170.

it. For the university-educated physician (whether by allegiance a 'Galenist', a Cartesian mechanist or anything else), the curability of a fever depended on the physician understanding *how the human body works*, and how *drugs work* in the human body; and to treat any case of fever (or anything else) successfully, the physician needed to know the constitution of the *individual* patient affected by the fever. For the conventional physician, called to his patient, the fever was what the patient *complained of* – it was the disease itself, or more properly, it was the consequences of the disease itself. And even if, in the course of an epidemic, a physician should see several cases of the same fever, they were to him *different* instances because they were in different, individual, patients. Within this way of seeing patients and conceptualizing fever, there was no room for a long-term programme of 'research', for every case urgently demanded cure, and every case was by definition different from any other case. There was neither room nor occasion for *experimenta*; nor was there any particular need for *new* observations; and given the large numbers of practical medical books with authoritative guidance in them that one could consult, there was specially no need for *one's own* observations! The *theory* one had been brought up on told one all one needed to know. Finally, because it was assumed that the fever was the consequence of the disease itself, so the adherents of each medical approach had come up with explanations of what prior events in the human body constituted *the cause* of the fever, whose rectification was necessary for a successful cure (let it be an imbalance of the humours, or a blockage of the tiny pores in the glands).

If we now visualize Thomas Sydenham engaged in earning a living as a physician amongst the poor, we can see that he would have had much more freedom of action than he would have had as a rich man's physician. For, if his cure failed on the poor, then his reputation would not be at such risk as if it failed on the rich. Thus some at least of the *urgency* of treatment would be gone – and hence the need to stick to conventional formulas of treatment. If we add to this what we know of Sydenham's resolve to *improve* medical practice – encapsulated in his reported statement that when he started to practise he resolved 'to act directly contrary in all cases to the common method then in fashion among the most eminent physicians' – then we know that he was also *giving himself* a certain freedom of action and vision. We can thus appreciate that Sydenham was making it possible for himself to see something different from what his fellow physicians were seeing.

Now, the first such thing which Sydenham could see from his

unusual position was that most of the poor who fell ill were not treated by physicians; yet most of them survived their illnesses, they even survived epidemic fevers. Hence with smallpox, for instance, a disease conventional physicians thought of as one of the most dangerous, he came to believe 'that if no mischief be done either by physician or nurse, it is the most slight and safe of all . . . diseases'.[17] The body left to itself cures itself: 'Nature by herself determines diseases, and is of herself sufficient in all things against all of them',[18] as he put it, quoting Hippocrates. The second thing he was able to see was that in the course of an epidemic, everyone who suffers is suffering from the *same* disease. In his own words: 'the order of Nature in producing diseases is so equable and everywhere so self-consistent, that in different (human) bodies, the same symptoms of a given disease may be usually found; and the very symptoms which were observed in the ill Socrates can generally be applied to any person whatever suffering from the same disease'.[19] Hence he could appreciate that each epidemic fever has its own set of manifestations, its own course, it own period, its own form of crisis. Because of his unique position, Sydenham's attention was on the fever, not on the body-of-the-suffering-patient and its workings: his attention was not on how the body 'fights' the disease – he was not seeing diseases 'as only the confused and disordered effects of Nature ill defending herself and thrown down from her customary state'[20] (as he characterized the common view). Finally, putting two and two together, he could see that since most people recover without the doctor's intervention, then the fever must itself be the *process of cure* – rather than being (the consequences of) the disease itself. 'A disease . . . is nothing else than the effort of Nature endeavouring with might and main the expulsion of the morbific matter for the health of the sick person.' [*Morbum . . . nihil esse aliud quam Naturae conamen, materiae morbificae exterminationem in aegri salutem omni ope molientis'.*][21]

These are the grand novelties of Sydenham's view: he looks at

---

[17] Letter Sydenham to Boyle, 1668, as printed in *ibid.*, p. 163. In a draft essay on smallpox (1669) he also wrote, 'Here by the way, from what hath been now said, it will appear how easy it is to solve that common doubt how it comes to pass that in the smallpox so few die amongst the common people, in comparison of the rich, which cannot be thought referable to any other cause, than that they are deprived through the narrowness of their fortunes and their rude way of living of the opportunities of hurting themselves with a more precise and tender keeping' (*Ibid.*, p. 116).

[18] Latham's translation, (*The Works of Thomas Sydenham*, p. 17) from the Greek of Hippocrates, as cited by Sydenham.

[19] My translation from the Latin in Greenhill, *Sydenham . . . Opera Omnia*, p. 13.

[20] My translation from the Latin *ibid.*, p. 13.

[21] My translation from the Latin *ibid.*, p. 26.

epidemic fevers while they fall on the demos; he sees that the unattended patient usually recovers; he sees the fever as the process of cure – as the solution, not the problem. He now set himself the task of improving nature's success rate, and of finding a new role for the physician – not interfering, but constructively helping nature, 'joining hands with Nature'.[22]

To return for a moment to Sydenham's choice of epidemic fever as a topic to investigate. Rather than locating it as a chance consequence of his enforced way of earning a living, should we instead be crediting the Honourable and ubiquitous Robert Boyle with first suggesting it to Sydenham? For Sydenham not only dedicated the first and second edition of his book to Boyle (and appears to have had Boyle's blessing for doing so),[23] reprinting this dedication again in the third edition, but in this dedication he also credited Boyle with having steered his attentions to this *provincia*.

> I confess that I have thus offered this Treatise to Your patronage, most especially because in the same way I took up this subject [*hanc provinciam suscepi*] by Your persuasion and instigation, so also You Yourself, the most sufficient witness, by [Your own] experience on several occasions have seen proven the truth and efficacy of the matters which are here related, when You several times so kindly condescended (a mark of Your outstanding goodness to others) to Yourself join me as a companion in visiting my patients.
> . . . Besides Fever (here to be considered), there are very many other diseases about which – since the core of their practice appears less satisfactory, and to be supported on a method contrary to reason (at least to mine) – I ought to have written as well, to have fulfilled the trust given by You to me.

But what does *provincia* refer to here? Given that Sydenham says that 'to have fulfilled the trust' given to him by Boyle he should have worked on many other diseases too, it would seem that it refers not to 'fevers' but more generally to the search for new cures for diseases.

Boyle's persistent concern with the improvement of medicine is well known. Not only was he always experimenting in dosing himself, but as early as 1649 he wrote a piece (to be published under Hartlib's auspices) on the need for people to make their secret remedies public. Again, in 1650 he is reported to have said that 'physicians hitherto had achieved better skill to know and discern diseases than to cure them', and that he wanted to improve this situation.[24] Equally well known is Boyle's life-long concern with the pursuit of 'useful knowledge' and its pertinence to the welfare of all mankind. Yet Boyle's own hopes for the

---

[22] My translation from the Latin *ibid.*, p. 18.
[23] Letter Sydenham to Boyle, 1668; printed in Dewhurst, *Dr Thomas Sydenham*, pp. 162–4.
[24] Marie Boas, *Robert Boyle and Seventeenth-Century Chemistry* (Cambridge, 1958), p. 18.

improvement of medicine lay down the chemical path; and this Sydenham most certainly was not interested in. So the debt Sydenham owed Boyle was an inspiration not from Boyle's own work, but from his general approach to natural philosophy. As we shall now see, the inspiration of Boyle is indeed the key to just how Sydenham went about his mission to improve medicine.

MEDICAL PRACTICE: HOW?

If we are right in assuming that Boyle was intellectually the senior, then Sydenham's debt to Boyle is everywhere apparent in his work. We have seen that epidemic fevers was Sydenham's primary choice of a topic for 'research'. In the pursuit of this he had of course to settle on some more precise objects of study – that is, certain particular things he was looking at and for. He also had to adopt certain ways of looking. It is for the specification of his initial set of these that he was indebted to Boyle. They constitute the core of the 'how' of Sydenham's clinical medicine. They are items 2, 2a and 2b as we numbered them above:

2. ...that only (i) accurate *observation of natural phenomena of diseases*... will enable one to improve medicine.

2a. The route is by performing *experimenta*.

2b. The work should be built on *one's own* observations.

These matters are of such consequence for our understanding of Sydenham's medical work, that it is important to establish that Sydenham knew and could have met Boyle often, that their relationship was (at least for a period) more than that just of an author seeking a non-controversial dedicatee. In fact, their paths could hardly have been closer. First, the man who introduced Sydenham to medicine, Thomas Coxe, was Boyle's physician; like Boyle himself, Coxe was part of Hartlib's retinue. Second, the house in which Boyle lived, experimented and wrote in Oxford (1654–)was opposite All Souls College, where Sydenham lived until 1656. Third, in London Sydenham lived next door to Boyle's sister, the one Boyle often visited and with whom he was eventually to live. And, if these did not provide enough opportunities for them to meet, there was always Dorset, where Sydenham's family lived, and where Boyle had his estate. Finally, their brothers (Lord Broghill and William Sydenham) not only knew each other and had similar political outlooks, but they also worked together for a while in the Protectorate government. Even their enemies knew that Boyle and Sydenham shared an approach in common. As Henry Stubbe wrote, in attacking Boyle:

I know what any physician may, as the mode is, tell you to your face, but except it be such as Dr Sydenham and young Coxe, I believe not one lives that doth not condemn your experimental philosophy.[25]

It has recently been shown that from his earliest investigations Boyle was concerned with (i) the doing of experiments, and with arguing for their crucial and indispensable role in natural philosophy; and (ii) with establishing certain things as 'matters of fact' – things on which one could gain general assent – and with arguing that discussion in natural philosophy ought properly to be limited to such 'matters of fact', and that there should be close restrictions on giving physical explanations to 'matters of fact'. The authors of this view of Boyle are Steve Shapin and Simon Schaffer, and rather than trying to summarize their recent book,[26] let me just make a little patchwork for my own purposes from some of their salient sentences about the practices of Robert Boyle. They are talking about work and writing Boyle was carrying out in his Oxford lodgings in the late 1650s, concerned with the air-pump:

Boyle proposed that matters of fact be established by the aggregation of individuals' *beliefs* . . . Matters of fact were the outcome of the process of having an empirical experience, warranting it to oneself, and assuring others that grounds for their belief were adequate. (p. 25)

His [Boyle's] overarching concern was to protect the matter of fact by separating it from various items of causal knowledge, and he repeatedly urged caution in moving from experimental matters of fact to their physical explanation. (p. 49)

There was a strong boundary placed between speech about the spring [of the air] as an explanation about matters of fact and speech about explanations of spring. Thus, in the first of the *New Experiments* Boyle claimed that his 'business [was] not . . . to assign the adequate cause of the spring of the air, but only to manifest that the air hath a spring, and to relate to some of its effects'. (p. 51)

Boyle's desire to revamp the defining criteria of natural philosophy may have derived from his eirenical inclinations, which have been pointed out recently by Jim Jacob.[27] For both in religion and in natural philosophy Boyle was looking constantly for the common ground of opinion that men shared, so that agreement could be built upon it in the interests of social stability and peace. In his life-long pursuit of this goal, the civil war gave Boyle early doubt about the possibility of reaching such common ground through the traditional route, the use of reason. Experiment is what Boyle found to stand in place of it. Quite why experiment was an alternative candidate (and why it became such a popular route to certainty among some people at this time) is still an

---

[25] 1670; as quoted by Dewhurst, *Dr Thomas Sydenham*, p. 63. I assume that 'young Coxe' is the son of Thomas Coxe.   [26] See note 14 above.

[27] J.R. Jacob, *Robert Boyle and the English Revolution* (New York, 1977).

open question. But experiment is what Boyle chose. Shapin and Schaffer have shown how he tried to claim that only 'matters of fact' brought into existence and visibility by the artifice of experiment, were to count: thus for Boyle it was primarily experiment which *created* 'matters of fact'. And such 'matters of fact' of course, consisted of 'natural phenomena' (to use a phrase of Sydenham's); and they had to be objects of perceptual experience.[28] Establishing what the natural phenomena produced by experiment are, is a matter of (i) seeing for oneself (personal experience); and (ii) having confirmation of what one sees by the complementary experience of witnesses. As Boyle himself warned, people should not believe even (chemical) experiments 'unless he, that delivers that, mentions his doing it upon his own particular knowledge, or upon the relation of some credible person, avowing it upon his own experience'.[29]

We can see here Sydenham's indebtedness to Boyle for vocabulary and concepts. Indeed Sydenham's dedication of his book to Boyle is itself couched in just this 'boyleian' language: experience, tested, sufficient witness. He even claims Boyle himself as his primary witness for the truth of what he is saying: of, that is, the accuracy of his own observation of the natural phenomena. And 'matter of fact' keeps appearing in Sydenham's published writings and in his letters; it is a phrase he constantly uses. As he wrote, for instance, to his friend Mapletoft in 1687, looking back over his life's work:

I can only say this, that as I have been very careful to write nothing but what was the product of faithful observation, so when the scandal of my person shall be laid aside and I in my grave, it will appear that I neither suffered myself to be deceived by indulging to idle speculations, nor have deceived others by obtruding anything upon them but downright matter of fact.[30]

Experiment was (as Boyle was to find) suspect amongst the natural philosophers. Amongst the physicians, however, it had an even lower reputation. For experiment was exactly what *empirics* did and relied on. Authentic (university-educated) physicians rely on reason–sense–experience, a quite different handful. As Gideon Harvey put it in an attack on Sydenham and others: 'it's necessary we should make a Transit to the reason of the thing, without resting satisfied, like Empiricks, in experiment only'.[31] When Sydenham took up and advocated experiment in medical practice, he was thus letting himself be

[28] Shapin and Schaffer, *Leviathan and the Air-Pump*, p. 36.
[29] *Sceptical Chymist*, quoted by Shapin and Schaffer *Leviathan and the Air-Pump*, p. 58.
[30] Printed in Dewhurst, *Dr Thomas Sydenham*, p. 174.
[31] Gideon Harvey, *The Conclave of Physicians, detecting their Intrigues, Frauds, and Plots against their Patients* (London, 1683), p. 149.

allied with the empirics, the enemies to good learning and good medicine. It would seem that it was Boyle who first convinced Sydenham of the proper role of experiment. And after the restoration, Sydenham may well have been pleased to continue to look like an empiric in contrast to those members of the College of Physicians who had (in his eyes) sold out politically.

The final area in which we may here point out the debt Sydenham owed to Boyle lies in the *air*. In all the experiments with his pump Boyle's object of study was the air (the very thing he was trying to exclude from his engine) and its properties. The air – rather than the human body – is where Sydenham was to locate the source of fevers; in changes in the air he was to find the origin of changes in the nature and kind of prevailing fever.

It was such beliefs about knowledge and experimental practice, absorbed from Boyle, that prompted Sydenham to resort to the clinic. Or, as he put it to the young Hans Sloane, 'You must go the bedside, it is there alone you can learn disease.' Only at the bedside can you make 'accurate observation of the natural phenomena of diseases' – one's own observations, confirmed (if possible) by reliable witnesses such as Boyle or Locke. We can now see why Sydenham would have described his method, on the title-page of his book, as being 'built on my own observations', and we can also now recognize the force of the title he chose for the revised edition: 'Medical Observations'. In a moment we shall turn to the nature of the *experiments* Sydenham performed at the bedside. Thus shall we see how he acquired the 'true and genuine curative indications' that he needed for the proper improvement of medicine.

For his first set of intellectual tools Sydenham was thus indebted to Boyle. But Sydenham parted company intellectually with Boyle after the restoration, for now their political aims were at odds. Boyle personally led the campaign to make experimental natural philosophy serve the new political arrangements. As Shapin and Schaffer put it:

producers of knowledge could find a place in Restoration society if they could supply weapons for the weakened churchmen. Boyle reckoned that experimental philosophy did provide such weapons. During the 1660s he instructed others on the way these weapons could be made and the right means of using them.[32]

Sydenham, however, was made of sterner political stuff. So much so that he was to be open to an account which dispensed altogether with 'physiological' explanations. He was to find it in the companionship of

---

[32] Shapin and Schaffer, *Leviathan and the Air-Pump*, p. 298.

discussion, thought and writing with John Locke as Locke moved politically nearer and nearer to supporting the Good Old Cause over the years.[33] It is through Locke's intellectual collaboration (and political sympathizing) that by 1676 Sydenham had come to the view that in medicine there can be no useful discussion at all about cause – and hence about the inner working of the human body. Cause was simply not the physician's concern. This is the most radical difference between Sydenham's books of 1666 and 1676 (the first and 'third' editions): the first is a product of his relationship with the politically moderate Boyle, the other of his encounter with the politically more radical Locke.

HISTORY AND METHOD

How did Thomas Sydenham actually perform his *experimenta* at the bedside – what did they look like? – and thus reach the *true and genuine curative indications* that he so earnestly sought? What was Sydenham's *method* (as he himself termed it) of discovering cures for diseases?

We will recall that Sydenham's practice was, perforce, primarily amongst the poor. In this circumstance he was able uninterruptedly to build what he regarded as the first requirement: *a history* of the disease. This is an account of the course of the disease (without intervention), of the symptoms it shows in the body, of the day and nature of its crisis, of its natural mortality (or recovery rate). None drawn up by Sydenham himself appear to survive, but here is one drawn up by his close friend John Locke in his role as physician and, curiously enough, concerning an illness of Sydenham's own son:

Measles. March 7th. 1670.
W. Sydenham, a boy of 11 years old, with only a delicate constitution by nature, rather weak lungs and very prone to cough, was seized by a shivering and rigor, followed by a slight praeternatural heat. Defluxion of the nose, drowziness, cough, anorexia.
4th day. The rigor and shivering and all these symptoms increased daily to the fourth day, the tongue on the fourth day also white and dry, bowels natural but the fever increased, and breathing difficult around the first two days, more frequent than usual, and vomiting.
5th day. In the evening there appeared on the forehead and cheeks some small red spots like flea bites, and all the other symptoms increased, especially the drowziness; the defluxion ceased . . .[34]

[33] On Locke's political opinions see, for instance, John Dunn, *Locke* (Past Masters Series) (Oxford, 1984); on Locke's relations with Sydenham see Patrick Romanell, *John Locke and Medicine: A New Key to Locke* (New York, 1984).

[34] My translation from the Latin as given by E.T. Withington, 'John Locke's "Observationes Medicae"', *Janus* IV, (1899), 579–81. For other individual case histories à la Sydenham (and again by a follower of Sydenham), see those printed by Andrew Brown at the end of his *De Febribus Continuis Tentamen Theoretico-Practicum seu nova febrium hypothesis mechanica audacta ex principiis Bellini constructa* (Edinburgh, 1695).

And so on, day by day, to the fourteenth day, when everything was much better. Such things are what you list. They are all *observable* phenomena, and (supposedly) *only* observable phenomena. Locke's journal has many such cases in it. They remind us, of course, of Hippocrates, of the so-called 'case-histories' of the books of *Epidemics*. That Sydenham (and thus Locke) took such histories may well be due to a desire to follow a direction for the improvement of medicine pointed out by the promoter of the Great Instauration himself, Francis Bacon:

The first [deficiency of current medicine] is the discontinuance of the ancient and serious diligence of Hippocrates, which used to set down a narrative of the special cases of his patients, and how they proceeded, and how they were judged by recovery or death . . . This continuance of medicinal history I find deficient.[35]

To go back to Hippocrates and to emulate his diligence was of course, for both Bacon and Sydenham, to go back beyond Galen and the medicine of the Schools.

But taking such histories of *individual* cases is not all you do, if you are Thomas Sydenham. For these are simply the raw material for making an *induction*, and thus reaching an account (a history) of that disease shorn of any of the restrictions, limitations, or special conditions that it shows in its occurrence in each individual it affects. You end up, that is, with a *general account* of the disease: that is Sydenham's own usage of the term 'history', and the 'histories' of diseases he published in his books are like this. After giving such a history of smallpox, for instance, he writes:

This is the natural history of the smallpox as comprehending the true and genuine phenomena belonging to them [i.e. the pocks] as they are in their own nature; but other anomalous accidents there are which attend the disease when unduly managed.[36]

The history deals with the disease as it is in its own nature, not in its particular manifestations. You do not explore any supposed 'causes' of the disease (there is no room for such speculation in a 'history'); you do not assume you know what has gone wrong in the body itself. You might perhaps (as Sydenham came to do) chart the seasons and weather onto your history, but that is about all. This, then is how you reach a proper account of a particular disease; and the history is itself your evidence that there exist *constant* diseases. It is also your way of distinguishing between diseases, as they are distinguished *in nature itself*.

Now, what about the cure? Just as the 'history' of the disease that

---

[35] Francis Bacon, *The Advancement of Learning* (1605), book 2, section X. 4.
[36] Dewhurst, *Dr Thomas Sydenham*, pp. 111–12.

Sydenham sought was a general history of that disease, so the cure too had to be a *general* cure. As he wrote in a 1669 (manuscript) revision of his book:

About the following observations I desire to premise these things – First, that these observations are not, as others commonly published, the history of particular cases, but (together with the nature and phenomena of several species of diseases) established practices and methods of curing, collected from a careful observation of a great number of instances in each disease.[37]

Here, in the finding of cures, we reach Sydenham's deployment of his *experimenta*. The 'history' of a disease has laid bare the 'true and genuine phenomena belonging to them as they are in their own nature'; the 'experiment' will make visible 'the true and genuine curative indications'. To Sydenham an experiment at the bedside consisted of nothing more sophisticated than *trial and error* in the administration of drugs and of medical techniques such as bleeding or purging. If a treatment seemed to be working, it was continued with, if not it was dropped. *Experimentum* thus meant making up, or working out, a sequence of cure as one went along! We can conveniently follow Sydenham's working in his own words, taken from a defensive account he gave (in English) of how he found the cure for smallpox:

it was not a forward affectation of novelty and opposition that made me run headlong upon this condemned course . . . but finding that notwithstanding the most diligent attendance and careful use of the choisest cordials many, of not only mine, but other physicians' patients miscarried in the smallpox . . . I first therefore bethought myself whether the sweats usually prescribed and prosecuted in the beginning might not be prejudicial in a disease the discharge whereof Nature had designed in another way, by pustules and little imposthumes. The first thing, therefore, that I decided on towards the curing of this disease was the cutting off and preventing those sweats, which attempts had no ill success but yet was not enough wholly to secure my patient. The next step therefore was to abate in great measure the plenty of cordials that used to be given in the smallpox, which I did but by slow degrees . . . my daily experience as I forebore their use gave me new encouragement the more to forebear them and that I found that those people always underwent this disease with least sickness and danger who took the least of them, and were most removed from hot-keeping either by clothes on their body or fire in their room. But finding that in the heat of summer in people of high and sanguine constitutions and a flox pox, bare abstaining from heat and hot medicines was not enough to preserve my patient . . . I had at last ventured to make them rise [out of bed] . . . But to perfect the cure there was one thing more yet required . . . actually to cool them not only by rising and wearing very thin clothes, but by giving them whey, which for several days together was the only physic or food they took. And by these slow steps has it been that after long and pensive thoughts about the smallpox I have in many years been able to perfect the cure of this disease . . .

If it were necessary to raise theories, and from thence draw probable arguments to prove that to be likely, which is already verified in matter of fact, and experience has justified to be true, the method I make use of in curing this disease is as capable of a fair

---

[37] As printed in Latham, *The Works Dr Thomas Sydenham*, I, lii.

defence and may be made out from as rational grounds as the cure of any disease whatsoever.[38]

Nothing could be more home-made than Sydenham's experiments; nothing could owe less to a millenium and a half of the authoritative and learned (Galenic) medical tradition, than Sydenham's method of improving medical practice. Here is how in the 1680s an enemy – quite accurately – described Sydenham's way of finding a cure. Amongst the London doctors Gideon Harvey is critizing there is

particularly the *Doctor of Contraries*, who with *Opium* and *Jesuits-Powder* shall make more various sorts of passes at Diseases, than even any *Roman* Gladiator with his *Weapon*; and these shall be hits, and do execution. As for instance, if the doctor is applyed unto, for his assistance against a continual *Feaver*, according to his last good or ill success in the like case, gives his direction for bleeding, or omits it; then with an unparallel'd assurance makes at the Distemper with an ample Dose of the *Jesuits-Powder*, pursuing this fierce on-set with a fresh supply of the same Bark every fourth hour; And finding the fiery adversary provoked, produces his other Champion (*Opium*) to encounter him, so between these two *Bravo's* frail Nature doth too oft lie down, and yield, and the Patient is brought to his *ultimum vale*.

The *Small Pox*, (a Distemper so unaccountable to most Physicians, and therefore Empirically treated, whence Nurses do equally vie with their Worships [i.e. with members of the College of Physicians of London] in the Cure) is by this *Generalissimo* (contrary to all sence and experience) countermined with Spirit of *Vitriol* and *Opium*, by which beyond all others he is infallible in procuring an *Euthanasia*. Good God, how the Universities do rob the Plough![39]

It was this way – 'according to his last good or ill success in the like case' – that Sydenham arrived at his unique treatments, such as 'the Bark and *Opium* in continual *Feavers*, and . . . *Spirit of Vitriol* and *Laudanum liquidum* in the *Small Pox*', as Harvey recorded with disgust.

What most outraged Gideon Harvey (and, we may assume, other of Sydenham's contemporaries) was the fact that Sydenham thought that this 'experimental' procedure qualified for the title of a 'method of healing' (*methodus medendi*): it was no accident that Galen's greatest book and Sydenham's book of 1666 both had the title of 'Method'. In Gideon Harvey's eyes, the last thing a proper method of healing was, was experimental! This is how he described a proper 'method of healing', for the edification of people like Sydenham. Harvey's usage of the term here is perfectly conventional, and perfectly Galenic:

---

[38] As printed in Dewhurst, *Dr Thomas Sydenham*, pp. 105–6.

[39] Harvey, *The Conclave of Physicians*, pp. 81–2. Harvey had a grudge against all the physicians of the London College of Physicians, and is here attacking Sydenham as just another member of the College. The language of this passage is sarcastic and full of political allusions; in calling Sydenham a 'Generalissmo', for instance, Harvey is probably making reference to the rule of the Major-Generals in the 1650s. (I am indebted to Julian Martin for pointing this out to me.)

Concerning the Description of the true Method of Physick, whereby a Physician is distinguish'd from a Quack:
. . . The Method of Physick is a rational order of Remedies to be applyed to the body of man, to cure Internal and External Diseases; which consists, first in the knowledge of the Distemper, and cause thereof, the Election or Choice of proper sufficient Remedies, and then reasoning within himself, which Remedy ought to be applyed first, when, and how, which Second, and Third, &c. How long each is to be continued, and when the First, Second, or Third is to be changed. So that here you see the *Method of Physick* implyes naturally two particulars; First, the knowing of the Distemper and cause; Secondly, adequate Remedies, that shall remove those Distempers, and their causes . . . 3dly, There is necessary (according to what my description of *Methodus* doth express) a Ratiocination in the Physician within himself, the result whereof makes a practical *Indication*, or conclusion: Thus, This Distemper is hot, therefore the Remedy must be cooling; it is caused by a Plethory, therefore Bleeding is the Remedy; or a Cacochymy, therefore purging is to be advised. So that the reasoning part is nothing else, but the apprehending rightly the nature of the Disease, and collecting thence, what Remedy it naturally doth point at, shew, or Indicate . . . 4ly, As then the *Methodus Medendi* doth imply the knowledge of all Diseases, Causes, Symptoms, and what parts they are inherent in, and likewise the knowledge and experience of all Remedies, so it doth consequently comprehend the sum, and substance, or total of the whole Art of Physick, and is the end and the ultimate point, whereto all the parts of Physick do tend, and therein concenter; so that only he, that is Master and Doctor in the Method of Physick, doth alone deserve the Name and Title of a Real Physician; all others that pretend to Physick, being Empiricks, Quacks, or Mountebancks.[40]

Sydenham's *methodus medendi* failed this test in every particular. He intended it to. For he was seeking to redefine what a medical method ought to be: it ought to be 'built on one's own observations', it ought to be a way of discovering *new* cures. Its 'history' revealed to the physician how a given disease is *in its nature*; given this, 'experiment' will show him its 'true and genuine curative indications'. This is to walk hand in hand with nature.

'AFTER MY DEATH, WHEN I AM BECOME AN AUTHORITY'

Little if any of the above is part of our familiar story about Sydenham. Our view of Sydenham is of an *empirical* practitioner (rather than an 'empiric'), sensibly rejecting Galenic theory while embracing anew the Hippocratic approach. 'Experiment' (like politics) seems to have been lost from our accounts of Sydenham. To us Sydenham is the restorer of clinical, bedside-medicine. For this we hold him to be a hero. Yet he was obviously not a hero to his contemporaries. Except of course to a handful, a handful of politically-motivated men, such as John Locke and Andrew Brown of Edinburgh.[41] When and how did Sydenham get

[40] *Ibid.*, pp. 96–100.
[41] See my 'Sydenham versus Newton: the Edinburgh fever dispute of the 1690s between Andrew Brown and Archibald Pitcairne', in W.F. Bynum and V. Nutton, (eds.), *Theories of Fever from Antiquity to the Enlightenment* (Supplement to *Medical History*, London, 1981), pp. 71–98.

sanitized and sanctified? Whence came that title of 'the English Hippocrates'? And whence indeed came the privileging of the matching reading of Hippocrates (the one we hold today) – of Hippocrates the bedside physician, the empiricist, working with nature, abjuring high-flown theory, practising expectant medicine?

It was in Leyden that Sydenham's reputation seems to have been laundered. While at home in England any defender of Sydenham was still virtually identifying himself as a supporter of the Good Old Cause (an unpopular position, especially after 1689) and the advocacy of Sydenhamian medicine was a politically loaded act, at the same time on the continent Sydenham was being presented as bland, safe, unpolitical, empirical, sensible. The transformation was performed above all by Hermann Boerhaave, who made Sydenham into one of the great masters of the history of medicine. In the early decades of the eighteenth century Boerhaave created a history whose peaks were Hippocrates, Bacon, Sydenham and Newton.[42] Hippocrates first practised proper medicine, Bacon pointed out the means for its restoration, Sydenham effected its restoration into practice, and Newton provided the (supposedly complementary) means to understand properly the workings of the body. Boerhaave also created a medical system which was (he claimed) constructed out of the eternal truths which had been brought to light primarily by the efforts of these four. And this medical system became the most widespread of any of the century, its influence being particularly felt in England and Scotland. Thus, some twenty years or so after Sydenham's death, a new image of Sydenham began to be re-imported into Britain by English-speaking students of Professor Boerhaave.

What is most remarkable about Boerhaave's list of heroes is how short it is: there is no-one significant between the fourth century B.C. and A.D. 1600. To put it another way, three of his four heroes come from the century immediately before his own (and all three of them from England, it may be noted). This is not the place to explore how and why Boerhaave created this particular pantheon. It will suffice for the moment to point out who, above all else, is missing from Boerhaave's list. It is of course Galen. Galen's domination of learned medicine for fifteen hundred years, Galen's interpretation of what Hippocrates was saying – they now count for nothing. It is indeed with and through the influence of Boerhaave's teaching that Galen's rule came to an end. What Boerhaave put into the vast space that Galen had occupied, was Thomas Sydenham. This Sydenham was restoring the

---

[42] For Boerhaave praising his heroes in his orations, see E. Kegel-Brinkgreve and A.M. Luyendijk-Elshout, *Boerhaave's Orations* (Leiden, 1983).

practice of Hippocrates. That Hippocrates, naturally enough, looked remarkably like Sydenham himself: making case histories and histories of disease, refraining from speculating about cause, working hand in hand with Nature.[43] And when Boerhaave had finished with them that Hippocrates and the Sydenham on whom he was modelled also sounded free of any interest but the impartial pursuit of truth in medicine for the good of mankind, inspired by a disinterested love of mankind. In other words, they began to sound like Boerhaave himself.

Finally we come back to the first Sydenham Society in the 1830s and 40s. In the lives of the people who ran and patronized this, clinical medicine was all the rage. That clinical medicine was, however, Parisian in origin, and hospital-based. In a roundabout way *an* image of Sydenham had indeed helped inspire this clinical medicine, but the way was very roundabout. *En route* the image of the true Thomas Sydenham, experimental physician of the Good Old Cause, had unfortunately been lost.

[43] Sydenham himself did quite a good job of characterizing Hippocrates as an early-day Sydenham:

> by these means and helps, the excellent Hippocrates arrived at the top of Physic, who laid this solid foundation for building the Art of Physic upon, viz. *Nature cures Diseases*. And he delivered plainly the *Phaenomena* of every disease, without pressing any Hypothesis for his service, as may be seen in his books of Diseases, Affections, and the like. He also delivered some Rules gathered from the Observation of the Method that Nature uses in promoting and removing Diseases; such are his *Praenotiones*, his *Aphorisms*, and the like. And of these things consisted the Theory of the Divine Old Man, which was not drawn from a vain and lascivious Fancy, like the Dreams of sick men, but it exhibited a legitimate History of those Operations of Nature, which she produces in the diseases of men. And now seeing this Theory was nothing else but an exquisite Description of Nature, it was very reasonable that in practice, his only aim should be to relieve her when she was oppressed, by the best means he could; and therefore he allowed no other Province for [medical] Art, than the succouring of Nature when she was weak, the restraining her when she was outrageous, and the reducing her to order, and to do all this in that way and manner whereby Nature endeavours to expel Diseases; for the sagacious Man [Hippocrates] perceives that Nature judges Diseases, and does in all, being helped by a few simple forms of remedies, and sometimes without any.

(Pechey's 1696 translation of part of the Preface to the *Medical Observations*). Though the fact may seem very strange to us now, this reading of Hippocrates was not a common one at Sydenham's time – even among men who were actually trying to model their own medical practice on that of Hippocrates; see Iain M. Lonie, 'Hippocrates the Iatromechanist', *Medical History*, XXV (1981), 113–50.

IX

Transforming plague
The laboratory and the identity of infectious disease

Introduction

The coming of the laboratory radically transformed the identity of infectious disease. This is one of the least appreciated – and, indeed, least obvious – of the changes of thinking and practice brought about in medicine by the coming of the laboratory. To show what this transformation comprised and how it happened, I shall take as my example the traditionally most feared, and in that sense the most important, of all infectious diseases: plague. These are the things I hope to show: one, that following the advent of laboratory medicine, infectious diseases are now necessarily and exclusively defined by the laboratory and thus receive their identity from the laboratory; two, that the laboratory concept of disease – with each disease having a single unique material cause, a cause which is identifiable in, and only in, the laboratory – is different from previous concepts of disease (and not merely a development of previous concepts); and three, that the dominance of the laboratory concept of disease has had a significant effect on our understanding of many *pre*-laboratory diseases – leading us to read them as if they were laboratory diseases; hence the coming of the laboratory has led to the *past* of medicine being rewritten to accord with the laboratory model of disease, and it has thereby been misunderstood.

The history of medicine as conventionally written is based on the assumption of a simple continuity in the identity of diseases, and thus tends to make invisible the issues which are actually involved in speaking of the 'identity' of a disease. For instance, a demographic historian has recently written that

Acknowledgements: My thanks to Dr Yoko Mitsui and Dr Perry Williams for their most helpful and sympathetic criticisms and guidance.

The most obviously difficult task for the historian is to identify the diseases at work in periods of high mortality and hence to try to form reliable hypotheses about the way in which they spread and the effects they might be expected to produce. He wants to be able to name the epidemic and to use modern experience of it as an aid to interpreting the partial evidence which survives for past outbreaks.[1]

This desire to name past diseases is quite common amongst historians of medicine as well as historical demographers, and it underlies much work in both disciplines. It is a claim about *identity*: that disease X in the past *was the same as* (or was identical to) disease Y in the present. The success of the venture of identifying past diseases depends on us having authentic means to make this claim of sameness, this claim of identity, about diseases. In pursuit of a more refined and reliable means of making such identifications, demographic and medical historians have increasingly turned to refining their own under-standing of the modern disease: if they have a full and proper understanding of disease Y in the present then, they believe, they will be on the firmest ground possible in saying that disease X in the past was (or was not) the same as disease Y in the present. That is, they have moved to acquiring greater technical medical knowledge. But this assumes that the making of such claims about sameness is in principle non-problematic, and simply a matter of having sufficient technical information at one's command. What really needs to be asked, however, is this: *what conditions would need to be satisfied* for the claim of sameness, of identity, to stick? And this is a philosophical and historiographical question, not a technical medical one. The historian of demography or of medicine customarily takes the familiar (modern medicine) and then applies it to the unfamiliar (past disease), seeking thereby to convert the unfamiliar into the familiar; hence, he feels, he can explain what a particular past disease 'really' was, how it was 'really' transmitted, what the past people 'really' saw, and what they 'missed'. What I shall be doing here, by contrast, is seeking to make an aspect of *modern* medicine look unfamiliar, and thereby explore how it was *constructed*; with respect to plague this will involve looking at the role of the laboratory in transforming the identity of plague. Then we should be able to see how *pre*-laboratory plague thereby became alien and unfamiliar to us.

Before proceeding with the argument, however, it will be useful to have some image of plague in our minds. In our world, the laboratory world, the 'medical facts' about it are usually taken, in brief, to be as

[1] Paul Slack, 'Introduction', in *The Plague Reconsidered: A New Look at its Origins and Effects in 16th and 17th Century England*, supplement to *Local Population Studies* (Matlock, Derbyshire), 1977, p. 6.

follows.[2] Plague is caused by a micro-organism, a bacillus, usually known as *Pasteurella pestis*. Plague is a disease of the rat. It is endemic in rat populations and periodically appears in a more virulent form and then becomes epidemic. The vector of the causative micro-organism of plague is the rat flea, whose proventriculus becomes 'blocked' with large quantities of the bacillus which it therefore cannot avoid regurgitating into rats as it tries to suck their blood. The rats thereby get plague. As the rats begin to die in great numbers from the plague so their fleas have to look for new hosts and hence they land on humans and transmit the bacillus into the bloodstreams of their human hosts. The humans are thereby given plague. There are three major forms of plague in the human, all of which have an extremely high mortality rate: (1) 'bubonic', spread directly by the rat flea, and whose typical lesions are 'buboes' which appear in the groin, armpit and neck; (2) the even more lethal variation of this, 'septicaemic', where the patient dies of sudden blood-poisoning before the buboes have time to appear; and (3) 'pneumonic', a complication of the bubonic and which can spread from person to person by infected droplets in the breath, and which does not therefore need the assistance of the rat flea in order to spread. These are the basic 'medical facts' about plague, as currently accepted.

Identifying disease

What constitutes the identity of a disease? This is not a topic which has received much discussion from doctors or philosophers or even medical historians, so there is little in the current literature to help us.[3]

---

[2] I have chosen to take this account from Leslie Bradley, 'Some medical aspects of plague' and Jean-Noel Biraben, 'Current medical and epidemiological views on plague', in *The Plague Reconsidered*, pp. 11–23 and 25–36, respectively, where the expression 'medical facts' is used, though without inverted commas. The accounts given in this work are completely conventional in modern terms, and the definition given in the text could have been derived from any modern handbook on plague or infectious diseases.

[3] There seem to be only two kinds of recent concern with it. The question has arisen over whether certain conditions are 'real' diseases, and in particular whether madness is one. The model of somatic disease against which comparison is often made is infectious disease, for here, it is assumed, the criteria of identity are the strictest. The other kind of discussion seems to be restricted to those who want medicine to continue to be treated as an art as well as a science. Here the treatment of the historical concept of disease has been unsatisfactory, being built around a supposed eternal dichotomy between 'ontological' and 'physiological' approaches. Yet this applies very poorly to the pre-laboratory age and anyway, as a dichotomy, seems to date from the emergence of laboratory medicine itself. Those historians who have chosen to write on the history of the concept of disease seem to be concerned to deny the validity of a strict 'ontological' concept, and to promote a form of 'physiological' concept. This is true even for Owsei Temkin; see his articles 'The scientific approach to

But we can certainly say that disease does not seem to be a 'natural kind'. Rather, a 'disease-entity' is a mental construct made up of experiences of pain, distress and debilitation, the outward visible appearances that accompany these experiences, the succession of all these over time together with the outcome (recovery, disablement death), the changes that the pathologist can find in the parts of the body, together with peoples' thoughts about the origin and reasons for what is happening and why it turns out as it does. A 'disease' is constituted by all these taken together. Two people undergoing the same set of these are usually judged to be undergoing the same (identical) disease.

Such mental groupings of experiences and natural phenomena, these 'disease-entities', do not necessarily correspond from one culture to another, and they have not necessarily been constant over time even within one particular culture. For instance, for a very long period in the history of our own western medical tradition it was the case that each 'disease' was thought of as being unique to each sufferer. Equally, the way a disease-entity is built up – the set of elements constituting its identity – may be changed over time. With the advent of French hospital medicine in the years round 1800, for instance, the localised pathological changes which happen within the body during the course of diseases came to be thought of as essential elements of the identity of those diseases, and disease nomenclature came to reflect this.[4] Again, some diseases in the past – diseases which were experienced, suffered from, treated, cured – ceased to be regarded as diseases at all; the most celebrated such ex-disease is perhaps chlorosis, from which thousands of young women suffered, especially in the nineteenth century, but which it is impossible to

disease: specific entity and individual sickness' (1963), and 'Health and disease' (1973), both reprinted in his *Double Face of Janus* (Baltimore, 1977), pp. 441–55 and 419–40; see also Sir Henry Cohen, 'The evolution of the concept of disease', *Proceedings of the Royal Society of Medicine*, 48 (1955), pp. 155–60. For a brief historical account of objections to the 'ontological' views of the bacteriologists, see Knud Faber, *Nosography in Modern Internal Medicine* (Oxford, 1923), pp. 186–94. For a recent example of the second type of concern, see Eric J. Cassell, 'Ideas in conflict: the rise and fall (and rise and fall) of new views of disease', *Daedalus* 1986 (= *Proceedings of the American Academy of Arts and Sciences*, 115), pp. 19–41. For examples of both types of concern see the entries by Guenter B. Risse and H. Tristram Engelhardt, Jr, under 'Health and Disease' in *Encyclopaedia of Bioethics*, ed. Warren T. Reich, 4 vols. (New York, 1978), vol. II. F. Kräupl Taylor, *The Concepts of Illness, Disease and Morbus* (Cambridge, 1979), gives a sophisticated philosophical treatment of the issues, but again assumes a long historical lineage for the 'ontological' concept.

[4] See Erwin H. Ackerknecht, *Medicine at the Paris Hospital, 1794–1848* (Baltimore, 1967); Michel Foucault, *The Birth of the Clinic: An Archaeology of Medical Perception* (London, 1973; original French edition, 1963); on the change in nomenclature see Faber, *Nosography*.

suffer from today, not because it has been eliminated but because people have ceased making identifications of it as a disease.[5]

Our actual concept of any disease, on the basis of which we may make claims that disease X is 'the same as' (identical to) disease Y, is constituted by what we may call its 'operational definition'. That is to say, by the questions people ask and the operations people actually engage in when finding or checking the 'identity' of any disease. It all comes down to the answers people give to certain questions – 'Is so-and-so ill?', 'What disease is it?', 'Is it cholera (or whatever)?' – and to the procedures people apply in reaching answers that satisfy at any given moment in history.[6]

Plague too is like this: it is defined by – that is, its identity derives from – the questions we ask and the activities we undertake in making the identification. How then do we today answer, to our own satisfaction, the question 'Is this a case of plague?' A modern expert at plague diagnosis explains how one starts:

The essential first step in the diagnosis of plague is to suspect the diagnosis of any person with a fever who lives in a known endemic area of the world or who has visited a known endemic area within the last few days. The diagnosis should be more strongly suspected if the patient with fever also has a painful bubo, cough, or signs of meningitis. Once the diagnosis is suspected, a physician should proceed to establish the diagnosis by bacteriological methods.[7]

That is the 'clinical diagnosis'. But all it can do is *suspect*. The only way a suspicion of plague can be confirmed or *established* is 'by bacteriological methods'; in other words, by a laboratory. The staff of the laboratory make this authentication by running tests to discover whether the pertinent bacterium is present or not: 'The absolute confirmation of plague infection in human beings, rodents, or fleas requires the isolation and identification of the plague bacillus,

[5] On chlorosis see Karl Figlio, 'Chlorosis and chronic disease in nineteenth-century Britain: the social constitution of somatic illness in a capitalist society', *Social History*, 3 (1978), pp. 167–197; but compare Irvin Loudon, 'The disease called chlorosis', *Psychological Medicine*, 14 (1984), pp. 27–36. For another instance, this time of the twentieth century, see Robert P. Hudson, 'Theory and therapy: ptosis, stasis, and autointoxication', *Bulletin of the History of Medicine*, 63 (1989), pp. 392–413.

[6] Although this particular formulation seems to be original with me, Charles Rosenberg has recently written 'In some ways disease does not exist until we have agreed that it does – by perceiving, naming, and responding to it', 'Disease in history: frames and framers', in *Framing Disease: The Creation and Negotiation of Explanatory Schemes*, Supplement 1 to *Milbank Quarterly*, 67 (1989) guest-edited by Charles Rosenberg and Janet Golden, pp. 1–16, at pp. 1–2. Similarly, though he is primarily concerned with the involvement of value judgements in the act of ascribing disease-status, Lawrie Reznek argues that disease-status 'is a division that is invented by our adoption of one descriptive definition of disease rather than another', and concludes that 'disease judgements, like moral judgements, are not factual ones', *The Nature of Disease* (London, 1987), pp. 80 and 213.

[7] Thomas Butler, *Plague and Other Yersinia Infections* (New York, c. 1983), pp. 163–4.

ISOLATION AND IDENTIFICATION OF *YERSINIA*
(PASTEURELLA) PESTIS

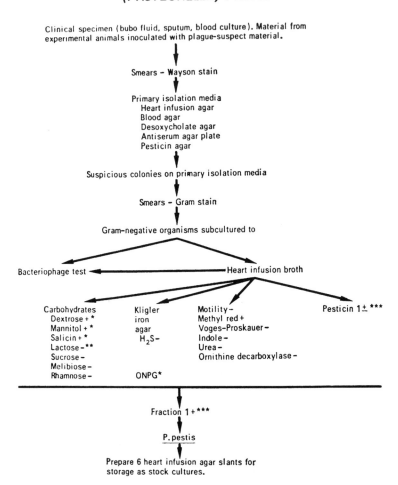

Clinical specimen (bubo fluid, sputum, blood culture). Material from experimental animals inoculated with plague-suspect material.

Smears – Wayson stain

Primary isolation media
Heart infusion agar
Blood agar
Desoxycholate agar
Antiserum agar plate
Pesticin agar

Suspicious colonies on primary isolation media

Smears – Gram stain

Gram-negative organisms subcultured to

Bacteriophage test ← Heart infusion broth

Carbohydrates	Kligler	Motility –	Pesticin 1± ***
Dextrose + *	iron	Methyl red +	
Mannitol + *	agar	Voges-Proskauer –	
Salicin + *	H₂S –	Indole –	
Lactose – **		Urea –	
Sucrose –		Ornithine decarboxylase –	
Melibiose –			
Rhamnose –	ONPG*		

Fraction 1 + ***

P. pestis

Prepare 6 heart infusion agar slants for storage as stock cultures.

* No gas formed.   ** No reaction 48 hours.   *** Most strains.

Fig. 6. How plague is identified today, according to laboratory criteria. This flowchart comes from the *Plague Manual* of the World Health Organisation. It should be noted that the internationally accepted name of the plague bacillus is now *Yersinia pestis*.

Yersinia (Pasteurella) pestis.[8] The necessary procedure to do this is detailed in Fig. 6. Clinical specimens are brought into the laboratory; they are then put through various treatments and tests, in a sequence

[8] M. Bahmanyar and D. C. Cavanaugh (eds.), *Plague Manual* (Geneva, 1976), p. 14.

shown by the arrows. If they pass all these tests (that is, if a particular, predictable result is obtained in each test), then we have identified *P. pestis*. These conditions must be satisfied, for they reveal either the presence or the absence of the plague bacillus. And if the bacillus is present, then this disease *is* plague.

It can be seen from this sequence of operations which we use today to identify plague that conclusive identification of plague can be made only by finding the bacillus. For we see this as the *cause* of the disease. Indeed we commonly refer to the bacillus as 'the causative micro-organism of plague'. Every modern account of plague that we might read today starts from this: from the bacillus as the *cause* of plague. But this is not only the case in medical texts, where we might expect it, but also in historical writings about plague, right at the beginning when the writer is about to make the identification of past plague with modern plague.[9] This seems so natural today that it scarcely bears remark. We simply take it for granted that the plague bacillus is the cause of plague, and that therefore the specification of the identity of plague, whether in the present or in the past, must start from the bacillus, the cause. Indeed, this is what I did above, when asking what are the 'medical facts' about plague today. But, as we shall see, this identity of plague derives from the laboratory, and was new with the introduction of the laboratory.

The term 'cause' is being used here in a specific sense. The sense in which we mean that this bacillus, *P. pestis*, is the cause of plague is specified by what are known as 'Koch's postulates'. In discussing the necessary and sufficient conditions that would have to be satisfied for a micro-organism to be accepted as the direct cause of a particular disease, Robert Koch wrote that

complete proof of the causal relationship demands, not merely a demonstration of the coincidence of the parasites with the disease, but, beyond this, it must be shown that the parasites directly produce the disease. To obtain this proof, it is necessary to isolate the parasites completely from the diseased organism, and from all the products of the disease to which any pathogenic influence could be ascribed; then to excite anew the disease with all its special characteristics by the introduction of the parasites alone into a healthy organism.[10]

[9] See, for instance, *The Plague Reconsidered*, or Folke Henschen, *The History of Diseases*, translated by Joan Tate (London, 1966; original Swedish edition, 1962), who treats every disease like this.

[10] Robert Koch, 'Die Aetiologie der Tuberkulose', *Mittheilungen aus dem Kaiserlichen Gesundheitsamte*, 2 (1884), pp. 1–88, at pp. 3–4. The translation is by Stanley Boyd, as given in W. Watson Cheyne (ed.), *Recent Essays by Various Authors on Bacteria in Relation to Disease* (London, 1886), pp. 70–1. There is, unfortunately, no one final statement of these postulates by Koch. For the development of Koch's arguments on the causal relationship, see the important recent series of articles by K. Codell Carter: 'Koch's postulates in relation to

These criteria date from 1884. The micro-organism must be constantly present in cases of the disease; it must be absent from other diseases (i.e. it must be unique to this disease); it must be possible to experimentally induce the disease in a healthy susceptible animal (if one exists) using *just* the micro-organism; and the micro-organism must then be found to have multiplied in that now-sick animal. After giving an example in the case of anthrax, Koch had concluded that

in the face of these facts, it is impossible to come to any other conclusion than that the splenic fever [= anthrax] bacillus is *the cause* of the disease, and not merely an accompaniment of it... These conclusions are so unanswerable that no one now opposes them, and science universally accepts the bacillus anthraxis as the cause both of the common typical splenic fever we are familiar with in our domestic animals, and also of the clinically different forms of the disease which occur in man.[11]

These are the criteria of 'cause' that we are applying to *P. pestis* when referring to it as the cause of plague. It follows from this that for infectious diseases in general, without the pertinent micro-organism present there can be no instance of the disease! For the micro-organism initiates, produces, brings into existence the whole sequence of inner changes in the body which constitute the disease. For instance the micro-organism might produce a toxin, which in turn damages certain cells, which leads in turn to malfunction, pain or death. In the case of plague the bacillus *P. pestis* is the cause which originates and produces all the bodily and mental experiences which constitute the disease of plague, right through to the production of buboes, and beyond to death. In this sense the micro-organism *P. pestis* is the essential initiator and cause of all the pathology, all the symptoms, all the experience of the disease of plague. Although today plague may be clinically *suspected* it is never *proved* (that is, never identified) without isolating its *cause*, the micro-organism. For a hundred years now this has been the case, and every outbreak of suspected plague in that hundred years which has been conclusively diagnosed as being plague has been diagnosed (identified) as plague only by finding the micro-organism *P. pestis* present.

What is more, because we take a particular micro-organism to be the *cause* of the disease, all the other features of the modern disease-entity of plague – that is, all the other features of its identity – follow from this 'cause'. Our understanding of how plague *spreads* depends

the work of Jacob Henle and Edwin Klebs', *Medical History*, 29 (1985), pp. 353–74; 'Edwin Klebs' criteria for disease causality', *Medizinhistorisches Journal*, 22 (1987), pp. 80–9; 'The Koch–Pasteur dispute on establishing the cause of anthrax', *Bulletin of the History of Medicine*, 62 (1988), pp. 42–57; see also William Coleman, 'Koch's comma bacillus: the first year', *Bulletin of the History of Medicine*, 61 (1987), pp. 315–42.

[11] Cheyne (ed.), *Recent Essays*, p. 72; emphasis as in Boyd's translation.

on seeing the micro-organism as the cause, for it is the micro-organism itself of which the rat flea (and, under the flea, the rat) are the vectors: the very cause is transmitted from one creature to another. Similarly our understanding of the different *forms* that plague takes comes from seeing the micro-organism as the cause: bubonic, septicaemic and pneumonic plague, although they present clinically in such different ways are, for us, all forms of the *same* disease because we find they all have the same micro-organism present as cause. As a modern expert has written:

Clinically, the two chief types [bubonic and pneumonic] are so distinct that they would rank as different diseases if they were not known to have a common origin and to be linked by intermediate types. At the bedside, nothing could be more unlike than cases of true pneumonic and uncomplicated bubonic plague; in the one the brunt of the infection falls upon the lungs, in the other on the lymphatic system.[12]

Again, our understanding of the *epidemiology* of plague is built on seeing the micro-organism as the cause: properly-identified cases of plague (i.e. identified in a laboratory) have a large-scale incidence of a certain distinctive form; thus if a yet-unidentified disease has this particular incidence, then it must be plague (and the bacillus will be there as its cause). The same is true of the *symptoms* of plague, and of its *pathology*: the 'true' symptoms and the pathological changes typical of plague are those regularly seen in cases where the 'causal micro-organism of plague' has been established as present. This way of working outwards *from* the cause *to* the symptoms and pathology (rather than in the other direction) has had significant effects on the classification of diseases – that is, on what conditions count as what disease. As a recent commentator has pointed out, the search for and discovery of causal micro-organisms of diseases has

led to the redefinition and reclassification of many disease entities... With the discovery of the tubercle bacillus and its role in disease, for instance, what had been designated phthisis was reordered into a number of conditions, only some forms of which were tuberculosis. The forms assigned as tuberculosis were those in which the bacillus could be demonstrated by staining and grown in culture...[13]

[12] L. F. Hirst, *The Conquest of Plague: A Study of the Evolution of Epidemiology* (Oxford, 1953), p. 28.

[13] Mervyn Susser, *Causal Thinking in the Health Sciences: Concepts and Strategies of Epidemiology* (New York, 1973), p. 23. For another instance, concerning the discovery of the bacillus of diphtheria in 1891, see Charles-Edward Amory Winslow, *The Conquest of Epidemic Disease: A Chapter in the History of Ideas* (Madison, Wis., 1971; original edition, 1943), p. 341: 'Thus, it appeared that between one-quarter and one-third of the cases of "clinical diphtheria" were not the true disease. On the other hand, 80 per cent of a series of cases of "membranous croup" proved on culture to be true laryngeal diphtheria. The value of the bacteriological criterion was further demonstrated by epidemiological studies, which showed the fatality of culturally-proved diphtheria to be 27 per cent, of "false diphtheria" to be less than 3 per cent.'

218

By seeing specific micro-organisms as the 'causes' of certain diseases, it has thus become possible to think in terms of *specific* infectious diseases: each such specific disease has a specific cause, a particular micro-organism. This sense of 'cause', as a specific causal agent which alone brings about the disease, is fundamental to our modern meaning for the term 'aetiology', the discipline where we discuss the causes of diseases.

The role of the laboratory in all this is absolutely crucial. The laboratory today holds total authority on the authentication of plague, for the final diagnosis – the identification – is impossible without the laboratory. So much so is this the case that even if a patient appears to have all the symptoms of plague, yet they cannot be said to have plague until the laboratory has spoken. As the eminent bacteriologist E. E. Klein wrote in 1906:

It is admitted on all sides that the *Bacillus pestis* is the real and essential cause of Oriental or bubonic plague, and consequently that the presence of this microbe in any material derived from a human or an animal being denotes the disease plague in such a being. It is likewise admitted that a patient, although exhibiting one or more symptoms suspicious of the disease plague – e.g. fever with swollen and inflamed subcutaneous lymph glands in one or other region of the body, cervical, axillary, inguinal, or femoral, – need not necessarily be affected with bubonic plague, notwithstanding that such person might have been indirectly exposed to plague infection. Should, however, in such swollen inflamed glands the *B. pestis* be demonstrated, epidemiologists and physicians would accept such a case unquestionably as true plague...[14]

And of course the presence of *B. pestis* can only be established in the laboratory, using all the proper tests. That tests of this kind are conducted in the laboratory is not a matter of mere convenience and coincidence – it is not the case that the laboratory just happens to be the best or most convenient and well-equipped place in which to make such tests – for such tests *are what constitute the central, the defining, activities of a medical microbiological laboratory*. Without it they are, quite literally, unthinkable. The laboratory makes the tests, but equally the tests make the laboratory.

The laboratory has this same role as the unique authenticator in the case of all laboratory-defined diseases, all those diseases where we now believe a micro-organism to be the unique cause; and therefore in all these cases too the laboratory has transformed the identity of disease. Indeed it is the laboratory which defines which diseases count as members of the category of 'infectious diseases', for infectious

[14] E. E. Klein, *Studies in the Bacteriology and Etiology of Oriental Plague* (London, 1906), p. xiii. On Klein see the extensive obituary by William Bulloch in *Journal of Pathology*, 28 (1925), pp. 684–97.

diseases today are those which have transmittable micro-organisms as their cause, whether a bacterium, a virus or other tiny parasite. This is a considerable proportion of all conditions known as 'disease' (even in the loosest formulation of the category of 'disease'), since infectious diseases constitute about half of all the cases that present to a general doctor. To this total we can add all those diseases where a micro-organism is believed to be the causative agent but which is still being hunted down in the laboratory. In most cases, of course, the doctor does not in practice invoke the laboratory before he offers a diagnosis of, for instance, influenza or chicken-pox, for he has been trained to trust in his skills in recognising and correlating symptoms and signs. But if the doctor has any doubt about his diagnosis in any particular case, then he does indeed send specimens and samples along to the laboratory for inspection so that a proper diagnosis can be made. Such behaviour on the doctor's part shows that all his diagnoses of infectious disease are only provisional until they have laboratory confirmation. This does not mean his diagnoses are necessarily inaccurate if he has not first consulted the laboratory, but it does mean that the laboratory is the final arbiter of the accuracy of the diagnoses the physician offers. It also shows that his diagnosis is based on a micro-organism as being the cause.

Old plague

How was plague identified – what was its identity – before the advent of the laboratory? While it would be misleading to claim that there was just a single picture of plague before the lab, since every outbreak revealed much disagreement and conflict amongst practitioners over the nature and treatment of plague,[15] yet the pre-lab identity of plague differed radically from its post-lab identity in a number of striking ways, which can be considered under two main heads.

In the first place, like all other diseases, plague before the laboratory was always identified by its *symptoms* and its course. Primarily this means it was identified by the presence of buboes. But pre-lab symptoms were many and complex, and even the buboes were not necessarily evident, nor were they even always taken to be the most important symptoms. Generally, as a late eighteenth-century investigator noted, buboes and carbuncles 'are equally diagnostics of

---

[15] On the complexities and varieties of concepts of plague before the lab see for instance Jon Arrizabalaga, 'Facing the Black Death: perceptions and reactions of university medical practitioners' in L. Garcia-Ballester *et al.* (eds.), *Practical Medicine from Salerno to the Black Death* (Cambridge, in press).

the true plague; their presence, separately or in conjunction, leaves the nature of the distemper unequivocal; but fatal has been the error of rashly, from their absence, pronouncing a distemper not to be the plague'.[16] If the buboes are present, then the disease is definitely plague; but if they are absent, it still might be plague. Dr Andrew White, a military surgeon with the British army, had to contain an outbreak of plague on the island of Corfu, and his account of the symptoms of plague (first published in 1845) hardly touches on the buboes:

The symptoms of the plague in Corfu, as collected from the medical officers employed on that occasion, were as follows:-

More or less fever, sometimes of a *remittent*, sometimes of an *intermittent* type; great prostration of strength; staggering like a drunken man; often violent headache; tremors; derangement of the stomach, with a sensation of burning heat; vomiting, sometimes of a yellow, at others of a blackish matter, like coffee-grounds; involuntary evacuations, both of urine and faeces at times, when the patients did not appear to be very ill, and which seemed the effect of fear, stupor, coma; often violent and sudden exacerbations of fever, which could not be said to belong to any type; a white, glossy tongue, the edges of which were generally clean, with a streak in the middle. The countenance exhibited an appearance of terror mixed with anxiety, and, as it were, claiming pity, which is difficult to describe, but which is well known to those who see plague patients, and is very characteristic of the disease. Sometimes the disease was ushered in with furious *delirium*, approaching to a state of *phrenitis*, with the eyes, as it were, ready to start from their sockets, and the face flushed, as if mad with the effects of drink and passion, so that for a time they became quite unmanageable. The duration of the paroxysm sometimes lasted for hours, after which they became calm and composed, and in some instances appeared to be quite rational. In some cases, these violent exacerbations were succeeded by cold *rigors*; these alternated, and the unhappy sufferer was carried off by them, sometimes without exhibiting those eruptions [i.e. the buboes] which are supposed necessary to form the character of plague.

Buboes and carbuncles were very common symptoms, particularly after the first ebullition of the disease was over...[17]

The buboes are very common, but not universal, symptoms, and other symptoms have to be recognised in order to be able to identify plague. Indeed, so important were these other symptoms that one night during this Corfu outbreak White himself 'went to bed with all the horrors of plague about me' because he believed his servant was showing severe symptoms of plague. What had happened in fact was that the servant had, quite untypically, got very drunk and was staggering around – and staggering was (as White's own report above shows) one of the prime symptoms of plague!

[16] Patrick Russell, *A Treatise of the Plague: Containing an Historical Journal, and a Medical Account, of the Plague, at Aleppo, in the Years 1760, 1761, and 1762* (London, 1791), p. 112.

[17] Andrew White, *A Treatise on the Plague, More Especially on the Police Management of That Disease* (London, 1846), p. 141.

This symptom-based identity of pre-lab plague meant that although medical men might claim to have supreme authority in making the identification, yet in practice theirs was neither the first nor the last word. Pre-lab plague could, in practice, be identified by anyone. If people thought there was an outbreak of plague amongst them, those who could afford to usually fled. They rarely waited for the doctors to tell them what they already knew about the identity of the disease. In the case of post-lab plague, by contrast, the identification of the disease cannot be made by lay-persons. Indeed it cannot even be made by clinical practitioners. It can only be made by the workers in a bacteriological laboratory.

The second major way in which pre-lab plague differed from post-lab plague, was with respect to *cause*. The term 'cause' was used in a different sense by pre-lab doctors, and had been for at least two millenia. 'Aetiology' also therefore meant something different. There was a whole hierarchy of causes, with many items at each level of the causal hierarchy.[18] These still prevailed in the nineteenth century, when generally there were taken to be four inter-related types of cause operative. There were *predisposing* causes, such as the particular constitution of the patient, the weather, the season of the year and the state of the soil. There were *external* (procatarctic or preceding) causes, such as the six 'non-naturals': the state of the air around a patient, and the nature, quality and quantity of his food and drink, his sleep and watch, his inanition and repletion, his movement and rest, and the passions of his mind. There were *antecedent* causes, such as some obstruction within the body. And there were *immediate* causes, such as a particular state of the blood. In plague, as in every other illness, these causes were operative. Some, such as the 'non-naturals', the physician could regulate to ward off or repel the disease; others he could not. Such causes were obviously not necessarily unique to a particular disease.

Moreover, pre-lab plague did not have a *specific causal agent*. As Vivian Nutton has written, about the concept of 'the seeds of disease' in the Ancient and Medieval periods:

A disease did not have an existence in its own right, but as a deviation from the norm within the patient, and although these authors accepted and wrote of such disease entities as fever and phthisis, they insisted on always taking into account 'the peculiar nature of each individual'. The nature of disease was to be found in man's temperament, the structure of his parts, his physiological and psychological dynamism, and could be defined very much in terms of impeded function. Set against

[18] See for instance the listings and explanations of them under 'Aetiology' in Robert Hooper, *Lexicon Medicum; or Medical Dictionary*, 6th edn (London, 1831).

this background, the seeds of disease act only as an initial cause: they are not the disease, any more than a blow to the head or a poisonous mushroom. They merely trigger a situation which eventually may lead to a humoral disorder and a bodily malfunction, and it is the latter which for Galen constituted disease and illness.[19]

These same attitudes described by Nutton for the Ancient and Medieval periods can still be seen in that great pre-laboratory work on plague in English, the 1791 *Treatise of the Plague* by Dr Patrick Russell, FRS, built on his experience of outbreaks of plague in Aleppo in the 1760s. In this Russell defines plague only by its symptoms. Of *cause*, in the sense in which the micro-organism is today seen as the cause of plague, as a unique material entity, specific to the disease, he naturally says nothing. Admittedly for Russell, as for many people of earlier centuries, there is indeed some invisible material entity present and active, a 'contagion': 'The plague is a contagious disease; that is, an emanation from a body diseased, passing into one which is sound, produces, in time, the same disease; and the person thus infected becomes in like manner capable of communicating the plague to others.'[20] But this 'contagion' is not the cause, but merely the *means of transmission* of the plague, acting only (as Nutton expresses it) to 'trigger' the disease. And this specific 'contagion' exists only under certain atmospheric conditions: it is brought into existence by a certain state of the air, and spreads only with an appropriate state of the air, and it has an effect on any given individual only if that individual is personally susceptible. As Russell wrote,

In what this particular constitution of the air consists, which in one case favours the spreading of the distemper, and in the other checks or extinguishes it; whether it operates by heightening the powers of the infectious effluvia, or by inducing an epidemical change on the human body, whereby it is rendered more or less susceptible of, or enabled to resist their influence, the effluvia remaining the same; are points involved in much obscurity. It seems in the mean while incontestible that without a concurrent state of the air, the plague will not become epidemical; and without a certain state of the body, the infection will not take effect.[21]

This 'contagion', originating in a particular state of the air, and spreading as a consequence of the state of the air, is quite different from the post-laboratory concept of a causative micro-organism which exists independently of any 'epidemic constitution'.

However, even though the pre-lab concept of 'cause' differed from the post-lab concept, it was nevertheless possible, as William Coleman has pointed out, for medical men in the early to mid nineteenth

---

[19] Vivian Nutton, 'The seeds of disease: an explanation of contagion and infection from the Greeks to the Renaissance', *Medical History*, 27 (1983), pp. 1–34, at p. 15.

[20] Russell, *Treatise*, p. 296.   [21] *Ibid.*, p. 261.

century to develop a science of epidemiology. They simply did not invoke aetiology. Coleman's study of work on yellow fever shows that clinicians could and did build pictures of the large-scale (epidemic) behaviour of yellow fever, without feeling the need to discuss the issue of cause at all.[22] Our modern-day discipline of epidemiology, by contrast, depends crucially upon the identification of the 'cause' of epidemic diseases: epidemic spread leads us in the first place to seek out a causal agent, and if we find a micro-organism candidate then thereafter our understanding of what is being spread epidemically is built on our view of the micro-organism which we have identified as the cause.

Not only did a pre-lab disease not have a specific causal agent, but it was possible to have 'mixed' diseases. It was also the case that diseases were considered capable of transforming into other diseases during their course, and that the 'morbific matter' could move around within the body, thus altering the locus of disease ('metastasis') and hence the form and nature of it.[23] This applied to plague as much as any other disease: a fever of one kind could, perhaps, turn into plague through a change in climatic conditions. Pre-lab diseases were simply not as fixed or constant as those post-lab ones whose identity is built on the isolation of a specific material causal agent.

All the differences between the identity of pre-lab plague and post-lab plague are encapsulated in one word: 'pathogen'. Although it is classical Greek in form, this word was new in the late nineteenth century. In the *Oxford English Dictionary*, which is a historical dictionary, the first recorded occurrence of 'pathogen' in English is dated as 1880, and defined as 'a micrococcus or bacterium that produces disease'.[24] 'Pathogenic', 'pathogenetic' and other forms are recorded as occurring before this date, but not the substantive 'pathogen': 'disease-causing entity'. A 'pathogen' is a specific material agent, which is itself the cause of the disease.

---

[22] William Coleman, *Yellow Fever in the North: The Methods of Early Epidemiology* (Madison, Wis., 1977). He writes, 'It is an untoward result of the triumph of the germ theory of disease that the development of epidemiology has come to be viewed primarily through the spectacles of that theory and of etiological reasoning in general. This is misleading, even false', p. xiii.

[23] See, for example, Malcolm Nicolson, 'The metastatic theory of pathogenesis and the professional interests of the eighteenth-century physician', *Medical History*, 32 (1988), pp. 277–300.

[24] The term does not appear, for instance, in the 1874 edition of Robley Dunglison's *Medical Lexicon: A Dictionary of Medical Science* (Philadelphia).

The transformation of plague

How was plague transformed from a disease whose identity was symptom-based into one whose identity was cause-based? The transformation of the identity of plague took place in the laboratory, and from that moment on plague would only be identifiable in the laboratory. It is common to see this event as a simple 'unmasking' or a 'drawing back the veil' on what had been known all along to be there but which had hitherto simply evaded the light of science, and that is the language that was also used by many contemporaries about the event just after it had happened, and which has been regularly used since. But it was not a simple 'unmasking'. Instead, the new view of the disease, its new identity, was a *construction* since it involved, and depended totally on, a new way of thinking and seeing, the laboratory way of thinking and seeing. As we have seen, post-lab plague is defined by and from its *cause*, and as Bruno Latour has correctly remarked, 'a cause is always the *consequence* of a long work of composition and a long struggle to attribute responsibility to some actors'.[25] The laboratory was the instrument used to attribute responsibility to micro-organisms. Yet the laboratory is never a mere instrument: it is also a *practice* which defines, limits and governs ways of thinking and seeing. Plague acquired its new identity from this new activity, this new practice. Therefore the laboratory had to precede, both in time and conceptually, the 'causative micro-organism' identity of the disease. It was necessary to take the laboratory to the disease, and then to take the disease through the laboratory.

The transformation of the identity of plague happened in Hong Kong in the summer of 1894. Plague broke out in early May, being immediately identified by the native population, who at once took to flight. The outbreak was of especial interest to those colonial powers with major interests in the region: Britain (the 'lessee' of Hong Kong), France and Japan. The first telegram report of it in *The Times* of London on 13 June expresses very clearly the precise nature of Britain's interests as the colonial master of this showpiece of Victorian commercial values:

Half native population Hong Kong left, numbering 100,000. Leaving by thousands daily; 1,500 deaths; several Europeans seized [by the disease], one died. Labour market paralyzed. Deaths nearly one hundred daily. Government anticipating failure of opium revenue; proposes taking over and destroying all unhealthy native quarters.

---

[25] Bruno Latour, *The Pasteurization of France*, translated by Alan Sheridan and John Law (Cambridge, Mass., 1988; original French edition, 1984), p. 258.

Fig. 7. Shibasaburo Kitasato and his teacher Robert Koch in Japan in 1908.
According to his biographer, 'Kitasato was a man of filial affection toward his
parents and a devoted follower of his teacher. While Koch was in Japan, Kitasato
always attended his teacher with the utmost care as though serving his own
father.' (Mikinosuke Miyajima, *British Medical Journal* (1939, 1), pp. 1141–2).
Kitasato was said to be Koch's favourite pupil.

The disease continued to rage over the summer. By 4 September, when Hong Kong was formally declared to be free from the plague, the official British government figure for those who had died was over 2,500, almost all of them Chinese, while unofficial reports put the figure at 'over 3,000, out of a normal native population of 150,000, reduced certainly to 100,000 by panic and flight'.[26]

The two investigators who (to use the customary phrase) 'discovered' the plague bacillus in Hong Kong in 1894 were Shibasaburo Kitasato and Alexandre Yersin.[27] The rivalry between Koch and Pasteur in Europe, on nationalist and scientific grounds, was continued here in Hong Kong by their volunteer champions: the German school of Koch was represented by Kitasato (Fig. 7), a Japanese, and the French school of Pasteur by Yersin (Fig. 8), a Swiss who had become a naturalised Frenchman. Kitasato was sent by the Japanese government, Yersin by the French Ministry of Colonies. They arrived within days of each other, they tried to work within the territory of the same plague hospital and they struggled for a monopoly of investigative facilities. In the event they worked separately, with Yersin choosing to follow Pasteur's advice on what to do when trying to defeat a Kochian rival: 'As much as possible, work

[26] The unofficial figure is from *The Times* of 28 August 1894, p. 6, col. 2. For the official figures, and the British Government's information on the epidemic, see *Correspondence Relative To the Outbreak of Bubonic Plague at Hong Kong*, and *Further Correspondence*, presented to both Houses of Parliament in July and August 1894 respectively (Command Papers 7461 and 7545).

[27] I have chosen to treat Kitasato's name in western style, that is with the given name first and the family name second. For information in English on the life and works of Kitasato see the obituaries of him in the *Proceedings of the Royal Society*, series B, 109 (1931–2), pp. xi–xvi, by William Bulloch, and in the *British Medical Journal* (1931, I), pp. 1141–2 by Mikinosuke Miyajima; the article in the *Dictionary of Scientific Biography* by Tsunesaburo Fujino. Both on Kitasato in particular and the general issue of Japanese involvement in late nineteenth century science in general, see the indispensable writings of James R. Bartholomew: 'Japanese culture and the problem of modern science' in Arnold Thackray and Everett Mendelsohn (eds), *Science and Values: Patterns of Tradition and Change* (New York, 1974), pp. 109–55; *The Formation of Science in Japan: Building a Research Tradition* (New Haven, 1989); and 'The acculturation of science in Japan: Kitasato Shibasaburo and the Japanese bacteriological community, 1885–1920', unpublished Ph.D dissertation (Stanford University, 1971). For Yersin see Paul Hauduroy (ed.), *Yersin et la peste: ouvrage publié pour la cinquantenaire de la découverte du microbe de la peste* (Lausanne, 1944); and now see the excellent biography by Henri H. Mollaret and Jacqueline Brossollet, *Alexandre Yersin ou le Vainqueur de la Peste* (Paris, 1985). Much of the literature on Kitasato and Yersin deals with the question of whether they in fact discovered the 'same' bacillus, and which of them therefore has the right to the credit for its discovery. This is not a concern of my paper, but on it see E. Lagrange, 'Concerning the discovery of the plague bacillus', *Journal of Tropical Medicine and Hygiene*, 29 (1926), pp. 299–303; Norman Howard-Jones, 'Kitasato, Yersin, and the plague bacillus', *Clio Medica*, 10 (1975), pp. 23–7; and David J. Bibel and T. H. Chen, 'Diagnosis of plague: an analysis of the Yersin–Kitasato controversy', *Bacteriological Reviews*, 40 (1976), pp. 633–51.

Fig. 8. Yersin in the first course of microbiological technique at the Pasteur Institute; the class was run by Dr Roux and Yersin was his *préparateur*. Yersin is seated in the front row, third from right, and to the left of him (on his right) sit Metchnikoff, Roux and Laveran.

by yourself. Keep your cadavers to yourself.'[28] Yet their roles in 'discovering' the plague bacillus and transforming the identity of plague, had a great deal in common. Let us see how the transformation was effected by their activities.

As Kitasato was a bacteriologist and Yersin was a microbiologist (these were the preferred terms used in their respective German and French schools), on the very first day that they could, they each *looked for* a micro-organism. As Yersin wrote, 'It was obvious that the first thing to do was to see whether there was a microbe in the blood of the patients and in the pulp of the buboes.'[29] And in order for them to search for a causative micro-organism it was necessary for each of them to have available to them a *laboratory*: that is, a dedicated space (a room) with special equipment in it. Kitasato, arriving first, made friends with Dr Lowson of the Colonial Medical Service, who 'put everything needful at our disposal in the most friendly spirit. A room in the Kennedy Town Hospital (one of the plague establishments) was given to us, and there we began our work on June 14th.'[30] Yersin, however, was denied space in the Kennedy Town Hospital, apart from a little space on a gallery. But he made friends with a long-established Catholic priest, and was thus able to build a laboratory of his own. As he wrote to his mother on 24 June, 'After having stayed at the hotel for some days, I have built myself a straw hut near the hospital for the plague victims and there I have set up my living quarters and my laboratory.'[31] This straw hut (Fig. 9), in the grounds

[28] On this rivalry see Yersin's letter to his mother of 24 June 1894, as translated by Ingrid Ebner in Butler, *Plague*, pp. 15–16; and the account by Mollaret and Brossollet, *Yersin*, pp. 133–8. Pasteur's advice, 'Autant que possible, travaillez seuls. Ayez vos cadavres à vous', was given in a letter to Straus and Roux in 1884; see *Correspondance de Pasteur, 1840–1895*, ed. Pasteur Vallery-Radot, 4 vols. (Paris, 1940–51), vol. III, p. 430; translated in Thomas D. Brock, *Robert Koch: A Life in Medicine and Bacteriology* (Madison, Wis., 1988), p. 176.

[29] 'Il était tout indiqué de rechercher tout d'abord s'il existe un microbe dans le sang des malades et dans la pulpe des bubons'; Alexandre Yersin, 'La peste bubonique à Hong-Kong', *Annales de l'Institut Pasteur*, 8 (1894), pp. 662–7, my translation. Translated in Hubert A. Lechevalier and Morris Solotorovsky, *Three Centuries of Microbiology* (New York, 1974; 1st edn, 1965), pp. 152–6; all subsequent quotations from Yersin are from here, unless otherwise noted. This translation is reprinted in Butler, *Plague*, pp. 17–22.

[30] S. Kitasato, 'The bacillus of bubonic plague', *Lancet*, 25 August 1894, pp. 428–30. All subsequent quotations from Kitasato are from here. Kitasato wrote his paper in German, and this was translated into English by Dr Lowson; this English form, as published in the *Lancet*, is the definitive version. See Bartholomew 'The acculturation of science in Japan', p. 174.

[31] Letter to his mother, 24 June 1894, as translated in Butler, *Plague*, pp. 15–16; the French original is published by H. H. Mollaret, 'Alexandre Yersin tel qu'en lui même enfin... Les révélations d'une correspondance inédite échelonnée de 1884 à 1926', *La Nouvelle Presse Médicale*, 2 (1973), pp. 2575–80, at p. 2577, and reproduced in facsimile on p. 2578: 'Après être resté quelques jours à l'hôtel, je me suis fait construire une paillotte à la côté de l'hôpital des pestiférés et j'ai établi là mon domicile et mon laboratoire'.

Fig. 9. Yersin outside his straw-hut laboratory in Hong Kong, 1894. He took this photograph himself to send home to his mother.

of the Alice Memorial Hospital, was Yersin's laboratory, within which he discovered his plague bacillus, and he constantly referred to it as such. To make these enclosed areas into laboratories proper, it was necessary to install in them certain equipment. 'I settled with my laboratory equipment in a straw hut that I had built with the permission of the English government, in the grounds of the main hospital', as Yersin wrote.[32] This equipment to transform inner space

---

[32] 'Je m'installai avec mon materiel de laboratoire dans une cabane en paillotte que je fis construire', Yersin, 'La peste bubonique', p. 662; translation as printed in Butler, *Plague*, p. 17. The set of items which constitute the essential equipment of a microbiological laboratory for the investigation of plague could be carried in three hands, as is shown in the opening passage of the autobiography of a slightly later 'plague fighter' (as he called himself), Wu Lien-Teh: 'Late in the bitterly cold afternoon of December 24, 1910, there arrived at the large railway station of Harbin in North Manchuria a young Chinese doctor...accompanied by his assistant... The doctor had in his right hand a compact, medium-sized British-made Beck microscope fitted with all necessaries for bacteriological work, while the assistant carried a handy-sized rattan basket containing various stains, glass slides, cover-glasses, small bottles of alcohol, test-tubes, platinum loops, needles, dissecting forceps, scissors and such other paraphernalia as are needed for laboratory investigation. Another but smaller basket held three dozen tubes of agar media packed in an upright position and held in place by packets of cotton wool; these media were most important for the routine growth of bacteria, particularly plague bacteria.' Wu Lien-Teh, *Plague Fighter: The Autobiography of a Modern Chinese Physician* (Cambridge, 1959), p. 1.

into a laboratory each investigator had brought with him. The most important item in a bacteriologist's arsenal was his microscope; it was virtually his emblem of office: indeed when Yersin got off the boat he was carrying his microscope in one hand and his autoclave in the other.

Thus they both arrived, transformed certain areas into laboratories and established themselves inside them, and they both started looking for a causative micro-organism. On the very first day, at the very first autopsy, each of them found one. Kitasato reported:

On that day we were able to see a post-mortem examination performed by Professor Aoyama [one of the Japanese team]. I found numerous bacilli in the bubo (in this case a swelling of the inguinal glands), in the blood of the heart, in the lungs, liver, spleen, &c.

When Yersin eventually got started with his investigations he too immediately found a candidate micro-organism:

With the help of Father Vigano, I try to persuade some English sailors, whose duty it is to bury the dead from the city and the other hospitals, to let me take the buboes from the dead, before they are buried. A few dollars conveniently distributed and the promise of a good tip for every case have a striking effect. The bodies before they are carried to the cemetery are deposed for one or two hours in a cellar. They are already in their coffins in a bed of lime. The coffin is opened, I move the lime to clear the crural region. The bubo is exposed, within less than a minute I cut it away and run to my laboratory. A film is prepared and put under the microscope; at the first glance, I see a real mass of bacilli, all identical. They are very small rods, thick with rounded ends and lightly coloured.[33]

It was never a problem for Yersin to see the bacillus: 'I always find it; for me there is no doubt.'[34]

Looking down his microscope each man could see a micro-organism of a particular known form (rod-like, i.e. a bacillus) but

[33] 'J'essaye, avec le père Vigano, d'obtenir de quelques matelots anglais qui ont pour mission de faire enterrer les cadavres de la ville et des autres hôpitaux qu'ils me laissent enlever les bubons des morts, avant qu'on ne les porte en terre. Quelques piastres judicieusement distribuées et la promesse d'un bon pourboire pour chaque bubon que je pourrai enlever ont un effet immédiat. Les morts, avant d'être enterrés au cimetière, sont déposés pendant une heure ou deux dans une sorte de cave. Ils sont déjà dans leur cercueil et recouverts de chaux. On ouvre un des cercueils et j'enlève un peu de la chaux pour découvrir la région crurale. Le bubon est bien net, je l'enlève en moins d'une minute et je monte à mon laboratoire. Je fais rapidement une préparation et la mets sous le microscope. Au premier coup d'oeuil je reconnais une véritable purée de microbes, tous semblables. Ce sont de très petits bâtonnets trapus, à extremités arrondies et assez mal colorés'. Yersin's diary for 20 June 1894; reproduced in facsimile in Hauduroy (ed.) *Yersin et la peste*, and in Lagrange, 'Discovery of the plague bacillus'; as translated by Lagrange.

[34] Yersin's letter to his mother of 24 June 1894, as translated in N. Howard-Jones, *The Scientific Background of the International Sanitary Conferences, 1851–1938* (Geneva, 1975), p. 79; the French original is 'Je le retrouve toujours; pour moi, il n'y a pas de doute', see Mollaret, 'Alexandre Yersin tel qu'en lui-même enfin', p. 2577.

with its own unusual and distinctive properties. It could be uniquely characterised in the laboratory according to its motility, staining reactions and behaviour when cultivated, and these characteristics gave it its unique identity, making it distinguishable from all other micro-organisms. This unique identity was described by each investigator. Kitasato's micro-organisms

are rods with rounded ends, which are readily stained by the ordinary analine dyes, the poles being stained darker than the middle part, especially in blood preparations, and presenting a capsule sometimes well marked, sometimes indistinct ... I am at present unable to say whether or not 'Gram's double-staining method' can be employed ... The bacilli show very little movement, and those grown in the incubator, in beef-tea, make the medium somewhat cloudy.

Yersin's micro-organisms were

short, stubby bacilli which are rather easy to stain with analine dyes and are not stained by the method of Gram. The ends of the bacilli are colored more strongly than the center. Sometimes the bacilli seem to be surrounded by a capsule ... In broth, the bacillus has a very characteristic appearance resembling that of the erysipelas culture: clear liquid with lumps depositing on the walls and bottom of the tube.

Both investigators thought that the bacillus they had found was the causative agent of plague that they had come to Hong Kong to find. But before their rod-shaped micro-organisms, their bacilli, could be determined to be the causative agent of plague, they had to be put to and pass certain other tests, which both investigators employed. These were the tests to fulfil 'Koch's postulates'. Kitasato wrote, 'I still had doubts about the true significance of what I found; I therefore made a cultivation...'. This was the first test he applied: could his bacillus be cultivated artificially, would it grow in a pure form outside the human body? Kitasato found that it would:

The growth of the bacilli is strongest on blood serum at the normal temperature of the human body (37 °C): under these conditions they develop luxuriantly and are moist in consistence and of a yellowish grey colour; they do not liquefy the serum. On agar-agar jelly (the best is good glycerine agar) they also grow freely. The different colonies are of a whitish-grey colour and by a reflected light have a bluish appearance; under the microscope they appear moist and in rounded patches with uneven edges ... If a cover-glass preparation is made from a cultivation on agar-agar, and, after having been stained, is observed under the microscope long threads of bacilli are seen.

Yersin too had success in making his bacillus grow in pure cultures outside the body:

The pulp of buboes, seeded on agar, gives rise to transparent, white colonies, with margins that are iridescent when examined with reflected light. Growth is even better if glycerol is incorporated into the agar. The bacillus also grows on coagulated serum ... Microscopical examination of the cultures reveals true chains of short bacilli interspersed with larger spherical bodies.

Once each bacillus had been successfully cultivated there was a second test to be made, the test on live animals. Kitasato mainly used the experimental animal which Koch had made indispensable, the white mouse:

The mice, which were inoculated on the first day with a piece of spleen and some blood from the finger-tips [of the first corpse post-mortemed by Professor Aoyama], died in two days' time, and at the post-mortem examination upon them I found oedema round the place of inoculation, and the same bacilli in the blood, in the internal organs, and in the oedematous part around the place of inoculation. All animals which had been inoculated with the cultivations (pigeons excepted) died after periods extending from one to four days, according to the size of the animal. The same state of the organs after death and the same bacteriological observations always obtained as in the case of the mice.

Yersin too turned to animals to test whether his bacillus was the true causative micro-organism of plague. He found it was.

If one inoculates mice, rats or guinea pigs with the pulp from buboes, they die, and at autopsy one can note the characteristic lesions as well as numerous bacilli in the lymph nodes, spleen and blood. Guinea pigs die in 2 to 5 days, mice in 1 to 3 days.

So when the bacillus was inoculated into healthy animals it produced 'the characteristic lesions'; it killed the animals in a few days; it produced 'the same state of organs after death', and 'the same bacilli' were found in the blood and in the internal organs of the dead animals.

The third test each bacillus had to pass was whether it was present in all cases of the disease. Kitasato found that his bacillus was indeed always present (well, almost always):

Every day I took blood from many plague patients and examined it, and almost every time I found the bacilli as above described, sometimes in great numbers, sometimes only few in number... On the other hand, these same bacilli were to be found at every post-mortem examination (of which we had upwards of fifteen) in great quantity in the bubonic swellings, in the spleen, the lungs, the liver, in the blood contained in the heart, in the brain, intestines – in fact, in all internal organs without exception – and every cultivation from any particle of these parts invariably produced the same bacilli.

Yersin too found his bacillus everywhere present in the body of humans or animals suffering from the plague: 'The pulp of the buboes always contains masses of short, stubby, bacilli... One can find them in large numbers in the buboes and the lymph nodes of the diseased persons', and in inoculated animals 'at autopsy, one recovers the bacillus from the blood, the liver, the spleen and the lymph nodes'.

Yersin, as a true Pasteurian, subjected his bacillus to one further question: did it exist naturally in, or could it be cultivated into, forms

THE LANCET,]   THE PLAGUE AT HONG-KONG.   [AUGUST 11, 1894.   325

Madeira; rainless and cloudless days are more frequent, and the temptations to invalids to overdo their strength are consequently greater. There is also more wind and more dust." Regarding Puerto Orotava, the author says : "In a good winter there is but little cold weather, but what cold there may be is felt, as it is accompanied by damp. From the middle of January to the end of February is the worst time, and in a bad year may be disagreeable, the sky being overcast day after day, and the sun being obscured by the thick mantle of clouds which will then envelop the island. Such winters are, however, the exception ...... a day when there is no sunshine, or when one cannot be out of doors for at least three or four hours, is almost unknown."

THE PLAGUE AT HONG-KONG.

WE have received the following notes from Dr. J. A. Lowson of Hong-Kong, who has forwarded a number of preparations of the plague bacillus, some of them prepared for him by Professor Kitasato, others prepared by himself, of which we give several representations. The organism—which is a bacterium resembling the bacilli found in the hæmorrhagic septicæmias, except that the ends are somewhat more rounded—when stained lightly appears almost like an encapsuled diplococcus, but when more deeply stained it has the appearance of an ovoid bacillus, with a somewhat lighter centre, especially when not accurately focussed. When, however, it is focussed more accurately it is still possible to make out the diplococcus form. It is quite possible that the capsule has been produced artificially,

FIG. 1.

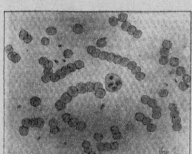

Ob. ₁/₁₂ homog. + oc. 3 compens.; length of tube 140 mm. Bacilli and blood from case of plague. Illustration of preparation made by Professor Kitasato and forwarded by Dr. Lowson.

though in Fig. 3 this does not appear to be the case. The positions in which it is most frequently met with—sometimes apparently in almost pure cultures—are the glandular enlargements which occur in the groin, in the axilla, and in the neck, though these enlargements are not always met with in the rapidly fatal cases. These enlarged glands are intensely congested, or rather they appear to be infiltrated with blood. In this blood, which is in a state of disintegration, mixed with the elements of the glandular tissue which are also broken down, the organisms are exceedingly numerous. They are also met with in considerable numbers in the spleen and in the other organs in those positions where there is a slowing of the circulation as it passes through capillary networks or sinuses. The

organisms are also found even in the blood in the heart and large bloodvessels, as seen in Fig. 1. Dr. Lowson hopes shortly to be able to send an account of the disease and also

FIG. 2.

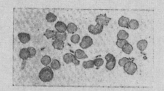

Ob. ₁/₁₂ homog. + oc. 3 compens.; length of tube 180 mm. Bacilli in blood of mouse. Illustration of preparation made by Professor Kitasato and forwarded by Dr. Lowson.

of the appearance of the micro-organism when cultivated outside the body; but he says : "I have recently been so

FIG. 3.

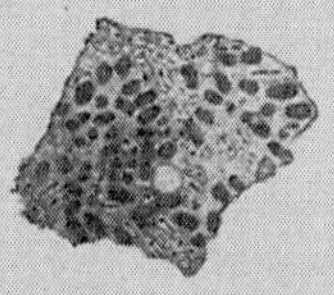

Ob. ₁/₁₂ homog. + oc. 3 compens.; length of tube 180 mm. Illustration of preparation of splenic pulp in case of plague made by Dr. Lowson.

engaged in looking after the sick, organising hospital work, inspecting insanitary houses, and looking after the disposal

FIG. 4.

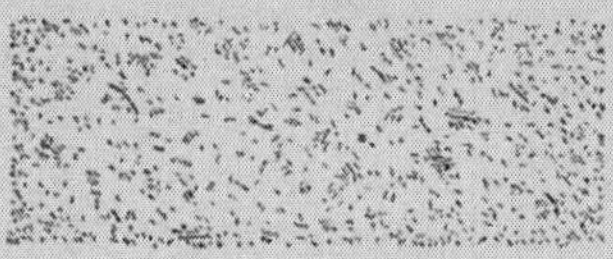

Ob. ₁/₁₂ homog. + oc. 3 compens.; length of tube 140 mm. Illustration of plague bacillus prepared by Professor Kitasato and sent by Dr. Lowson.

of the dead that I have been unable to find time to do more than send you these few notes and specimens, which, however, I thought might be of interest to some of your readers."

EXHIBITION OF AMBULANCE WORK AT CHUD-LEIGH.—On July 26th the annual exhibition of the Chudleigh Cottage Garden Society was held in a field close to Chudleigh Rocks. In the evening the Ambulance and Field Stretcher Bearer Company, under the direction of Surgeon-Captain C. L. Cunningham, gave an excellent exhibition, in which they showed that they had fully availed themselves of the lessons in ambulance work received from their director. A sham fight was organised between a supposed party of Arabs and a company of Engineers; the wounded were brought out of action and were promptly attended to with all proper detail. The men in the course of the sham fight availed themselves for signalling purposes of an ingenious method of signalling invented by Surgeon-Captain Cunningham for use in the Soudan campaign.

Fig. 10. Plague as identified by Kitasato. Note that most of the article announcing Kitasato's discovery is taken up by reproductions of the microscopic image of the plague bacillus.

with lesser virulence, and which could thus be used to give animals and humans immunity against the plague? Kitasato, as a true Kochian, asked and answered a different final question: 'What means are to be employed against the plague? – preventive measures, general hygiene, good drainage, perfect water-supply, cleanliness in dwelling-houses, and cleanliness in the streets.'[35]

The bacillus that Kitasato found and the bacillus that Yersin found passed all the tests of Koch. 'From this evidence', wrote Kitasato, 'we must come to the conclusion that this bacillus is the cause of the disease known as the bubonic plague; therefore the bubonic plague is an infectious disease produced by a specific bacillus'. Yersin concluded similarly that 'Plague is thus a contagious and transmissible disease', whose cause is the bacillus he had found.

As soon as they were each certain that they had discovered the causal micro-organism of plague, each of them got into print as quickly as possible. Yersin, typically for a Pasteurian, placed his report in the *Annales de l'Institut Pasteur*. Kitasato, however, had been befriended by the British Dr Lowson, who encouraged him to let him translate his original German text into English and to send it to London to be published in *The Lancet*. Yersin and Kitasato sent not just descriptions of their successful hunt for 'the causative micro-organism' of plague but, most importantly, pictures (Fig. 10 and 11). What was portrayed in these pictures was not the *symptoms* – the patients suffering the disease – but the *microbe*, a thing which could only be seen down the microscope. And the message of these pictures is this: 'here is the micro-organism = here is the disease (plague)'. They had taken into their laboratories a disease whose identity was constituted by symptoms; they had emerged with a disease whose identity was constituted by its causal agent.

After Kitasato and Yersin had each found their plague bacillus, the tiny micro-organism was given its new scientific name. That name was in Latin. Hitherto what had been named in Latin or Greek was the disease (*pestis*); henceforth it was the micro-organism. The name given to the micro-organism directly expresses its causal relation to the disease. Its first name was *Bacterium pestis*, the bacterium of the plague. In 1900 it was renamed *Bacillus pestis* the bacillus of the plague. From 1923 it was called *Pasteurella pestis*, which is short for 'the Pasteur-genus causative micro-organism of the plague'. A new genus, *Yersinia*, was proposed in 1954, and today the bacillus is

---

[35] For this contrast between the styles of Pasteurians and Kochians, see Brock, *Robert Koch*, p. 177.

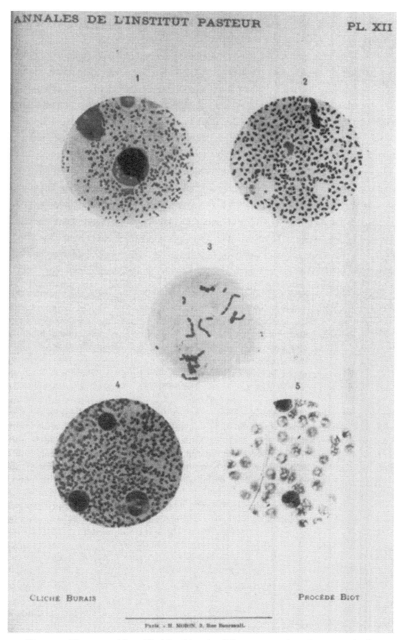

Fig. 11. Plague as identified by Yersin. These were the pictures which
accompanied his announcement of his discovery of a plague bacillus in the
Annales of the Pasteur Institute.

increasingly referred to as *Yersinia pestis*, the 'Yersin-genus causative micro-organism of plague'.[36] (Evidently the Pasteurians won the naming contest over the Kochians.)

Over the next decade other investigators, most notably perhaps P.-L. Simond working in his tent laboratory in Bombay in 1897, worked out the details about how this plague bacillus was transmitted from creature to creature by the rat flea, and the circumstances under which man becomes affected. The transformation of the identity of plague was now complete: the modern identity of plague was now established.[37]

Having looked at the central moment of identification of plague by Kitasato and Yersin, we have seen the essential role of the laboratory in it. Now we can look briefly at a subsequent moment of identification of plague to see how the laboratory thereafter was and is always crucial. Here, for instance, are the procedures adopted by Dr Wu Lien-Teh in the Manchuria epidemic of 1910–11 before he could diagnose (identify) as plague the disease he encountered:

The opportunity to perform a post-mortem upon a patient came on the morning of December 27, when a telephone message informed the office that a Japanese woman inn-keeper, married to a Chinese at Fuchiaten, had died during the night showing symptoms of cough and spitting of blood. Dr Wu and his assistant brought out their case containing all the necessary instruments and apparatus for exactly such an emergency, and at once drove to a small house lying in a poor quarter of the town. A female corpse, dressed in a cheap cotton-padded kimono, was lying on a soiled *tatami* laid on some planks raised two feet from the earthen floor. The room was dark and untidy, but sufficient water was available for a limited postmortem to be made. After the cartilaginous portion of the chest had been removed, a thick-bored syringe needle was plunged into the right auricle and sufficient blood was removed for culture in two agar tubes and for thin films on slides. Next, the surface of a lung and the spleen were scarified, and a platinum needle was inserted into the substance of each organ and the necessary cultures and films made. Pieces of the affected lungs, spleen and liver, each two inches by two inches, were removed and placed in glass jars containing 10 per cent formalin...

... The small party was glad to return to their quarters, and since no proper laboratory had yet been established, they had to work temporarily in a room allotted to them in the Chamber of Commerce. After simple staining with Loeffler's methylene blue, all the specimens from blood, heart, lungs, liver and spleen, when seen under a high-power microscope, showed swarms of the characteristic oval-shaped plague bacilli with bipolar staining at the ends. Further confirmation of plague was established in the growths in the agar tubes. After these had been left three days in the

---

[36] Butler, *Plague*, p. 25; William Bulloch, *The History of Bacteriology* (Oxford, 1938; reprinted 1960), chapter 8, 'Classification of bacteria'.

[37] Establishing the role of the rat and of the rat flea as the vector of the plague bacillus was not done without opposition; for the stories see Hirst, *Conquest of Plague*, pp. 130–74. For the key works in the medical literature, see Arthur L. Bloomfield, *A Bibliography of Internal Medicine: Communicable Diseases* (Chicago, 1958), pp. 47–60.

Fig. 12. Wu Lien-Teh in his temporary plague laboratory, Harbin 1911, photographed by himself.

ordinary temperature of the living room, small pin-point translucent colonies appeared in profusion. Films taken from one of these colonies showed the characteristic plague organisms. The cultures from the heart-blood and the spleen were quite pure, that is, they were not contaminated with other organisms; only the lung cultures showed slight contamination. This discovery was announced both to the local officials and the higher ones in Peking. The Taotai, the magistrate and the Chief of Police – all laymen – were invited to look down the microscope and be convinced, if possible, of the true cause of the suspicious deaths, but it was not always easy to convince persons who lack the foundations of modern knowledge and of science...[38]

The patient showed certain symptoms (including that most striking one of all, sudden death); these symptoms lead the investigators to suspect plague. 'But is it plague?', they ask. These are the procedures that Dr Wu follows in order to answer this question with certainty, either in the affirmative or the negative. He carries out a post-mortem in order to acquire samples of some of the deceased's organs and blood, 'and the necessary cultures and films' are made. The

[38] Wu, *Plague Fighter*, pp. 11–12.

investigators then return to a room which is their makeshift *laboratory* (Fig. 12). First confirmation that the disease is plague comes from staining the specimens and then inspecting them under a microscope: the characteristic plague bacilli are observed to be present. 'Further confirmation of plague was established' by inspecting some time later the growths in the agar tubes, where again the characteristic plague organisms were readily seen in artificial culture. No attempts were made on this occasion by Dr Wu to test whether experimental animals would show the symptoms of the disease after inoculation, but he seems to have thought that the evidence in front of his eyes was sufficient, for he invited the laymen present to look down the microscope at 'the true cause of the suspicious deaths'; but they could not readily be convinced that this microscopically small thing could be 'the true cause' of a disease. Wu was one of those who wished to modernise (that is, westernise) traditional China, as his autobiography makes clear. He was right to attribute the unwillingness to believe exhibited by the local Chinese officials to their ignorance of the foundations of modern (= western) knowledge and science. In particular they lacked the laboratory way of looking and thinking. Indeed this moment of incomprehension in Manchuria, this clash between cultures, sums up the transformation of the identity of plague; for, in order to see a micro-organism as 'the true cause' of a disease one needs to be already in the laboratory thought-world. To be able to see 'the true cause' of a disease down a microscope, one needs to bring to it eyes and mind familiar with the laboratory way of seeing.

Plague transformed

Thus, as we can see here in the case of plague, the coming of the laboratory created a watershed in the identity of infectious diseases. Indeed, one could be even more precise, and actually specify the date on which they acquired their new identities: for plague it was 14–24 June 1894.[39] But in speaking of 1894 as the watershed in the definition and identity of plague, I do not mean to imply that everyone immediately in 1894 adopted this new 'bacteriologic' (as I call it) understanding. We have already seen how difficult a way of thinking this was for Chinese officials to adopt. But it was also very difficult for western medical men. There was a long battle between on the one hand the bacteriologists, and on the other the 'epidemiologists' or

[39] Kitasato found his bacillus on 14 June, Yersin his on 24 June; there is continuing debate over whether they were the same one; see note 27.

'localists' who believed some 'epidemic constitution' or special state of the soil was necessary before an epidemic could occur.[40] This battle between Kochians and Pettenkoferians, between as it were Berlin and Munich, between bacteriologists and hygienists, continued a long time. Pettenkofer's great challenge to the germ theory is famous: in 1892 he drank a culture of the cholera vibrio and did not get cholera! Indeed, one might say that it was his stubborn resistance to the germ theory both in mind and in body which obliged him ultimately to commit suicide. Similarly a long resistance was put up by clinicians, epitomised by Professor Rosenbach and his book *Physician versus Bacteriologist* of 1903.[41] For a long time it was possible to say, with the eminent British abdominal surgeon Lawson Tait in the late 1890s, that the true *causa causans* of typhus fever was not a microbe but the terrible living conditions of the poor, and that 'the laboratory fails utterly': 'Human beings alive differ in their individual results from exactly similar conditions and in ways altogether irreconcilable with the laboratory facts of bacteriology.'[42]

The fight was probably not finally over until the 1930s.[43] But eventually the bacteriologists won. We are all bacteriologists now, and none of us would attempt to identify plague today without a laboratory. To oppose the claims of bacteriology is now not a rival view, nor an alternative view, nor even a dissident view. It is now a lunatic view. That indicates how comprehensive the victory of the bacteriologists has turned out to be. But while the fight was still on – while the bacteriologists were for the first time promoting their new

[40] See Winslow, *Conquest of Epidemic Disease*, chapter 15, 'Pettenkofer – the last stand'; Richard J. Evans, *Death in Hamburg: Society and Politics in the Cholera Years 1830–1910* (Oxford, 1987), esp. pp. 237–43 and 490–507.

[41] First published as *Artz contra Bakteriologe*. See Russell C. Maulitz, '"Physician versus Bacteriologist": the ideology of science in clinical medicine' in Morris J. Vogel and Charles E. Rosenberg (eds.), *The Therapeutic Revolution: Essays in the Social History of American Medicine* (Philadelphia, 1979), pp. 91–107.

[42] Tait talked of the true 'causa causans' of typhus fever in a letter to the *British Medical Journal* (1899, 1), p. 879. The other two quotations from him are from 'The evolution of the aseptic method in surgery', *Medical Press and Circular* (1898, 1), pp. 427–30, at p. 427, where he writes also: 'this mysterious "life" is the most perfect antiseptic we have, and this prime fact is that which makes all "cultures" and all laboratory experiments fail absolutely in giving results which can safely be applied to the living body, particularly to man... When we come to consider man we find all our facts at sixes and sevens, and all theories of no avail. The laboratory fails utterly; and whilst in the processes of his body we may see verisimilitudes with those of the lower animals, we are met at once by facts which show they are not...'. On this issue in general see also Lloyd G. Stevenson, 'Science down the drain: on the hostility of certain sanitarians to animal experimentation, bacteriology and immunology', *Bulletin of the History of Medicine*, 29 (1955), pp. 1–26.

[43] See Howard-Jones, *The Scientific Background of the International Sanitary Conferences*, p. 89.

way of seeing, their laboratory way of seeing, and claiming total truth for it – at that time it was an essential (and spontaneous) part of their strategy to claim that they, and only they, had mastered the causal secrets of the age-old scourges of mankind. This is also of course the way they experienced it. And this view – that theirs was the first *successful* understanding of plague and other terrible diseases, which replaced the old, unsuccessful and *misguided* attempts – was the basis of the way they now rewrote the history of man's life with disease. The bacteriologists presented the laboratory as the glorious weapon giving men victory at last in an age-old battle between men and microbes, hitherto fruitlessly fought with the wrong weapons by their enemies – the Pettenkofers of the past and the present. They now had new heroes whom they exhumed from old books to be early microbiologists and bacteriologists, men such as Fracastoro, Leeuwenhoek, Redi, Spallanzani and Semmelweis. They themselves were the successors to these far-sighted men whose fate had inevitably been not to have been appreciated in their own day.[44] This bacteriologic account of the history was presented as the story of the fight of evidence and common sense over theory and stupidity. And this is why, in writing their histories of the triumphs of laboratory medicine, bacteriologic historians have usually applauded Kitasato and Yersin as having correctly *focused on* the essential thing about plague, its causative bacillus, and not noticed that by deploying the laboratory in investigating plague, Kitasato and Yersin were *making* the bacillus the essential thing about plague.

This bacteriologic view of the past is still shaping our accounts of infectious diseases in the long past before the laboratory. To take just

[44] For instance, nowadays we tend to see Ignaz Semmelweis (1818–65) as an early worker in the cause of the germ theory, because of his views on childbed fever. But formal recognition of the supposed 'contribution' of Semmelweis to bacteriology and antisepsis was first made only in the 1880s, by Joseph Lister – that is, after Lister himself had received widespread praise and acceptance for his introduction of antisepsis. And even then the claims of Semmelweis to 'credit' were only recognised through the insistence of the Hungarian nationalist, Dr Duka, in virtually forcing Semmelweis on Lister's attention. As Lister's biographer wrote: '[Semmelweis'] work and almost his name were forgotten until this festival at Pesth was held in Lister's honour ... The history and fate of this great and unhappy man [Semmelweis] excited the liveliest interest in Lister. Here, at the completion of his antiseptic triumph, quite unexpectedly he came across a true but almost forgotten forerunner, who owing to the happy deduction of an original mind fought the unseen enemies of disease with weapons similar to Lister's, though unsupported by the scientific evidence which Pasteur had given to Lister', G. T. Wrench, *Lord Lister: His Life and Work* (London, 1913), pp. 347 and 342. Semmelweis was constructed as two heroes at once by Duka: as a Hungarian patriot and as a microbiological 'predecessor', and each facet received lustre from the other; see Liza H. Gold, 'Ignaz Philip Semmelweis and the reception of his work on puerperal fever in 19th century Britain', unpublished M.Phil. thesis (Wellcome Unit for the History of Medicine, Cambridge University, 1982).

one instance, it underlies Carlo Cipolla's account of the plague in Prato in 1629–30, such that Cipolla can write that the medical officers of Prato 'fought against an invisible enemy. They did not know what the enemy was, or how it struck. Medical knowledge was of no help and medical treatment was of no value.'[45] The only kind of knowledge about the 'enemy' that matters in Cipolla's eyes is knowledge about the bacillus! This attitude is completely typical of most historians' accounts and assumptions today. Yet these people in the past *did* know what plague was and how it struck: they just knew these things differently from how we know them, because the laboratory way of thinking did not exist.

In its co-option of the history of infectious diseases, this bacteriologic view is also the reason why the transformation in the identity of plague has gone unnoticed, not only by the historians but also by the very bacteriologists who made these new identities for infectious diseases. For Kitasato and Yersin themselves were at pains to point out that the disease whose cause they had discovered was indeed the same disease which had been a scourge for centuries. They both said (and, indeed, so did the newspapers at the time) that the disease whose cause they had discovered was 'ancient bubonic plague'. Kitasato, for instance, writes in his report:

History shows us that plague epidemics existed in the fourteenth century both in Asia and Europe, and thousands of human beings perished. Since then from time to time, now here, now there, an epidemic has appeared, and until lately the disease almost seemed to have vanished from the face of the earth. This, however, was not so. In China it has existed to this day... The recent outbreak has given us opportunity for studying this disease – a cause of mystery for centuries – with the means which modern science places in our hand...

And indeed they were right. But only at that moment. For at the moment Kitasato and Yersin decided to go into their respective laboratories carrying their blood and tissue specimens, they were working with ancient bubonic plague. But by the time they came out of their laboratories, they had given plague a new identity. From this moment onwards plague would never – *could* never – again be identified by symptoms alone, but by its causal agent. It is this detour through the laboratory which means that pre-laboratory plague and post-laboratory plague do not have one consistent and continuous identity.[46]

---

[45] Carlo N. Cipolla, *Cristofano and the Plague: A Study in the History of Public Health in the Age of Galileo* (London, 1973), p. 120.

[46] On the importance and significance of the detour through the laboratory for plague see Latour, *Pasteurization of France*, pp. 94–100.

The laboratory construction of plague means that there is an unbridgeable gap between past 'plague' and our plague. The identities of pre-1894 plague and post-1894 plague have become incommensurable. We are simply unable to say whether they were the same, since the criteria of 'sameness' have been changed. As I have been arguing, this is not a technical medical issue but a logical, philosophical and historiographic one. Nevertheless, historians and bacteriologists regularly put themselves through intellectual contortions in their determination to make identifications across this divide, and presumably will continue to do so, bizarre as their assertions sometimes are by their own usual standards of evidence and proof.[47]

The laboratory identity of modern plague and other infectious diseases is now such a great truth that it has become enshrined in that higher truth, modern fiction. Fictionalised plague pre-1894 can be seen in Daniel Defoe's *Journal of the Plague Year*, written in 1722 and posing as an account of the London plague of 1665. While Defoe himself, in the voice of the narrator, asserted his belief that the disease was spread by infection, 'that is to say, by some certain steams or fumes, which the physicians call effluvia... immediately penetrating

---

[47] What is typically done by such enthusiasts is to identify the disease in the past on the basis of symptoms alone, working from past descriptions of 'plague' – and despite the fact that the symptoms mentioned do not tally with the modern set of symptoms. Then an unconscious logical leap is taken to identify this symptom-identified disease with our modern cause-identified plague. Hence one infers the presence of the plague bacillus in these past diseases (and treats it as the cause of the disease) from an ill-fitting set of symptoms which had been recorded by people with a pre-lab way of seeing and a pre-lab set of diagnostic categories! For instance, Dr Wu claimed that 'It is undeniable that some of the symptoms of the Black Death [of 1348] are not frequently encountered in modern outbreaks of plague, while others are altogether absent... The fact that the Black Death does not quite correspond to the form of infection as it is known today cannot eliminate the ample evidence that it was plague. The description of both the bubonic and the pneumonic types, as given by contemporary observers, leave no room for doubt', *A Treatise on Pneumonic Plague* (Geneva, 1926), p. 3. But, apart from anything else, the bubonic/pneumonic distinction had not in fact been available to those 'contemporaries', since it derives from seeing the bacillus as the cause of both of these clinically different conditions (as discussed earlier). Again, it has been written by modern historians of medicine that 'Considering the extent of the disaster, we know surprisingly little of the Black Death; for instance, the very scanty descriptions of signs and symptoms do not so much as mention blood-stained sputum (though *vomiting* of blood is described), yet this is one of the cardinal signs of a virulent pneumonia. Nevertheless, the widespread and high mortality of the plague indicate that it must have been predominantly of the pneumonic type...', F. R. Cartwright in collaboration with M. D. Biddis, *Disease and History* (London, 1973), p. 38. Similarly, Hirst, discussing Emile Rocher's 1871 account of the Yunnan epidemic, writes: 'The detailed clinical description of the human cases leaves no room for doubt that the disease was bubonic plague. Rocher also states, however, that a number of domestic animals were affected, including cattle, sheep, and goats, and sometimes even farmyard birds. We now know that none of these animals is subject to true plague', Hirst, *Conquest of Plague*, pp. 101–2. Identifications like this require something of the eye of hope.

the vital parts' of sound persons, 'putting their blood into an immediate ferment', yet he looked 'with contempt' on the opinion of those

who talk of infection being carried on by the air only, by carrying with it vast numbers of insects and invisible creatures, who enter into the body with the breath, or even at the pores with the air, and there generate or emit most acute poisons, or poisonous ovae or eggs, which mingle themselves with the blood, and so infect the body.[48]

He condemned this as 'a discourse full of learned simplicity, and manifested to be so by universal experience'. So it is clear that Defoe's plague was transmitted by the 'seeds of disease' but not by a unique and specific material cause. We may contrast this with H. G. Wells' famous story, 'The stolen bacillus'.[49] This was published on 21 June 1894, which was by extraordinary chance in the few days interval between Kitasato's (14 June) and Yersin's (24 June) discoveries of their plague bacilli in Hong Kong. In this story, the absent-minded Bacteriologist shows round his laboratory a mysterious and evil visitor ('Certainly the man was not a Teutonic type nor a common Latin one. "A morbid product, anyhow, I'm afraid", said the Bacteriologist to himself'). The Bacteriologist takes up a sealed tube, saying 'Here is the living thing. This is a cultivation of the actual living disease bacteria ... Bottled cholera, so to speak.' His visitor, as is obvious from the description of him, is an Anarchist, who steals the tube and with it intends to poison the whole city and thus advance his nefarious political aims. Wells' story begins that popular view of the mighty power of the microbe in the laboratory tube: the deadly single material cause of untold misery. Just tip it into the water supply and the cholera is abroad: 'those little particles, those mere atomies, might multiply and devastate a city! Wonderful!', as the Bacteriologist exults.

The power of the laboratory in controlling the identification of infectious disease is perhaps best shown in the most famous appearance of plague in modern fiction, Camus' *La Peste*, about a fictional outbreak in an Algerian town in the 1940s.[50] Part of the early drama revolves around the identification of the disease. The laboratory is therefore at the centre of things, but it has not yet spoken. Castel, an older practitioner, comes to see the hero Dr Rieux.

---

[48] Daniel Defoe, *A Journal of the Plague Year* (first published 1722; Harmondsworth, 1966), pp. 92–3.
[49] First published in the *Pall Mall Budget*; quoted from *Selected Short Stories* (Harmondsworth, 1958; reprinted 1981), pp. 149 and 145.
[50] Albert Camus, *La Peste* (first published 1947; Paris, 1972; reprinted 1988), pp. 39–41 and 45; my translation.

'Naturally', he says, 'you know what this is, Rieux?'

'I am awaiting the result of the analyses.'

'Well, I know what it is. And I don't need any analyses. I spent part of my career in China, and I saw some cases in Paris twenty years ago. It was just that one didn't dare give it a name at that moment. Public opinion is sacred: no panic, above all no panic. And, then, as a colleague said, "It's impossible, everyone knows that it has disappeared in the West." Yes, everyone knew it, except the corpses. Come on, Rieux, you know as well as I do what it is.'

There is a pause while Rieux reflects and looks out of the window.

'Yes, Castel', he said, 'it's hardly believable. But it really appears as if it could be plague'

...

The word 'plague' had just been uttered for the first time.

A little later Rieux is presented with the mortality statistics by a clerk. 'We ought to make up our mind to call this disease by its name', Rieux says. 'Up to now we have just been shilly-shallying'. But Rieux cannot in fact bring himself to mention the name to the clerk. Not yet. For the analyses have not been completed – the laboratory has not yet spoken. Instead he turns to the clerk and says, 'But come with me. I must go to the laboratory.'

X

IDENTIFYING DISEASE IN THE PAST:
CUTTING THE GORDIAN KNOT

I: DISEASE IDENTITY

How should we historians approach the issue of the identity of disease in the past? Can we legitimately identify past diseases? Can we legitimately identify past diseases with present diseases? Can we legitimately identify particular epidemics in the past? Can we legitimately talk of the evolution of diseases or pathogens? Can we legitimately reach past people's experience of disease by identifying what they were suffering from? Are our attempts at retrospective diagnosis legitimate? Is retrospective diagnosis either possible or desirable?

These questions —surprisingly— virtually never crop up in the work of historians of medicine, who for the most part assume that the identification of past diseases is simply not a problem, since they assume the continuous identity of past diseases with modern diseases. They just get on with identifying past diseases in modern terms – whether it is logically, philosophically or historiographically possible or not. Hence my use of the term 'legitimately' in my questions. Certainly we *can* make such identifications, and we do. But do they mean anything? Do they mean what we want them to mean? Are they logical, sensible, and coherent things for us to do? Most important of all, do they tell us anything at all about *the past*, or are they simply projections backwards of present-day issues and concerns?

Before we can explore these issues we have to ask a fundamental question: what is disease? There has been a lot of confusion on this issue, based on misplaced and un-inspected assumptions. What we can say is that, at its most fundamental, disease is (1) an *experience* – an experience of debilitation, pain, suffering, together with (2) the spontaneous *appearance of non-customary phenomena* with respect to the body, such as spots, vomiting, sweating, aches, and (3) with *outcomes* of recovery, death or disability. In this, disease is something that humans and animals have in common. But there is one big difference. For unlike other animals, humans seem to insist on seeking reasons or *causes* for disease: for its incidence, its origin, its course, its outcome. Some of the reasons given in the past and in other present-day societies seem to us to be either unreasonable or irrational – something which prejudices them in our eyes and could exclude them from our considerations. So I prefer to speak in terms of *causes*, since this is a more inclusive term than *reasons* or rationales. It is this 'cause' dimension in human disease which means that *disease is always experienced socially*, that it is not just a biological phenomenon but just as much a social phenomenon. It is not just that we have to use language, itself essentially social, as our only means to think and express ourselves about disease. It is rather that we seem to be unable to talk about or even conceptualise disease without invoking *cause* in one sense or another. Different societies, separated culturally by space or time, will have different views as to what states constitute disease and what its causes are. But at all times and places, in all societies, disease identity, and especially the cause dimension of it, is going to be an expression of how people in that society think the world functions. Cause, with respect to disease, is thus expressed in many ways and thought to be many different kinds of thing, from djins and the evil eye, to humoral imbalance, to germs and vitamin deficiency, to poor DNA and social disadvantage. For we can only think about our experience of disease —as of anything else— in the terms and categories of whichever particular society we are in.

For almost 150 years the dominant model of disease we have had in Western developed societies has been that many diseases are caused by some *encounter* of the human body with some dangerous element (or a *failed* encounter with some necessary element) 'out there' in Nature. This model applies primarily to our categories of infectious disease and deficiency disease, but it is pertinent also to a large extent also to degenerative diseases, which occupy a large part of our present-day disease landscape. As we also believe —since this is part of our modern scientific world-view— that Nature is pretty consistent in its behaviour, and the elements in it (whether good or bad with respect to humans) are pretty constant in their behaviour, so it feels obvious to us that these encounters with Nature or the environment are —and always have been— constant too. Thus when talking about disease identity we tend to make a little logical leap at this point, and assume that disease identity has been pretty constant over time, and that diseases are themselves a-cultural and a-social, that they have an identity quite separate from the social circumstances in which they are experienced. It seems obvious to us, looking through our scientific medicine spectacles, that of course social *interpretations* of disease (and what 'counts as' a disease) do and have varied from society to society, but it seems to us that these just express greater or less success in coping with the underlying constant disease reality 'out there' in Nature. As we assume that our own success in coping with disease has been the greatest, we naturally take *our* models of disease identity as the final, and thereby the only legitimate, models. In particular, for those diseases where we believe some minute material living entity to be the cause, we believe we in effect encapsulate the whole disease *experience* by talking about the *encounter* between microbe and human organism. So when we come to doing the history, when we come to trying to identify past outbreaks of plague for instance, we assume that what we need is the best *modern* thinking about the disease and its manifestations. Armed with this supreme form of knowledge we are able, we believe, to *correctly* identify outbreaks of plague in the past, even down to pronouncing on the presence or absence of the bacillus, and we correct the people of the past in their identifications of plague, telling them when they were right and when they were wrong, since our form of knowledge is clearly superior to theirs.

Yet in fact this is just our society's way of thinking: true for us and our world, but not necessarily true for other societies and other times.

Some of the large issues about the problems in making retrospective diagnosis have been raised very ably in a recent extended critique of other historians' approaches[1]. Using Ludwig Fleck's analysis of the history of the concept of syphilis, it is there made clear that, through all their many changes, disease concepts are *always* social products. But, even more strikingly, it is shown that the maintenance of the *stability* of disease concepts is *also* a social phenomenon, achieved by social reinforcement. As a necessary consequence of this social reinforcement, 'the historicity of [disease] concepts is necessarily eliminated'. This elimination of the historicity of disease concepts is 'a necessary by-product, within «popular science» of the modern concept itself'. This is how diseases are attributed eternal reality by us: 'the modern concept is extended backwards in time: the disease as presently conceived is seen as a permanent entity, and it is assumed that it can be diagnosed retrospectively'. In the light of this account of the social construction of disease identity *and* of the permanent being of diseases, it can now be quite easily understood why the assumption that retrospective diagnosis is both possible and desirable, has so dominated the history of medicine. Medical historians have turned to modern medical knowledge to help their investigations of the past, and modern doctors have felt themselves specially well-placed to make retrospective diagnosis themselves. But the assumption of the persistence of disease identity is simply that: an assumption – and one which is not open to proof or disproof, because of the incommensurability of old disease concepts with new disease concepts.

Once we understand the sources of our unjustified assumption about the validity of retrospective diagnosis, we can stop trying to do it. But where should we go from there? The author of this particular critique recommends that instead of concentrating on *diseases* (as real entities), we should instead concentrate on disease *concepts* (as thought entities), and write the history of these[2]. However, this alternate approach —that we should be studying the history of disease concepts— is also subject to fatal criticism, primarily that concepts, as things thought, don't actually have histories. Concepts are

---

[1] WILSON, A. (2000), «On the History of Disease Concepts: The Case of Pleurisy», *History of Science* 38, 271-319; the quotations are from page 275.

[2] Ibid., especially the conclusions, pp. 303-6.

the product or outcome or perhaps the elements of thinking; they do not have their own histories separate from the thinking act. This thinking act, by contrast, certainly does have a history because it is a human activity[3].

The main thrust of the present paper is to follow this line and to offer an alternative approach by cutting through the Gordian knot of disease history. It will involve turning our attention away both from *diseases* (the old way) and also away from disease *concepts* (the proposed new way), and turning it instead towards *how diagnosis happens*. In other words, to *people thinking and acting* in particular cultures, situations and times. It will be noted that looking at *people thinking* is not the same as looking at their *mental concepts*, although of course mental concepts are involved. Looking at *people thinking and acting* will, I believe, give us a properly historical view of disease history, placing past disease firmly in the past, and interpreting that past experience of disease in such a way that people of the present may empathise with that past experience, but not turning it into some early version of modern disease and hence of modern experience. By making *how diagnosis happens* central to our historical investigations, we are using the only sure thing we have, the only thing which we can rely on. For it is by the act of diagnosis that disease identity is given or established. The operations that humans perform in making diagnosis are not just the key to disease identity, but the source of disease identity. The only identity disease has is this *operational* identity. I see this as the equivalent of Alexander the Great's solution to the complexities of the Gordian knot. He simply sliced through it, rather than seeking to disentangle it. If we concentrate on *how diagnosis happens*, we need no longer worry about disentangling disease entities, disease concepts, linguistic and conceptual incommensurability, germ evolution or anything else.

Diagnosis always proceeds operatively or operationally, by people asking and answering one or more specific questions about the patient and his or her affliction. But these sequences of operations differ from one medical system to another. And while these operative sequences are what actually *give* the disease identity at all times and in all cultures, we all tend to think that they are merely procedures enabling us to *recognise* the disease, with its (supposed) pre-existing identity.

Here I shall make two points about the identity of disease as it affects the practice of the historian, and which are built on this principle that *how diagnosis happens* is the source and key to disease identity at any time and in any society.

The first of these two points is extremely simple. It is that *the identity of any disease is made up of a compound of elements, of which the biological or medical is only one,*

---

[3] This criticism of the pursuit of a history of *concepts* of disease, I hope to justify at length in the larger work of which the present essay is a preliminary part. A second problem with trying to write a history of disease concepts is that when we ask 'What is a disease concept a concept of?' the answer still has to be *disease*. So disease —the very category which is problematic here— remains the focus in this approach, and its ontological status still goes unexplored, or appears to be begged.

and sometimes the least important one. This is to reiterate the point that disease is always experienced socially.

The second point is as simple as the first one. It is that *you die of what your doctor says you die of.* Your cause-of-death certificate is not negotiable. While this might seem a reasonable thing to say about people dying today, I want to argue that it also applies to everyone in the past. *They* died from what *their* doctors said they died of. *Their* cause-of-death certificates (as it were, for of course such certificates are very modern and very western) are equally not negotiable, neither by the modern medic, whether clinician, pathologist, epidemiologist or psychiatrist, nor by the modern historian.

Of course, in the matter of specifying cause of death, the doctor or other practitioner (even a witchdoctor) is simply a bystander who has been given or conceded special authority by the other bystanders to speak, and whose pronouncements are thus accepted as locally definitive. So for situations and societies where there are no acknowledged doctors or other practitioners with equivalent special authority, it is the bystanders, whoever they may be —those whom John Graunt in 1672 called 'the generality of the World'— who define cause of death and thus determine the array of available diseases in any particular time and culture. The more general form of my point that *you die of what your doctor says you die of* can therefore be reformulated as *People die of what their bystanders say they die of,* with it being understood that there is often a special class of bystander (the doctor) who makes claims to, or is ascribed, special authority based on specialised knowledge. But in general the specification of cause of death requires no medical expertise: if there is no medical practitioner present, then anyone can do it, and does, and what they say goes. No matter how ignorant we may regard such bystanders as being, nevertheless the cause of death they ascribe is not negotiable afterwards.

Only if a doctor is involved does specification of the cause of death require *medical* expertise. But in all cases, including those where a doctor is involved, what the specification of the cause of death really requires is *social* expertise. That is to say, it requires full immersion in and acquaintance with the mores and beliefs of the pertinent society. This is what everyone brought up in a given society possesses as second nature, and it is precisely what the outsider to that society, whether anthropologist, missionary, visitor from Mars or even historian, does *not* possess. What the outsider does, spontaneously, is *translate* what he or she sees or hears from the bystanders into his or her own language and culture. And as we all know, *traduttore traditore.*

To illustrate these points I need a modern moment of death, a volunteer from the audience as it were. The particular volunteer I have chosen is my own father, partly I suppose as a personal *memento mori.* My father died at about midnight on the longest day of the year in 1987, at home in Swansea in Wales. He had been ill for some months, and had spent a period in hospital where he had undergone abdominal surgery. He was 76 years old, but did not die of old age because, although it was once a regular cause of death, it is a very difficult thing to die of these days. The illness from which he died is what might be called 'the disease which dare not speak its name' (if I may paraphrase Oscar Wilde). In my father's case this was quite literally so. For my mother reports that just before he died, and fully aware that these were his last moments, my father in his very weakened voice told her he loved her, spoke of the last matters that needed to be attended to when he'd gone, and then said, 'Is it —(pause)— cancer?' My mother said, yes, it was. Then he died.

So my father had experienced a long illness which had made him progressively weaker and which over a number of months had caused him to waste away in front of his own eyes; he had been into hospital for what was probably only the second time in his life, and he had received major invasive surgery for this illness; he had discussed his condition with his doctors, his wife, his family and visitors over a period of months; and yet no-one had felt able to tell him what he was suffering from, and nor had he asked. He was not a shy man about such matters. If he had wanted to know, he would certainly have asked. But he didn't want to know. He really didn't. He waited until the very moment of his death to have his worst fears confirmed – and they were. Because it *was* cancer. In fact it was cancer of the bowel.

How do we know this? Why can we trust this? Well, we take the doctors' word for it (at least I did). But how did the doctors themselves know what my father died of? They did so by following their training as modern doctors. How else? Thus, my father complained of certain problems, which happened to be rectal bleeding, severe abdominal pains and weight loss. This, together with my father's age, raised bowel cancer as a possible clinical diagnosis in their minds, something they had learned at medical school. They then went and tested the provisional diagnosis, by using the modern methods of diagnosis they had been taught. To distinguish with certainty between a benign and a malignant (cancerous) tumour or growth, a biopsy was performed. These tests were at the cellular level, because this is the primary level of medical understanding today, that organic diseases are cellular phenomena. Thus carcinomas are nowadays defined as products of abnormal cell activity, or as someone has said, cancer is 'the misguided cell'[4]. With this evidence from the laboratory, the doctors could be certain of the identity of the disease in biological terms: they now knew what my father was suffering from. As far as treatment was concerned, the doctors again naturally turned to what they had been taught when they were trained. Thus, as the tumour seemed to them to be so far confined only to a portion of the bowel, so their next move was to submit my father to surgery to remove the affected

---

[4] PRESCOTT, D. M. and FLEXER, A. S. (1986), *Cancer: the Misguided Cell*, Sunderland, Massachusetts, Sinauer Associates Inc. Another author calls cancer 'the wayward cell'; see Richards, V. (1978), *The Wayward Cell: Cancer, its Origins, Nature, and Treatment*, Berkeley, Los Angeles, University of California Press. GRAHAM, R. M. (1963), *The Cytologic Diagnosis of Cancer*, Philadelphia, W. B. Saunders Company, notes that methods for securely distinguishing cancerous from non-cancerous cells have been developed only since the 1940s (see Preface).

section. In this way the doctors hoped to stop the cancer spreading, either by direct extension from the primary site, or by the cells entering the vascular system by invading the lymphatics or blood vessels. So a large length of bowel was removed, and the healthy ends of the intestine were joined up. My father did not respond as hoped. So the doctors concluded that the cancerous cells had spread and that there was nothing practical more that they could do for him, especially given his age. So my father went home, where he grew weaker and gradually faded away, all his actions and responses slowing down until even watching him eat, dress or speak required great patience on everyone's part[5]. Eventually he took to his bed and never got up again.

So in the present example, my father died of cancer of the bowel because this is what his doctors said he died of. And that's that. His cause-of-death certificate is not negotiable, either today or in the future. The diagnosis was offered, tested and confirmed, through all the procedures that modern Western hospital-based medicine requires. The cells had been found guilty. On the basis of the diagnosis the treatment was given. The sequence of diagnostic steps tells you that this is the answer, that this *has to be* the answer. There is no alternative diagnosis possible in this case, nor any other possible cause of death available.

With respect to *how diagnosis happens* today, and that *you die of what your doctor says you die of*, this must suffice for now. But what of the other point, that *the identity of any disease is made up of a compound of elements, of which the biological or medical is only one, and sometimes the least important one*? Of all modern disease categories, cancer is perhaps the one most straightforward for making this point about how far the social element makes up the disease's identity. It is clear from my father's own reaction —and this is something widely shared in our modern society— that cancer has a very special place in the modern disease spectrum. It is the most feared of modern diseases, in most of its forms. The social meaning with which we load cancer, and lead each other to load it with – seems to be the reason why cancer has this special position of being unmentionable in our social experience of disease in today's society. Susan Sontag, the celebrated modern commentator on the social dimensions of cancer, has remarked that no-one is embarrassed to say that they have had a heart attack if they've had one. Similarly, no-one will hide the fact of their heart surgery if they have had heart surgery. But few people want other people to know that they have cancer. In many cases, my father's included, they themselves do not want to know that they have cancer. There is something about our attitude to this disease which makes it different. It comes with overtones of dirt and shame, and of blame and punishment, which few other diseases have. Lester and Devra Breslow

---

[5] All this was rationalised by my father in a way which made cancer irrelevant as a possible cause. As he knew sections of his bowel had been removed (though did not ask why) it was obvious to him that there was not enough intestine left to detain and absorb the food long enough to nourish him. As he noticed that his food just 'fell straight through' him, he was not at all surprised he was fading away. He explained all this to me. Very slowly.

have written that 'Cancer was and is still perceived largely as a disease that attacks individuals one at a time, each uniquely, rather than as a public burden requiring large-scale public efforts'[6]. This focus on the individual as responsible for his or her cancer has, according to the Breslows, been inadvertently maintained, at least in the U.S.A., by the action of interested parties such as the private medical business, the private industrial business and the biomedical research establishment. So strong is this social element in the identity of cancer in our society today, that it over-rides and renders nugatory the medical view that it is just a few cells misbehaving, and that five-year survival rates from cancer treatment are improving all the time. My father did not die from a disease whose identity is purely biological. His experience (demonstrated precisely by his refusal to voice his concerns), together with the experience of the doctors around him, the experience of his family, and the experience of all the other bystanders, was that he was dying from a shameful disease, the disease which dare not speak its name. All this human experience was part of the identity of the disease.

The social component of the identity of cancer may be the most easy to point out, thanks to the work of sociologists and other commentators over the last few decades. But every disease has its social component, and the social component (like the medical component) of a disease can and often does change over time. This applies equally to all conditions labelled as diseases in all societies. I hope to make that case at length elsewhere sometime.

I have here only given one case history tracing the steps of diagnosis and treatment appropriate to the late 20th century. I chose it because of its familiarity: we all know how the suspected cancer patient is treated today. Now imagine tracing the steps of diagnosis and treatment of any episode of disease of the past. Here we will be on unfamiliar territory. The conditions will be different, the content of diagnosis and treatment will be different, the thought patterns guiding diagnosis and treatment will be different. And yet, the same conclusion will have to be reached by us: that the patient died of what his doctor said he died of. Because there is nothing else in the encounter. There is no 'real' disease, with an identity separate from its sufferers at any given time, which can be separated out as a timeless entity for us to give our modern labels to, years —centuries— after the events.

II: THE ARRAY OF AVAILABLE DISEASES

The present conference paper is part of a longer argument about disease identity in the past that I am currently engaged on, and there can be room here only to discuss one particular phase of the longer argument. So to support my major claims about the

---

[6] BRESLOW, L. and D. M. BRESLOW (1982), «Historical Perspectives [on cancer epidemiology and prevention]», *Cancer Epidemiology and Prevention,* Schottenfeld, D. and J. F. Fraumeni, eds., Philadelphia, W. B. Saunders Co. 1039-1048, 1040.

identity of disease, and in particular the claim that *you die of what your doctor says you die of*, I shall look here only at what I shall call *the array of available diseases* in any given historic or modern society, that is to say at what set of diseases is thought to exist at a given time in a particular society. Other questions that arise naturally from this, such as how diseases are added to or subtracted from these arrays, or how individual cases of disease are mapped onto these arrays in any particular historic or present society, as also with most questions about the operative steps involved in diagnosis, I will have to leave for another occasion. For convenience I shall use here only examples of diseases that are or were believed to kill people, but the analysis could in principle be deployed also for all non-fatal diseases too.

Different societies and different periods in any one society have different sets or arrays of 'available diseases'. Such arrays are different both in what they include and also —and this is even more important— in the theoretical underpinnings of what constitutes the identity of a particular disease. Hence what you can die of at any particular time differs. To illustrate what I mean by this I shall compare three cause-of-death analyses drawn up at intervals of about 160 years from each other. They are

(1) the London Bills of Mortality, kept from the 1590s to 1849, as analysed by John Graunt in 1672;
(2) the first comprehensive report on causes of death in England, viz. the statistics drawn up by the Registrar-General in 1839, as commented on by the Compiler of Abstracts, William Farr; and
(3) what is at the time of writing the most recent edition of *Health Statistics Quarterly* for England and Wales, the issue covering the year 1999.

At first glance the three documents and the information they contain seem reasonably comparable, given that they cover much the same area and society (England, London), are in the same language (English), were drawn up for much the same reasons (to track contemporary causes of death), were drawn up by people with a strong interest in 'political arithmetic' or statistics as a means of assessing the state of the State, and were drawn up within a period of less than four hundred years. That is, they seem like successive attempts to solve the same problem, and indeed this is how they have been generally treated.[7] But they are not comparable at all because, given the changes in disease conceptualisation that had occurred between the first and the second and between the second and the third of them, they were, in effect, drawn up in different societies and cultures. In a word, they are incommensurable. Thus causes of death ascribed in the first two are not amenable to subsequent diagnosis by the modern medic or historian. And the causes of death ascribed in the most recent one, in its turn, may well not be amenable to later re-diagnosis by future medics or historians.

---

[7] See for instance GREENWOOD, M. (1948), *Medical Statistics from Graunt to Farr*, Cambridge, Cambridge University Press.

1. The London Bills of Mortality

The Bills of Mortality were kept in London sporadically from the 1592 plague, and regularly from the 1603 plague. They are believed to have been started in order to inform the royal court and the rich residents of London when plague or other epidemics had reached such a state that flight from town was necessary. Outside plague times, however, they were consulted by Londoners 'so as they might take the same as a *Text* to talk upon, in the next Company', that is as dinner-table conversation[8]. They were drawn up by parish clerks, as part of their duties, and the records were kept in their guild headquarters, Parish-Clerks Hall. They were printed every week on Thursdays, and a general account of each year was printed on the Thursday before Christmas Day. They covered the 97 parishes within the walls of the city of London, plus the 16 parishes in the liberties but outside the walls (together with their pest-house), plus the nine adjoining parishes. A few other parishes were added in later years. Because of the way they were conducted, they recorded only the births and deaths of members of the Church of England, omitting Catholics and Dissenters[9].

How were the causes of death gathered? This is what John Graunt says about the procedure:

> When anyone dies, then, either by tolling, or ringing of a Bell, or by bespeaking of a Grave of the Sexton, the same is known to the Searchers, corresponding with the said Sexton. The Searchers hereupon (who are ancient Matrons, sworn to their Office) repair to the place, where the dead Corps lies, and by view of the same, and by other enquiries, they examine by what Disease, or Casualty the Corps died. Hereupon they make their Report to the Parish-Clerk, and he, every Tuesday night, carries in an Account of all the Burials, and Christenings, happening that Week, to the Clerk of the [Parish-Clerks'] Hall[10].

---

[8] GRAUNT, J. (1672), *Natural and Political Observations Mentioned in a following Index, and made upon the Bills of Mortality. By John Graunt, Citizen of London. With Reference to the Government, Religion, Trade, Growth, Ayre, Diseases, and the several Changes of the said City*, London, John Martin, James Allestry and Tho. Dicas, 1.

[9] For an extensive series of them, and information about their history, see [Heberden, W., Ed.] (1759), *A Collection of the Yearly Bills of Mortality from 1657 to 1758 inclusive. Together with several other Bills of an earlier Date. To which are subjoined 1. Natural and Political Observations on the bills of mortality: by Capt. John Graunt, F. R.S. reprinted from the sixth edition, in 1676. II. Another essay in political arithmetic, concerning the growth of London; with measures, periods, causes, and consequences thereof. By Sir William Petty, Kt. F.R.S. reprinted from the edition printed at London in 1683. III. Observations on the past growth and present state of the city of London; reprinted from the edition printed at London in 1751; with a continuation of the tables to the end of the year 1757. By Corbyn Morris Esq; F.R.S. IV. A comparative view of the diseases and ages, and a table of the probabilities of life, for the last thirty years. By J. P. esq; F.R.S.*, London, A. Millar.

[10] GRAUNT (1672), (note 8 above), 11. Here and in all other quotations from 17th century sources I have silently modernised spellings.

For Graunt's purposes in making his natural and political observations, it mattered little 'whether the Disease were exactly the same, as Physicians define it in their Books' (p. 14). As Graunt points out, what was sometimes reported was the predominant symptom, rather than what a doctor would count as the disease proper. Sometimes the two old women 'after the mist of a Cup of Ale, and the bribe of a two-groat fee' could perhaps be persuaded to report all deaths from emaciation as consumption, and not specify whether it was a phthisis, a hectic fever or an atrophy. But even so, in general, Graunt concluded, these diagnoses of cause of death could be pretty well trusted:

> To conclude, In many of these cases the Searchers are able to report the Opinion of the Physician, who was with the Patient, as they receive the same from the Friends of the Defunct, and in very many cases, such as Drowning, Scalding, Bleeding, Vomiting, making-away themselves, Lunatics, Sores, Small-Pox, &c. their own senses are sufficient, and the generality of the World, are able pretty well to distinguish the Gout, Stone, Dropsy, Falling-Sickness, Palsy, Agues, Plurisy, Rickets, &c. one from another. (pp. 14-15)

Graunt reported that the Bills of Mortality revealed that some new diseases had appeared in the early 17th century in London. Rickets first appeared in 1634, and Graunt believed it was a new disease, not just a hitherto unreported one. The 'stopping of the stomach' first appeared in 1636.

What is particularly striking about the Bills of Mortality is that the establishing of causes of death was barely at all a medical affair. The whole business was a concern of church administration (the parish clerk system), since it was something which had simply been added on to the arrangements for funerals and burials in the parish. It was not the concern of any medical institution. Neither the College of Physicians, nor the Barber-Surgeons Company of London, nor the Society of Apothecaries were involved. Nor were even individual doctors directly involved. The opinions of the bystanders at the bedside of the dead person, and those of the two old women themselves, about cause of death were at least as significant as the opinions of any doctors.

The Table (TABLE 1) shows what diseases were effectively available to die from in London in a randomly chosen sample year, 1632: that is to say, they are the diseases which constituted the only official current form of listing of causes of death. The diseases are listed alphabetically, and no-one now seems to know who first chose this set of available diseases, nor how the list came in time to include one or two new diseases such as stopping of the stomach and rickets. Presumably the listing was systematised by the parish clerks, so that in practice the Searchers were effectively limited to choosing a candidate disease from those already on the list.

Quite a number of these available diseases have no modern equivalent whatever as possible causes of death, even in name. Fright, ague, bloody flux, canker (a spreading sore, not to be confused with cancer), consumption, fever, grief, fallen jaw, the King's Evil, lethargy, livergrown, piles, planet, purples, quinsy, rising of the lights, suddenness, surfeit and worms, are none of them recognised possible causes of death in

(9)

The Diseases, and Casualties this year being 1632.

Abortive, and Stilborn —— 445
Affrighted ———————— 1
Aged ———————————— 628
Ague —————————————— 43
Apoplex, and Meagrom —— 17
Bit with a mad dog ———— 1
Bleeding ———————————— 3
Bloody flux, scowring, and flux 348
Brused, Issues, sores, and ulcers, 28
Burnt, and Scalded ———— 5
Burst, and Rupture ———— 9
Cancer, and Wolf ———— 10
Canker ———————————— 1
Childbed ———————————— 171
Chrisomes, and Infants —— 2268
Cold, and Cough ———— 55
Colick, Stone, and Strangury — 56
Consumption ———————— 1797
Convulsion ———————— 241
Cut of the Stone ———— 5
Dead in the street, and starved — 6
Droplie, and Swelling —— 267
Drowned ———————————— 34
Executed, and prest to death — 18
Falling Sickness ———— 7
Fever ———————————— 1108
Fistula ———————————— 13
Flocks, and small Pox —— 531
French Pox ———————— 12
Gangrene ———————————— 5
Gout ———————————— 4
Grief ———————————— 11

Jaundies ———————————— 43
Jawfaln ———————————— 8
Impostume ———————— 74
Kil'd by several accidents—— 46
King's Evil ———————— 38
Lethargie ———————————— 2
Livergrown ———————— 87
Lunatique ———————————— 5
Made away themselves ———— 15
Measles ———————————— 80
Murthered ———————————— 7
Over-laid, and starved at nurse — 7
Pallie ———————————— 25
Piles ———————————— 1
Plague ———————————— 8
Planet ———————————— 13
Pleurisie, and Spleen———— 36
Purples, and spotted Feaver — 38
Quinsie ———————————— 7
Rising of the Lights ———— 98
Sciatica ———————————— 1
Scurvey, and Itch ———— 9
Suddenly ———————————— 62
Surfet ———————————— 86
Swine Pox ———————— 6
Teeth ———————————— 470
Thrush, and Sore mouth —— 40
Tympany ———————————— 13
Tissick ———————————— 34
Vomiting ———————————— 1
Worms ———————————— 27

	Christened			Buried			
	Males	4994		Males	4932		Whereof,
	Females	4590		Females	4603		of the
	In all	9584		In all	9535		Plague 8

Increased in the Burials in the 122 Parishes, and at the Pesthouse this year   993
Decreased of the Plague in the 122 Parishes, and at the Pesthouse this year,   266

C   7 In

Courtesy of the University Library, Cambridge.

the early twenty-first century in the West[11]. Yet it seems that, according to the Searchers, in London in 1632 no less than 3,783 people —over a third of all the deaths that year— died of one or other of these diseases. Of the four major kinds of disease which the majority of people die from today in England and Wales —the degenerative diseases: cancer, heart disease, respiratory disease and cerebrovascular disease— the name of only one of them (cancer) is even present in this list. The big causes of death in 1632 London, by contrast, were 1. 'chrisomes and infants', that is children dying within the first month of life or shortly after, the chrism being the baptismal robe in which the dead infant would also be buried (2,268 cases); 2. 'consumption' (1,797 cases); 3. 'fever' (1,108 cases). The next biggest causes of death were 4. 'old age' (628 cases); 5. 'flocks and small pox' (531 cases); and 6. 'teeth' (470 cases).

This system of drawing up the weekly Bills of Mortality continued into the early 19th century, without substantial change[12]. In time it came to be replaced by our next kind of report.

2. 19th century Registrar-General reports

We now move on one hundred and sixty-seven years, to the first governmental listing of causes of death throughout England[13]. The office of Registrar-General of Births, Deaths and Marriages in England, was set up by the Registration Act of 1836 and its first annual report appeared in 1839[14]. The Benthamite sanitary reformer, Edwin Chadwick, claimed that it was at his suggestion that the Registrars had to collect causes of death, and he had even hoped that the Registrars themselves would be medical men[15]. The point of collecting causes of death, according to the Registrar-General, was that 'if the cause of death be correctly inserted [in the registers], there

---

[11] I appreciate that some of these are conventionally taken to have been renamed later, and thus are conventionally taken to correspond to modern diseases (for instance, both consumption and the King's Evil were supposedly later reclassified as tuberculosis). I shall deal in the larger version of this paper with what happens in such reclassifications, and show that as a consequence of how reclassification takes place, the earlier disease does not in fact correspond with the modern one.

[12] For as complete a series of Bills as could be assembled in 1749, see [Heberden, W., Ed.] (1759), (note 9 above).

[13] For some of the interests in counting people and their causes of death in Britain in the interval between Graunt and the Registrar-General, see Greenwood (1948), (note 7 above), and Glass, D. V. (1973, repr. 1978), *Numbering the People. The Eighteenth-Century Population Controversy and the Development of Census and Vital Statistics in Britain*, London, Gordon and Cremonesi.

[14] (1839), *First Annual Report of the Registrar-General of Births, Deaths, and Marriages in England.*, London, Longman, Orme, Brown, Green, and Longmans. For a comparable attempt to register and tabulate causes of death in Ireland, see (1843), *Report of the Commissioners appointed to take the Census of Ireland for the Year 1841. Presented to Both Houses of Parliament by Command of Her Majesty*, Dublin, Her Majesty's Stationery Office.

[15] GLASS (1973), (note 13 above), 149.

will exist thenceforward public documents, from whence may be derived a more accurate knowledge, not only of the comparative prevalence of various mortal diseases, as regards the whole of England and Wales, but also of the *localities* in which they respectively prevail, and the *sex, age,* and *condition of life* which each principally affects'. The Act itself had specified only that 'some person present at the death, or in attendance during the last illness' should give this information to the Registrar. But to avoid reliance on 'persons ignorant of medicine, and of the names and natures of diseases', the Registrar-General 'earnestly recommended that every practising member of any branch of the medical profession who may have been present at the death, or in attendance during the last illness of any person, shall, immediately after such death' place in the hands of the person who would report to the Registrar 'written statements of the cause of death'. For London, the Presidents of the Royal College of Physicians and of the Royal College of Surgeons, together with the Master of the Worshipful Society of Apothecaries, were prevailed upon to pledge themselves 'to give, in every instance which may fall under our care, an authentic name of the fatal disease', and, they said, they entreated 'all authorized practitioners throughout the country to follow our example, and adopt the same practice'. The popular or common name of the disease was to be preferred to the technical name known only to medical men[16].

Thus the causes of death collected by the Registrars were intended to furnish new knowledge about the incidence of mortal diseases, and the co-operation of medical men of all kinds meant that the causes given should be highly reliable for statistical analysis.

After the information on causes of death had been collected each year, William Farr, who held the post of Compiler of Abstracts from 1838 to 1880, wrote a long letter to the Registrar-General analysing the material, and this letter was printed in the Registrar-General's annual report, and came to form the largest section of it throughout Farr's forty-two years in the job. Farr himself was a Paris- and London-trained medical man, and by the time of his appointment at the age of 32 he was also a fairly accomplished statistician. As a liberal reformer who was also a medical man, his concerns were with promoting the sciences in medicine, especially hygiene, and with investigating 'all the laws of vitality capable of being observed in masses of men, expressed in numbers'[17].

So Farr had to construct a nosology suitable for statistical purposes, and based on some medical principles of the day. (TABLE 2) A nosology is any classification of diseases, usually presented in tabulated form. That is to say, it is a list of 'available diseases' of a particular period and place, drawn up by an individual or organisation, established on the basis of some medical principles or other, with the diseases ar-

16 (1839), *First Annual Report,* (note 14 above), 77-9.

17 EYLER, J. M. (1979), *Victorian Social Medicine. The Ideas and Methods of William Farr,* Baltimore, Johns Hopkins University Press, 8, quoting Farr's own aims for his journal *British Annals of Medicine.*

IDENTIFYING DISEASE IN THE PAST: CUTTING THE GORDIAN KNOT

TABLE A.—ENGLAND AND WALES.

ABSTRACT of the CAUSES of DEATH registered in England and Wales, from 1st July to 31st December, 1837, both inclusive.

Area in Square Miles.	Population according to Census of 1831.	Families in 1831.			
		Employed chiefly in Agriculture.	Chiefly in Trade, Manufactures, and Handicraft.	Other Families.	Total.
57,805	13,897,187	834,543	1,227,614	849,717	2,911,974

		Diseases.	M.	F.	Tot.		Diseases.	M.	F.	Tot
		Cholera	246	214	460		Nephritis	37	23	60
		Influenza	220	264	484		Ischuria	49	4	53
		Small-pox	3050	2761	5811	Of the Urinary Organs.	Diabetes (17)	68	27	96
		Measles	2340	2392	4732		Granular Diseases	2	1	3
		Scarlatina	1238	1282	2520		Cystitis	61	9	70
Epidemic, Endemic, and Contagious Diseases.		Hooping Cough	1277	1767	3044		Stone (18)	161	19	180
		Croup	879	776	1655		Stricture	43	3	46
		Thrush (1)	381	326	707		Disease	262	47	309
		Diarrhœa	1451	1304	2755					
		Dysentery	350	325	675		Total	683	133	816
		Ague	39	37	76					
		Typhus (2)	4439	4608	9047					
		Erysipelas (3)	237	245	482	Of the Organs of Generation.	Childbed (19)	..	1265	1265
		Syphilis	30	43	73		Paramenia	..	49	49
		Hydrophobia	13	3	16		Ovarian Dropsy	..	21	21
							Disease (20)	13	150	163
		Total	6190	16347	32537					
							Total	13	1485	1498
		Cephalitis (4)	567	454	1021					
		Hydrocephalus	1933	1637	3570	Of the Organs of Locomotion	Arthritis	7	8	15
		Apoplexy (5)	1447	1264	2711		Rheumatism (21)	221	216	437
Of the Nervous System.		Paralysis	987	1052	2039		Disease (22)	277	200	477
		Convulsions (6)	5798	4931	10729					
		Tetanus	45	11	56		Total	505	424	929
		Chorea (7)	3	9	12					
		Epilepsy (8)	278	292	570					
		Insanity (9)	147	138	285					
		Delirium Tremens	86	9	95					
		Disease (10)	438	326	764		Carbuncle	14	5	19
						Of the integumentary System.	Phlegmon (23)	29	17	46
		Total	11729	10123	21852		Ulcer	37	45	82
		Laryngitis	11	13	24		Fistula	39	12	51
		Quinsey	141	148	289		Disease (24)	39	27	66
		Bronchitis	248	212	460					
Of the Respiratory Organs.		Pleurisy	140	96	236		Total	158	106	264
		Pneumonia	3187	2637	5824					
		Hydrothorax	557	438	995					
		Asthma	1020	744	1764					
		Consumption (11)	9494	10753	20247					
		Decline	3474	4033	7507		Inflammation	1201	1136	2337
		Disease (12)	653	523	1176		Hæmorrhage (25)	369	213	582
							Dropsy	2445	3139	5584
		Total	18925	19597	38522		Abscess	247	217	464
		Pericarditis	31	31	6.		Mortification (26)	305	276	581
Of the Organs of Circulation.		Aneurism	37	15	52		Scrofula (27)	286	255	541
		Disease (13)	834	648	1482	Of Uncertain Seat.	Carcinoma (28)	355	873	1228
							Tumor	48	81	199
		Total	902	694	1596		Gout	67	12	79
							Intemperance (29)	70	15	85
		Teething	998	905	1903		Atrophy	478	481	959
		Gastro-Enteritis	1710	1686	3396		Debility	1328	1078	2406
		Peritonitis	35	47	92		Starvation	34	29	63
		Tabes Mesenterica	228	209	437		Malformations (30)	75	41	116
Of the Digestive Organs.	Intestinal Canal.	Ascites	28	23	51		Sudden Deaths	634	419	1053
		Ulceration (14)	96	74	170					
		Hernia	150	102	252		Total	7942	8265	16207
		Colic	39	19	58					
		Constipation (15)	253	208	461					
		Worms	119	145	264					
		Disease (16)	437	416	853					
	Pancreas	Disease	..	2	2		Old Age	5674	7017	12691
		Hepatitis	91	92	183					
	Liver	Jaundice (15*)	211	194	405	Violent Deaths (31)		3605	1240	4845
		Disease (16*)	716	605	1321					
	Spleen	Disease	4	8	12	Causes not specified		3718	3376	7094
		Total	5115	4735	9850		Total	75159	73542	148701

Courtesy of the University Library, Cambridge.

ranged usually according to some theory about cause, and where that is not possible or desirable, by some other supposedly meaningful classification, such as anatomical location in the body. The term 'nosology' was popularised in 1763 by Boissier de Sauvages for his tables of diseases[18], and the term was rapidly adopted by other medical teachers in the century, and it is still in use, though the expression 'classification of disease' is more common today. Nosologies are not usually very stable. If someone came along who had a different view of the range of possible diseases, of the relative importance of different diseases, of the natural relations of diseases, or of the causal relationships of different diseases, then he could simply draw up his own nosology if he felt strongly enough about it. By 1838 there were many possible existing nosologies that Farr could choose from if he wished – he himself mentioned those of Pinel, Richerand, Bichat, Parr, Young and Mason Good[19]. However, Farr had his own medical principles and a strong view about the doctor's role. In this respect he was particularly concerned with the diseases which he believed were most amenable to prevention. So of course he had to draw up his own, new, nosology[20]. This is what is employed in the present table.

The disease taking pride of place —the most sudden and spectacular killing disease of the time— is cholera, which had scourged Europe in 1832. It headed the 'first division' of diseases in Farr's table, which consisted of the epidemic, endemic, and contagious diseases. In Farr's view these all constituted a natural group, because he believed that they all had similar causation – what he was soon to call *zymotic*, a poisonous effluvium given off by decay and dirt[21]. This primary division is therefore a classification by *cause*: all these diseases, for Farr, had similar causes, and hence were amenable to similar management. Their 'exciting causes', were, for Farr, the insalubrity of the local conditions. Action could be taken to reduce their incidence, by undertaking engineering measures to introduce clean air, clean water, remove wastes, and so on. This belief on Farr's part in 'filth diseases' and their causation preceded his classificatory activity, and was as much a political belief —that is, about how society ought to be structured and run— as it was a medical one. In 1866 Farr himself stated that 'No variation in the health of the states of Europe is the result of chance; it is the direct result of the physical and political conditions in which nations live'[22]. Indeed, the Registry was for Farr in effect an instrument to advance this politico-medical agenda. He did not draw up the array of available diseases in an innocent manner and

---

[18] This was a development of his *Nouvelles Classes des Maladies*, first published in 1731. See Martin, J. (1990), «Sauvages's Nosology: Medical Enlightenment in Montpellier», *The Medical Enlightenment of the Eighteenth Century*, Cunningham, A. and R. French, eds., Cambridge, Cambridge University Press, 111-37.

[19] (1839). *First Annual Report* (note 14 above), 92.

[20] EYLER, (1979), (note 20 above), 53-60.

[21] (1842), *Fourth Annual Report of the Registrar-General of Births, Deaths, and Marriages in England*, London, Longman, Brown, Green, and Longmans, 199-205.

[22] Eyler (1979), (note 19 above), 199.

then make supposedly Baconian inferences from it. No: the necessary structure of the table and the array of diseases was given to him by his pre-existing beliefs about disease causation. And because Farr for forty-two years personally drew up the tables and statistics from the material he was sent, he continued to be the gate-keeper of 'what counted' as a disease, and the arbiter of which diseases were more important than others, even though his positions on these matters were actually controversial.

So what could one die from in William Farr's England in 1838-9? By far the single greatest cause of death was consumption (20, 247), followed at a great distance by convulsions (10, 729) and typhus (9,047) – the first in killing-power of the 'first division' diseases. Indeed, as a group, Farr's 'epidemic, endemic and contagious diseases', comprised just under a quarter of all deaths that year, with smallpox, measles, whooping-cough and diarrhoea following typhus in their mortality. But it needs to be noted that, in grouping this 'first division' together, Farr's statistics show that over 32,000 out of almost 150,000 deaths were actually caused by the zymotic products of 'filth'. Cancer, a disease classified as 'of uncertain seat', killed only 1,228 people. Meanwhile, debility killed more than cancer did, and atrophy, and intemperance (a morally defined disease) too claimed their victims as causes of death.

3. The 1999 statistics

This most recent table of causes of death reflects how internationalised medicine has become in recent decades. In Western countries it is no longer optional which nosology you use for cause of death if you are a medic. What you use is the *International Classification of Diseases* (hereafter ICD). First drawn up in the 1890s by the Chef des Travaux statistiques de la ville de Paris, Jacques Bertillon, at the initiative of the International Statistical Institute (a statisticians' pressure group), it was soon adopted by the registrars of the United States, Canada and Mexico. It has been regularly revised since, approximately once every ten years until recently, and the 1979 revision, the ninth (hereafter ICD9), though technically out of date, is the one currently in use[23]. Since 1946 the World Health Organisation has been entrusted with revising the classification and getting governments to adopt it. As the use of the code numbers reveals, it is the classification used in the U.K. statistics in TABLE 3[24].

---

[23] On the history of the International Classification, see (1977), *International Classification of Diseases. Manual of the International Statistical Classification of Diseases, Injuries, and Causes of Death. Based on the Recommendations of the Ninth Revision Conference, 1975, and Adopted by the Twenty-ninth World Health Assembly*, Geneva, World Health Organization, Introduction.

[24] «Report: Death Registrations 1999: cause England and Wales», *Health Statistics Quarterly* (2000), n. p. The table is five pages long, and only pages one to three are reproduced here.

Table 2 — Deaths by age, sex and underlying cause, 1999 registrations — *England and Wales*

ICD9 code	Causes of death *		All ages	Under 1	1–4	5–14	15–24	25–34	35–44	45–54	55–64	65–74	75–84	85 and over
	All causes, all ages	M	263,166	2,080	408	510	2,170	3,978	5,918	13,633	28,532	64,017	89,963	51,957
		F	290,366	1,555	308	387	883	1,707	3,773	8,999	17,949	44,958	93,360	116,487
	All causes, ages under 28 days	M	1,393	1,393	-	-	-	-	-	-	-	-	-	-
		F	1,046	1,046	-	-	-	-	-	-	-	-	-	-
	All causes, age 28 days and over	M	261,773	687	408	510	2,170	3,978	5,918	13,633	28,532	64,017	89,963	51,957
		F	289,320	509	308	387	883	1,707	3,773	8,999	17,949	44,958	93,360	116,487
001–139	Infectious and parasitic diseases	M	1,848	72	38	21	53	82	160	180	192	344	445	261
		F	1,763	68	25	18	45	39	58	86	142	277	483	522
001–009	Intestinal infectious diseases	M	166	17	-	3	2	-	2	5	12	21	49	55
		F	312	22	-	2	-	-	-	5	11	34	88	150
010–018	Tuberculosis	M	248	1	-	-	2	6	14	15	29	66	84	31
		F	139	-	-	-	3	4	6	7	10	34	42	33
010–012	Pulmonary and other respiratory tuberculosis	M	208	1	-	-	2	6	13	10	26	56	65	29
		F	95	-	-	-	1	3	4	4	6	26	32	19
033	Whooping cough	M	1	1	-	-	-	-	-	-	-	-	-	-
		F	2	2	-	-	-	-	-	-	-	-	-	-
034–035	Streptococcal sore throat, scarlatina and erysipelas	M	2	-	-	-	-	1	-	-	1	-	-	-
		F	-	-	-	-	-	-	-	-	-	-	-	-
036	Meningococcal infection	M	105	14	23	7	26	12	5	7	7	3	-	1
		F	111	17	16	9	23	7	4	9	6	7	10	3
038	Septicaemia	M	661	15	7	6	7	10	26	42	59	142	210	137
		F	730	13	5	2	6	5	18	23	53	118	239	248
042–044	HIV infection	M	128	-	1	-	2	21	54	28	15	6	1	-
		F	26	-	-	-	3	8	10	2	1	1	1	-
055	Measles	M	2	-	-	1	-	-	-	1	-	-	-	-
		F	-	-	-	-	-	-	-	-	-	-	-	-
084	Malaria	M	9	-	-	-	1	1	-	4	1	2	-	-
		F	2	-	-	-	-	-	-	1	-	1	-	-
137	Late effects of tuberculosis	M	32	-	-	-	-	-	-	-	2	13	15	4
		F	23	-	-	-	-	-	-	-	2	5	12	4
140–239	Neoplasms	M	70,259	9	63	122	188	395	1,051	4,370	10,806	21,797	23,094	8,364
		F	65,532	13	38	102	119	472	1,654	4,824	9,040	16,592	20,506	12,172
140–208	Malignant neoplasms	M	69,334	5	56	112	176	377	1,024	4,308	10,688	21,570	22,776	8,242
		F	64,415	8	33	89	109	453	1,627	4,753	8,947	16,365	20,103	11,928
140–149	Malignant neoplasm of lip, oral cavity and pharynx	M	1,079	-	-	1	3	11	43	199	264	277	198	83
		F	593	-	1	-	2	9	21	56	102	120	148	134
150–159	Malignant neoplasm of digestive organs and peritoneum	M	20,162	-	1	1	10	64	269	1,346	3,334	6,495	6,414	2,228
		F	17,064	1	-	2	9	46	196	759	1,810	4,114	6,000	4,127
150	Malignant neoplasm of oesophagus	M	3,722	-	-	-	-	11	56	336	708	1,181	1,110	320
		F	2,309	-	-	-	1	3	11	103	215	560	843	573
151	Malignant neoplasm of stomach	M	3,820	-	-	-	2	14	46	179	534	1,249	1,296	500
		F	2,313	-	-	-	1	11	22	78	213	519	855	614
153	Malignant neoplasm of colon	M	4,795	-	-	-	4	20	48	264	720	1,529	1,576	634
		F	5,085	-	-	-	1	13	62	206	530	1,216	1,742	1,315
154	Malignant neoplasm of rectum, rectosigmoid junction and anus	M	2,672	-	-	-	-	5	29	163	456	860	854	305
		F	2,012	-	-	-	3	6	34	116	207	450	667	529
157	Malignant neoplasm of pancreas	M	2,802	-	-	-	1	3	48	219	539	911	845	236
		F	3,124	-	-	-	-	3	33	150	376	819	1,119	624

* The figures for individual cause categories exclude deaths at ages under 28 days.

IDENTIFYING DISEASE IN THE PAST: CUTTING THE GORDIAN KNOT

Table 2 continued	Deaths by age, sex and underlying cause, 1999 registrations													England and Wales

								Age group						
ICD9 code	Causes of death *		All ages	Under 1	1–4	5–14	15–24	25–34	35–44	45–54	55–64	65–74	75–84	85 and over
161	Malignant neoplasm of larynx	M	580	-	-	-	-	2	8	47	138	179	153	53
		F	157	-	-	-	-	-	1	10	30	50	51	15
162	Malignant neoplasm of trachea, bronchus and lung	M	18,297	-	-	-	1	9	146	952	3,019	6,577	6,067	1,526
		F	11,109	-	-	-	1	9	126	672	1,663	3,742	3,758	1,138
172	Malignant melanoma of skin	M	761	-	-	-	6	31	60	127	151	179	159	48
		F	713	-	-	-	4	22	49	83	117	144	179	115
173	Other malignant neoplasm of skin	M	213	-	-	-	-	3	3	8	17	49	84	49
		F	157	-	-	-	-	-	-	8	7	12	51	79
174	Malignant neoplasm of female breast	F	11,548	-	-	-	1	123	603	1,534	1,998	2,466	2,745	2,078
179-189	Malignant neoplasm of genitourinary organs	M	13,200	-	1	1	4	24	72	349	1,244	3,472	5,266	2,767
		F	9,192	-	-	1	13	81	281	756	1,466	2,342	2,722	1,530
179	Malignant neoplasm of uterus, part unspecified	F	433	-	-	-	-	-	5	22	71	98	139	98
180	Malignant neoplasm of cervix uteri	F	1,106	-	-	-	6	51	144	178	153	199	269	106
182	Malignant neoplasm of body of uterus	F	800	-	-	-	-	1	7	30	125	227	248	162
183	Malignant neoplasm of ovary and other uterine adnexa	F	3,946	-	-	-	4	19	87	423	838	1,134	1,038	403
185	Malignant neoplasm of prostate	M	8,502	-	-	-	-	-	3	76	560	2,063	3,724	2,076
186	Malignant neoplasm of testis	M	71	-	-	1	3	14	22	14	6	6	5	-
188	Malignant neoplasm of bladder	M	2,849	-	-	-	1	1	11	64	299	828	1,093	552
		F	1,460	-	-	-	-	3	8	29	103	326	540	451
189	Malignant neoplasm of kidney and other and unspecified urinary organs	M	1,687	-	1	-	-	8	33	186	362	562	411	124
		F	1,030	-	-	1	3	4	24	61	142	289	342	164
191	Malignant neoplasm of brain	M	1,625	1	17	37	26	52	114	313	393	431	212	29
		F	1,171	2	9	26	16	37	77	157	244	333	218	52
200-208	Malignant neoplasm of lymphatic and haematopoietic tissue	M	5,294	2	17	47	83	114	174	422	829	1,524	1,539	543
		F	4,818	3	19	38	42	80	141	293	628	1,173	1,495	906
204-208	Leukaemia	M	1,946	2	17	41	51	58	64	129	247	546	556	235
		F	1,728	3	18	34	29	43	69	110	196	377	456	393
210-239	Benign, in situ, other and unspecified neoplasms	M	925	4	7	10	12	18	27	62	118	227	318	122
		F	1,117	5	5	13	10	19	27	71	93	227	403	244
240-279	Endocrine, nutritional and metabolic diseases and immunity disorders	M	3,471	19	18	13	41	85	95	190	429	913	1,083	585
		F	4,066	12	19	18	40	46	65	125	307	731	1,312	1,391
250	Diabetes mellitus	M	2,814	-	-	1	7	39	52	128	344	774	961	508
		F	3,135	-	-	2	8	13	34	71	230	584	1,080	1,113
260-269	Nutritional deficiencies	M	28	-	-	-	-	1	2	2	5	6	6	6
		F	32	-	-	-	-	-	1	1	2	6	6	16
280-289	Diseases of blood and blood-forming organs	M	853	4	7	5	5	19	13	27	62	180	317	214
		F	1,002	1	7	6	4	14	16	16	32	134	336	436
280-285	Anaemias	M	185	-	1	2	4	11	1	7	9	27	65	58
		F	400	-	2	2	3	5	5	6	10	43	119	205
290-319	Mental disorders	M	3,860	-	-	1	196	338	225	147	129	345	1,176	1,303
		F	7,285	-	-	3	45	53	64	60	91	321	2,112	4,536
290	Senile and presenile organic psychotic conditions	M	2,089	-	-	-	-	-	-	2	31	199	842	1,015
		F	5,516	-	-	-	-	-	1	4	28	210	1,596	3,677

* The figures for individual cause categories exclude deaths at ages under 28 days.

31

| Table 2 continued | Deaths by age, sex and underlying cause, 1999 registrations | | | | | | | | | | | | | *England and Wales* |

ICD9 code	Causes of death *		All ages	Under 1	1–4	5–14	15–24	25–34	35–44	45–54	55–64	65–74	75–84	85 and over
320-389	Diseases of the nervous	M	4,848	53	42	74	133	167	223	334	448	965	1,622	787
	system and sense organs	F	5,316	38	41	53	56	105	144	284	379	797	1,712	1,707
320-322	Meningitis	M	98	10	8	8	5	4	10	13	8	15	14	3
		F	86	6	1	3	4	8	9	3	15	20	13	4
332	Parkinson's disease	M	1,554	-	-	-	-	-	-	3	28	302	804	417
		F	1,220	-	-	-	-	-	-	1	15	122	579	503
340	Multiple sclerosis	M	279	-	-	-	2	9	25	70	84	62	24	3
		F	482	-	-	-	3	11	39	121	107	103	81	17
345	Epilepsy	M	505	2	6	11	31	99	109	95	49	46	45	12
		F	361	2	9	7	19	53	55	65	38	30	39	44
390-459	Diseases of the circulatory	M	104,650	29	34	26	81	330	1,355	4,616	11,211	27,134	39,201	20,633
	system	F	113,412	24	14	22	74	191	562	1,683	4,490	16,085	41,287	48,980
390-392†	Acute rheumatic fever	M	4	-	-	-	-	-	1	-	-	2	-	1
		F	4	-	-	-	-	-	-	1	1	2	-	-
393-398†	Chronic rheumatic heart	M	439	1	-	1	1	4	12	24	57	134	145	60
	disease	F	1,187	1	-	-	3	-	8	27	103	265	465	315
401-405	Hypertensive disease	M	1,464	-	-	-	1	5	39	99	177	395	527	221
		F	1,825	-	-	-	1	2	13	27	92	275	714	701
410-414	Ischaemic heart disease	M	62,996	-	-	2	6	106	813	3,196	7,980	17,790	22,817	10,286
		F	51,471	-	-	2	4	26	172	672	2,347	8,650	19,871	19,727
410	Acute myocardial infarction	M	30,865	-	-	1	3	54	415	1,629	4,114	9,111	11,050	4,488
		F	24,502	-	-	1	4	15	89	359	1,249	4,652	9,846	8,287
415-429	Diseases of pulmonary	M	10,295	14	29	18	48	121	218	479	864	2,019	3,648	2,837
	circulation and other forms	F	16,055	16	12	12	38	74	121	305	537	1,850	5,071	8,019
	of heart disease													
430-438	Cerebrovascular disease	M	20,653	11	3	5	23	71	208	600	1,433	4,316	8,387	5,596
		F	35,214	7	2	6	23	69	212	520	1,116	3,839	12,225	17,195
433-434	Cerebral infarction	M	1,838	2	-	-	1	4	17	69	192	439	691	423
		F	2,780	-	-	-	2	9	13	43	86	336	1,013	1,278
440	Atherosclerosis	M	462	-	-	-	-	-	-	6	16	51	190	199
		F	884	-	-	-	-	-	-	1	10	52	240	581
451-453	Phlebitis, thrombophlebitis,	M	621	1	-	-	-	2	21	51	88	161	202	95
	venous embolism and	F	1,076	-	-	-	2	9	14	58	87	211	391	304
	thrombosis													
460-519	Diseases of the respiratory	M	43,165	89	45	39	68	123	268	806	2,469	8,436	16,991	13,831
	system	F	53,288	44	37	25	53	92	222	565	1,769	6,431	16,828	27,222
480-486	Pneumonia	M	23,033	28	21	19	33	82	174	411	1,005	3,209	8,664	9,387
		F	35,416	25	15	10	22	48	126	236	681	2,530	10,303	21,420
487	Influenza	M	198	3	1	1	-	2	2	6	10	31	76	66
		F	363	-	1	1	-	1	2	4	12	27	101	214
490-496	Chronic obstructive pulmonary	M	15,532	4	5	13	26	19	46	291	1,174	4,310	6,618	3,026
	disease and allied conditions	F	12,400	1	3	7	21	23	60	249	914	3,340	4,894	2,888
490-491	Bronchitis, chronic and	M	1,031	2	1	1	-	-	3	35	82	267	433	207
	unspecified	F	640	-	2	1	-	3	6	13	41	130	228	216
493	Asthma	M	498	1	3	12	25	16	25	55	78	114	108	61
		F	858	-	1	5	21	18	38	64	97	179	232	203
496	Chronic airways obstruction,	M	12,514	1	1	-	1	-	10	150	839	3,483	5,486	2,543
	not elsewhere classified	F	9,917	-	-	1	-	-	12	130	660	2,756	4,050	2,308

* The figures for individual cause categories exclude deaths at ages under 28 days.
† Figures inconsistent with those published for 1998 in the corresponding Report in HSQ02, and with those published for 1993 to 1997 in Annual Reference Volumes. For more details see section 2.6 in volume DH2 no. 25 (1998).

According to the authors of the official U.S.A. clinical version of the ICD, it 'represents the best in contemporary thinking of clinicians, nosologists, epidemiologists, and statisticians from both the pubic and private sectors'[25]. All nosologies, of course, are intended to reflect the best in contemporary thinking, and this one therefore, like its predecessors, reflects perfectly the dominant medical thinking of its age. In particular it expresses the interests of the epidemiologists and statisticians in the management of public health, because the categories it uses are designed to be ones most useful for general statistical use. In fact they descend from Farr's distinction between general diseases on the one hand, and those localized to a particular organ or anatomical site on the other. This is probably because Farr was on the first committee![26] Most strikingly the great scare disease of the 1830s —cholera— still takes pride of place among the general diseases, being number 001. However, in other respects, how the diseases are distributed between these categories differs greatly from Farr's arrangement, especially with respect to the transfer, to the 'infectious' grouping, of several diseases which Farr had distributed according to anatomical seat.

What could one die of in 1999 in England and Wales? Of all the many differences between Farr's (first) version of 1839 and the tables of the year 1999 based on ICD9, the greatest is the enormous presence in the latter of 'neoplasms' which, like 'carcinomas', is a medical euphemism for cancers. The tables reveal that cancer now is the cause of death of 25% of the population of England and Wales. In 1632 in London, just ten deaths out of over nine thousand were attributed to 'cancer or the wolf'. In Farr's 1839 statistics 'carcinoma' was a disease 'of uncertain seat' and killed only 1228 out of a total of almost a hundred and fifty thousand deaths, which was less than 1%. Today cancer kills one person in four.

We could extend this analysis of the array of available diseases to other periods, other cultures: we could even take the practice of a witch-doctor. But for the moment I will limit the analysis to England over this four hundred year period.

We find people in these three periods able to die only of a limited, specifiable and specified, array of diseases. These arrays are different. In the second two, it is the doctors and epidemiologists who specify what diseases are possible, whereas in 1632 it was two old women in each parish and then the parish clerk. In 1632 and 1999 death from zymotic diseases was impossible, while in 1839 almost a quarter of the people who died were killed by filth diseases. Other comparisons will occur to each reader. While some historians may choose to read these three different tables and their disease categories in a direct and naive way, making the 1632 cancer and plague the same as the 1839 and 1999 cancer and plague, to most medical historians it will be self-evident that these three moments are separated not just in time but also in

---

[25] (1980), *ICD.9.CM The International Classification of Diseases 9th Revision Clinical Modification,* Washington, U.S. Department of Health and Human Sciences, vol. 1, Preface.
[26] EYLER (1979), (note 20 above). 58.

ways of thinking. For a host of innovations in medical thinking and transformations in social attitudes separate each of these different cause-of-death tables. The early 19th century views of pathology and disease causation were radically different from those obtaining in the mid-17th century. Similarly, the nosology of William Farr, though it seems to correspond more to the modern ICD classification, had to give ground to a new way of thinking about causation —the germ theory— in the latter decades of the 19th century. What counted as a disease and, more particularly, precisely how a disease was diagnosed, had changed beyond recognition[27]. That is to say, the disease identities changed. More properly, the disease identities *were changed*, for this is something which happens entirely through human action. Hence it remained and remains true, that in all these periods the people died of what their doctor (or their bystander) said they died of. And that's that.

---

[27] CUNNINGHAM, A. (1992), «Transforming Plague: The Laboratory and the Identity of Infectious Disease», *The Laboratory Revolution in Medicine,* Cunningham, A. and P. Williams, eds., Cambridge, Cambridge University Press, pp. 209-44.

INDEX

Printed and bound by CPI Group (UK) Ltd, Croydon, CR0 4YY

18/10/2024

01776245-0002